W9-BIJ-018

# ADVANCES IN CHEMICAL PHYSICS

VOLUME XXV

# Advances in CHEMICAL PHYSICS

EDITED BY

I. PRIGOGINE

University of Brussels,
Brussels, Belgium

AND

STUART A. RICE

Department of Chemistry
and
The James Franck Institute
The University of Chicago
Chicago, Illinois

VOLUME XXV

AN INTERSCIENCE® PUBLICATION

JOHN WILEY AND SONS

NEW YORK · LONDON · SYDNEY · TORONTO

AN INTERSCIENCE® PUBLICATION



***Library of Congress Catalog Card Number:* 58-9935**

ISBN 0-471-69930-6

Printed in the United States of America

10 9 8 7 6 5 4 3 2 1

# INTRODUCTION

In the last decades chemical physics has attracted an ever-increasing amount of interest. The variety of problems, such as those of chemical kinetics, molecular physics, molecular spectroscopy, transport processes, thermodynamics, the study of the state of matter, and the variety of experimental methods used, makes the great development of this field understandable. But the consequence of this breadth of subject matter has been the scattering of the relevant literature in a great number of publications.

Despite this variety and the implicit difficulty of exactly defining the topic of chemical physics, there are a certain number of basic problems that concern the properties of individual molecules and atoms as well as the behavior of statistical ensembles of molecules and atoms. This new series is devoted to this group of problems which are characteristic of modern chemical physics.

As a consequence of the enormous growth in the amount of information to be transmitted, the original papers, as published in the leading scientific journals, have of necessity been made as short as is compatible with a minimum of scientific clarity. They have, therefore, become increasingly difficult to follow for anyone who is not an expert in this specific field. In order to alleviate this situation, numerous publications have recently appeared which are devoted to review articles and which contain a more or less critical survey of the literature in a specific field.

An alternative way to improve the situation, however, is to ask an expert to write a comprehensive article in which he explains his view on a subject freely and without limitation of space. The emphasis in this case would be on the personal ideas of the author. This is the approach that has been attempted in this new series. We hope that as a consequence of this approach, the series may become especially stimulating for new research.

Finally, we hope that the style of this series will develop into something more personal and less academic than what has become the standard scientific style. Such a hope, however, is not likely to be completely realized until a certain degree of maturity has been attained—a process which normally requires a few years.

At present, we intend to publish one volume a year, and occasionally several volumes, but this schedule may be revised in the future.

In order to proceed to a more effective coverage of the different aspects of chemical physics, it has seemed appropriate to form an editorial board. I want to express to them my thanks for their cooperation.

I. PRIGOGINE

# CONTRIBUTORS TO VOLUME XXV

JOSEPH ABDALLAH, JR., Department of Chemistry, University of Minnesota, Minneapolis, Minnesota

W. H. FLYGARE, Noyes Chemical Laboratory and Materials Research Laboratory, University of Illinois, Urbana, Illinois

SHELDON GREEN, Institute for Space Studies, New York, New York

J. C. MCGURK, Noyes Chemical Laboratory and Materials Research Laboratory, University of Illinois, Urbana, Illinois

WILLIAM H. MILLER, Department of Chemistry, University of California, Berkeley, California

T. G. SCHMALZ, Noyes Chemical Laboratory and Materials Research Laboratory, University of Illinois, Urbana, Illinois

RICHARD L. SMITH, Department of Chemistry, University of Minnesota, Minneapolis, Minnesota

DONALD G. TRUHLAR, Department of Chemistry, University of Minnesota, Minneapolis, Minnesota

# CONTENTS

# ADVANCES IN CHEMICAL PHYSICS

VOLUME XXV

# A DENSITY MATRIX, BLOCH EQUATION DESCRIPTION OF INFRARED AND MICROWAVE TRANSIENT PHENOMENA*

J. C. McGURK, T. G. SCHMALZ, AND W. H. FLYGARE

*Noyes Chemical Laboratory and Materials Research Laboratory, University of Illinois, Urbana, Illinois*

## CONTENTS



## I. INTRODUCTION

Infrared and microwave transient phenomena are observed immediately after a sudden change in the equilibrium condition of the system. The transient phenomena are those time dependent changes in the system that occur as the system relaxes to the new equilibrium condition. This new equilibrium condition need not be the normal Boltzmann condition. It is evident that transient phenomena must be observed in periods that are short relative to the relaxation times of the molecules involved. The steps in observing the transient behavior are first, to prepare the system in an

* Supported in part by National Science Foundation Grants NSF GH-33634 and NSF GP-12382X3

equilibrium condition; second, to change the equilibrium condition in a period much shorter than the relaxation time; and third, to observe the system relax to the new equilibrium condition.

In this paper we describe the infrared and microwave transient phenomena obtained by using electromagnetic radiation to probe the vibrational and rotational energy levels of a molecule. The availability of infrared laser sources has opened up the coherent vibrational level probe that has been available before in microwave spectroscopy with tunable electronic oscillators. Of course, for the experiments described here we require radiation sources that can be tuned to molecular resonance frequencies and have sufficient power to drive a two-level system into a new equilibrium condition. A perusal of the current literature shows that new data on infrared and microwave transient effects are being generated by combining high-power coherent infrared and microwave frequency sources with Stark effect molecular energy level tuning. There appears to be some confusion in the literature about what is being observed and in some cases there is confusion about the molecular relaxation information that is being obtained. One reason for the confusion is that there is no general treatment of the theory of infrared and microwave transient phenomena in the literature. In this article we employ a semiclassical model using the density matrix formalism to solve the problem of a two-level quantum system interacting with a classical electromagnetic field. By introducing the molecular relaxation times phenomenologically, we obtain the electric field analogues of the Bloch equations of nuclear magnetic resonance (NMR). We then employ the analogy with NMR to provide a systematic interpretation of infrared-microwave transient effects.

Throughout this paper we draw attention to the analogy between NMR transient behavior and infrared and microwave transient phenomena. Nuclear magnetic transient effects are well characterized. In NMR, relaxation times are normally on the order of seconds. Therefore it is a relatively simple experimental problem to observe transient effects on a time scale of less than 1 sec. In rotational and vibrational spectroscopy, relaxation effects are normally dominated by rotational relaxation times that are typically on the order of $10^{-6}$ see · mtorr pressure. As sensitivity requirements force infrared and microwave observations to be made at pressures greater than 1 mtorr, the corresponding transient effects must be observed in times short relative to 1 $\mu$sec. This is a considerable experimental challenge and it is quite understandable why the observation of transient effects in the infrared and microwave spectra of gases has lagged so far behind the NMR development. In spite of the difference of $10^6$ in time scales, many of the equations governing NMR relaxation processes can be carried over directly into the discussion of vibrational and rotational relaxation processes. The major conceptual difference is that many times in the gas phase the Doppler effect gives rise to an additional convolution of the information that is obtained directly in the case of the effectively stationary nuclei observed in NMR.

In Section II of this article we describe the theoretical solution of the two level problem and derive the electric dipole analogue of the Bloch equations. In Section III we solve the Bloch equations for the steady state and relate these solutions to power saturation, Lamb dip, and saturation dip experiments. We discuss the measurement of $T_1$ and $T_2$ by steady-state line shape studies. Sections IV through VI present the time dependent solutions to the equations of motion and applications to the analogues of the common transient phenomena of nuclear magnetic resonance: transient nutation (transient absorption), free induction decay, and pulse echoes. We also discuss the possibility of employing Fourier transform techniques in microwave spectroscopy. Section VII discusses transient double resonance effects where a change in population difference between two levels is induced by pumping a connected transition with a high power pulse. We discuss both microwave-microwave and microwave-infrared experiments.

## II. THEORY

Let us consider an ensemble of nondegenerate two-level quantum systems interacting with a plane-polarized electromagnetic wave through the electric dipole interaction. We shall first consider the case where the positions of the two-level systems are fixed in space and where they do not interact with each other.

The Hamiltonian is

$$\mathscr{H} = \mathscr{H}_0 - 2\mu\mathscr{E} \cos(\omega t - kz) \tag{1}$$

where $\mathscr{H}_0$ is the time independent Hamiltonian of the two-level system with eigenvalues $E_a$ and $E_b$ corresponding to the lower and upper states, respectively. $\mu$ is the dipole moment operator and

$$E(t) = 2\mathscr{E} \cos(\omega t - kz) \tag{2}$$

is the electric field with $k = \omega/c$ for a plane wave in a vacuum. Note that $\mathscr{E}$ is in general a function of $z$, $t$. The resonance frequency of the two-level system is given by

$$\omega_0 = \frac{E_b - E_a}{\hbar} \tag{3}$$

which we suppose differs from the frequency $\omega$ of the applied field by

$$\Delta\omega = \omega_0 - \omega \tag{4}$$

The time dependence of this system is best followed by examining the time dependence of its density matrix.[1] If $\sigma$ is the density matrix corresponding to the Hamiltonian in (1), then its time development is determined by the equation

$$i\hbar \frac{\partial \sigma}{\partial t} = [\mathscr{H}, \sigma] = \mathscr{H}\sigma - \sigma\mathscr{H} \tag{5}$$

Equation (5) is most simply solved by transforming to the interaction representation. The density matrix in the interaction representation is defined by

$$\rho = \exp\left[\frac{iS}{\hbar}\left(t - \frac{z}{c}\right)\right] \sigma \exp\left[-\frac{iS}{\hbar}\left(t - \frac{z}{c}\right)\right] \tag{6}$$

where $S$ is a diagonal matrix with eigenvalues $E_a$ and $E_a + \hbar\omega$. Differentiating (6) with respect to $t$, we have

$$i\hbar \frac{\partial \rho}{\partial t} = \exp\left[\frac{iS}{\hbar}\left(t - \frac{z}{c}\right)\right] \left\{ i\hbar \frac{\partial \sigma}{\partial t} - [S, \sigma] \right\} \exp\left[-\frac{iS}{\hbar}\left(t - \frac{z}{c}\right)\right] \tag{7}$$

which with (1) and (5) becomes

$$i\hbar \frac{\partial \rho}{\partial t} = [\mathscr{H}_0 - S - 2\mu_s \mathscr{E} \cos(\omega t - kz), \rho] \tag{8}$$

where

$$\mu_s = \exp\left[\frac{iS}{\hbar}\left(t - \frac{z}{c}\right)\right] \mu \exp\left[-\frac{iS}{\hbar}\left(t - \frac{z}{c}\right)\right]$$

The matrix element of the interaction coupling states $a$ and $b$ is

$$\begin{aligned} -2\mathscr{E} \cos(\omega t - kz)\langle a|\,\mu_s\,|b\rangle &= -2\mathscr{E} \cos(\omega t - kz)\langle a|\,\mu\,|b\rangle e^{-i(\omega t - kz)} \\ &= -\mathscr{E}\langle a|\,\mu\,|b\rangle(1 + e^{-2i(\omega t - kz)}) \end{aligned} \tag{9}$$

where the last line is obtained by using the identity

$$2\cos(\omega t - kz) = e^{+i(\omega t - kz)} + e^{-i(\omega t - kz)} \tag{10}$$

We now make the rotating wave approximation which consists of neglecting the rapidly varying term $e^{-2i(\omega t - kz)}$ with respect to one (unity). Since the diagonal matrix elements of the interaction are zero or rapidly varying, we can now rewrite (8) as

$$i\hbar \frac{\partial \rho}{\partial t} = [\mathscr{H}_0 - S - \mathscr{E}\mu, \rho] \tag{11}$$

We note that the only nonvanishing matrix element of $\mathscr{H}_0 - S$ is

$$\langle b| \mathscr{H}_0 - S |b\rangle = \hbar\, \Delta\omega \tag{12}$$

The effect of an electric field on an ensemble of small dipoles is to align them along the direction of the field, thus inducing a macroscopic dipole moment. The polarization (per unit volume) $P$ of $N$ two-level systems is given by

$$P = N \operatorname{Tr} \{\mu\sigma\} \tag{13}$$

where Tr indicates the trace of the matrix representation of $\mu\sigma$. Substituting (6) into (13) gives

$$P = N \operatorname{Tr} \left\{\mu \exp\left[-\frac{iS}{\hbar}\left(t - \frac{z}{c}\right)\right] \rho \exp\left[\frac{iS}{\hbar}\left(t - \frac{z}{c}\right)\right]\right\} \tag{14}$$

Expanding the trace as the sum of the diagonal elements and again using the fact that the diagonal matrix elements of $\mu$ are neglected, we have

$$P = N\{\langle b| \mu |a\rangle\langle a| \rho |b\rangle e^{i(\omega t - kz)} + \langle a| \mu |b\rangle\langle b| \rho |a\rangle e^{-i(\omega t - kz)}\} \tag{15}$$

Noting that the second term is just the complex conjugate of the first allows us to write

$$P = (P_r + iP_i)e^{i(\omega t - kz)} + (P_r - iP_i)e^{-i(\omega t - kz)} \tag{16}$$

introducing a real and an imaginary component of the polarization. Notice, though, that the macroscopic polarization $P$, $P_r$, and $P_i$ are all real quantities.

We now determine the time dependence of $P_r$ and $P_i$. Taking matrix elements of (11), we have

$$i\hbar\langle a| \frac{\partial\rho}{\partial t} |b\rangle = \mathscr{E}\langle a| \mu |b\rangle[\langle a| \rho |a\rangle - \langle b| \rho |b\rangle] - \hbar\, \Delta\omega\langle a| \rho |b\rangle \tag{17}$$

$$i\hbar\langle a| \frac{\partial\rho}{\partial t} |a\rangle = -\mathscr{E}\langle a| \mu |b\rangle\langle b| \rho |a\rangle + \mathscr{E}\langle a| \rho |b\rangle\langle b| \mu |a\rangle \tag{18}$$

and

$$i\hbar\langle b| \frac{\partial\rho}{\partial t} |b\rangle = -\mathscr{E}\langle b| \mu |a\rangle\langle a| \rho |b\rangle + \mathscr{E}\langle b| \rho |a\rangle\langle a| \mu |b\rangle \tag{19}$$

From (15) and (17) and the definition in (16) we have

$$\begin{aligned} i\hbar\frac{\partial}{\partial t}(P_r + iP_i) &= i\hbar N\langle b| \mu |a\rangle\langle a| \frac{\partial\rho}{\partial t} |b\rangle \\ &= \mathscr{E}\langle a| \mu |b\rangle\langle b| \mu |a\rangle[N\langle a| \rho |a\rangle - N\langle b| \rho |b\rangle] \\ &\quad - \hbar\, \Delta\omega N\langle b| \mu |a\rangle\langle a| \rho |b\rangle \end{aligned} \tag{20}$$

$N\langle a|\,\rho\,|a\rangle$ and $N\langle b|\,\rho\,|b\rangle$ are just the number densities of molecules in the $a$ and $b$ states, respectively, so let

$$N\langle a|\,\rho\,|a\rangle = N_a$$

and

$$N\langle b|\,\rho\,|b\rangle = N_b \tag{21}$$

Equation (20) may now be rewritten as

$$i\hbar \frac{\partial}{\partial t}(P_r + iP_i) = \mathscr{E}\mu_{ab}{}^2\,\Delta N - \hbar\,\Delta\omega(P_r + iP_i) \tag{22}$$

where

$$\mu_{ab}{}^2 = |\langle a|\,\mu\,|b\rangle|^2 \tag{23}$$

and

$$\Delta N = N_a - N_b \tag{24}$$

For the time dependence of $\Delta N$ using (18) and (19) we have

$$i\hbar \frac{\partial\,\Delta N}{\partial t} = 2N\mathscr{E}[\langle a|\,\rho\,|b\rangle\langle b|\,\mu\,|a\rangle - \text{c.c.}] \tag{25}$$

where c.c. is the complex conjugate. From (15) and (16) we can rewrite (25) as

$$i\hbar \frac{\partial\,\Delta N}{\partial t} = 4i\mathscr{E}P_i \tag{26}$$

Separating the real and imaginary parts of (22) and combining with (26) gives a set of three coupled differential equations in the three unknowns $P_r$, $P_i$, and $\Delta N$.

$$\frac{\partial P_r}{\partial t} + \Delta\omega P_i = 0$$

$$\frac{\partial P_i}{dt} - \Delta\omega P_r + \kappa^2\mathscr{E}\left(\frac{\hbar\,\Delta N}{4}\right) = 0 \tag{27}$$

$$\frac{\partial}{\partial t}\left(\frac{\hbar\,\Delta N}{4}\right) - \mathscr{E}P_i = 0$$

where

$$\kappa = \frac{2\mu_{ab}}{\hbar} \tag{28}$$

Before discussing these equations further we now derive an expression for the power absorbed by the system from the field since this is one quantity that is experimentally accessible. Let the $N$ two-level systems be bounded by the planes $z = 0$, $z = l$. Let the traveling wave

$$E_+(z, t) = \mathscr{E}(z, t)e^{i(\omega t - kz)} \tag{29}$$

be the complex solution of the wave equation

$$\frac{\partial^2 E_+}{\partial z^2} = \frac{1}{c^2}\frac{\partial^2 E_+}{\partial t^2} + \frac{4\pi}{c^2}\frac{\partial^2 P_+}{\partial t^2} \tag{30}$$

where from (16)

$$P_+ = (P_r + iP_i)e^{i(\omega t - kz)} \tag{31}$$

Substituting (29) and (31) into (30) and neglecting the slowly varying terms in

$$\frac{\partial P_r}{\partial t}, \quad \frac{\partial^2 P_r}{\partial t^2}, \quad \frac{\partial P_i}{\partial t}, \quad \frac{\partial^2 P_i}{\partial t^2}, \quad \frac{\partial^2 \mathscr{E}}{\partial z^2}, \quad \frac{\partial^2 \mathscr{E}}{\partial t^2}$$

we obtain the equation

$$\frac{\partial \mathscr{E}}{\partial z} + \frac{1}{c}\frac{\partial \mathscr{E}}{\partial t} = \frac{2\pi\omega}{c} P_i \tag{32}$$

In general, then, a complete description of the system is given by the solutions to (27) and (32) subject to the boundary conditions

$$\begin{aligned} z < 0 \qquad & \mathscr{E}(z, t) = \mathscr{E}(t) \\ & P_r(z, t) = P_i(z, t) = \Delta N(z, t) = 0 \\ z > l \qquad & \mathscr{E}(z, t) = \mathscr{E}(l, t) \\ & P_r(z, t) = P_i(z, t) = \Delta N(z, t) = 0 \end{aligned} \tag{33}$$

The equations may be solved by an iteration technique. First assume a functional form for $\mathscr{E}(z, t)$ and solve (27) for $P_i(z, t)$. This value is then used in (32) to calculate a new value for $\mathscr{E}(z, t)$. The process is then repeated a number of times until the degree of accuracy required is attained. In this article we discuss only the case where the applied radiation is very weakly attenuated by the absorbing medium. In that case it is sufficient to solve (27) using for $\mathscr{E}(z, t)$ the applied field $\mathscr{E}(t)$. In this case we note that $P_r$, $P_i$, and $\Delta N$ are functions only of $t$. The partial derivatives with respect to $t$ appearing in (27) may then be replaced by total derivatives. The solution of $P_i(t)$ so obtained is then used in (32) to solve for $\mathscr{E}(z, t)$. In (32) the term $(1/c)(\partial\mathscr{E}/\partial t)$ is for our purposes very small compared with $\partial\mathscr{E}/\partial z$. This claim may be justified as follows. Anticipating our results for the moment, we may take a field of the general form $\mathscr{E}_0 e^{-\gamma z}$, with the absorption coefficient $\gamma$ of the form $\gamma_0 e^{-t/T}$ where $T$ is the relaxation time appropriate for the process under consideration. Performing the indicated differentiation gives

$$\frac{1}{c}\left|\frac{\partial \mathscr{E}}{\partial t}\right| = \frac{z}{cT}\left|\frac{\partial \mathscr{E}}{\partial z}\right| \tag{34}$$

For typical path lengths and relaxation times in microwave and infrared spectroscopy, $z/cT$ is on the order of $10^{-3}$ or less, and hence we may neglect the term $(1/c)(\partial\mathscr{E}/\partial t)$ and write

$$\frac{\partial\mathscr{E}}{\partial z} = \frac{2\pi\omega}{c} P_i \tag{35}$$

The average power per unit cross-sectional area for the electromagnetic radiation is given by

$$W = \langle E(t)^2 \rangle \frac{c}{4\pi} \tag{36}$$

where the brackets indicate here the time average. Using (2) gives

$$W = \frac{\mathscr{E}^2 c}{2\pi} \tag{37}$$

From (35)

$$\frac{d(\mathscr{E}^2)}{dz} = 2\mathscr{E}\frac{d\mathscr{E}}{dz} = \frac{4\pi\omega}{c}\mathscr{E}P_i$$

or

$$\frac{d(\mathscr{E}^2)}{\mathscr{E}^2} = \frac{4\pi\omega}{c}\frac{P_i}{\mathscr{E}}\, dz$$

which from (37) becomes

$$\frac{dW}{W} = \frac{4\pi\omega}{c}\frac{P_i}{\mathscr{E}}\, dz \tag{38}$$

Defining the absorption coefficient $\gamma$ by the equation

$$\frac{dW}{W} = -\gamma\, dz \tag{39}$$

we have, comparing with (38), the resultant absorption coefficient

$$\gamma = -\frac{4\pi\omega}{c}\frac{P_i}{\mathscr{E}} \tag{40}$$

Equation (40) may be related to the more common expression for the absorption coefficient[2] in the following way. Consider a wave of the form given in (29) propagating through a medium. The wave vector in the absorbing medium, $\tilde{k}$, may be related to the electric polarizability $\alpha$ by[3]

$$\tilde{k} = k(1 + 2\pi\alpha) \tag{41}$$

If $\alpha$ has an imaginary part, there will be a damping term in (29) and hence an attenuation of the field by the medium.

The absorption coefficient is defined by

$$I = I_0 e^{-\gamma z} \tag{42}$$

$I$ is proportional to $|\mathscr{E}|^2$ so

$$\frac{I}{I_0} = e^{2\tilde{k}_i z} = e^{(4\pi\omega/c)\alpha i} \tag{43}$$

where $\tilde{k}_i$ and $\alpha_i$ are the imaginary parts of $\tilde{k}$ and $\alpha$, respectively. Comparing (42) and (43) gives the standard result,

$$\gamma = -\frac{4\pi\omega}{c}\alpha_i \tag{44}$$

Comparison of (44) and (40) allows the identification of $\alpha_i$ with $P_i/\mathscr{E}$, which is consistent with the usual picture of $P$ being the induced dipole that results when a field is applied to a polarizable system,

$$P = \alpha\mathscr{E} \tag{45}$$

Equations (27) and (32) contain a complete description of the physical system under the very restrictive conditions with which we began. We now begin to relax these restrictions and show how the additional complications that arise may be handled.

In any real physical situation the two-level systems are not independent but are in constant interaction through collisions. These collisions induce relaxation of both the polarization and the population difference. We shall take these processes into account phenomenologically by assuming that when the radiation is switched off $P_r$, $P_i$, and $\Delta N$ decay to their equilibrium values of 0, 0, and $\Delta N_0$ by first-order relaxation with the characteristic relaxation times $T_2$, $T_2$, and $T_1$, respectively. In the presence of relaxation then we rewrite (27) in the form

$$\begin{aligned} \frac{dP_r}{dt} + \Delta\omega P_i + \frac{P_r}{T_2} &= 0 \\ \frac{dP_i}{dt} - \Delta\omega P_r + \kappa^2\mathscr{E}\left(\frac{\hbar\,\Delta N}{4}\right) + \frac{P_i}{T_2} &= 0 \\ \frac{d}{dt}\left(\frac{\hbar\,\Delta N}{4}\right) - \mathscr{E}P_i + \frac{\hbar}{4}\frac{(\Delta N - \Delta N_0)}{T_1} &= 0 \end{aligned} \tag{46}$$

Equations (46) show a marked resemblance to the Bloch equations used in nuclear magnetic resonance (NMR).[4] This is not surprising since with little modification the above analysis could have been applied equally well to a spin $\frac{1}{2}$ system in a constant magnetic field interacting with a coherent electromagnetic wave through the magnetic dipole interaction.[5] The Hamiltonian for the spin $\frac{1}{2}$ system in a magnetic field of the form

$$\mathbf{H} = H_1[\hat{\mathbf{i}} \cos(\omega t - kz) + \hat{\mathbf{j}} \sin(\omega t - kz)] + \hat{\mathbf{k}} H_z \tag{47}$$

is given by

$$\mathscr{H} = -\boldsymbol{\mu} \cdot \mathbf{H} = -\hbar\gamma \mathbf{I} \cdot \mathbf{H} \tag{48}$$

where $\hat{\mathbf{i}}$, $\hat{\mathbf{j}}$, and $\hat{\mathbf{k}}$ are unit vectors. We have used here a circularly polarized plane wave in (47), which is the usual case in NMR. $\gamma$ in (48) is the gyromagnetic ratio and $\mathbf{I}$ is the nuclear spin operator. From (47) and (48) we derive the following expression for the density matrix of the system

$$i\frac{\partial \sigma}{\partial t} = [\omega_0 I_z - \tfrac{1}{2}\gamma(I_+ H_- + I_- H_+)] \tag{49}$$

where

$$\omega_0 = -\gamma H_z$$

and

$$H_\pm = H_1 e^{\pm i(\omega t - kz)} \tag{50}$$

Transforming to the interaction representation defined by

$$\rho = e^{-iI_z(\omega t - kz)} \sigma e^{iI_z(\omega t - kz)} \tag{51}$$

we find for the time dependence of $\rho$ the equation

$$\frac{i\,\partial \rho}{\partial t} = [I_z \,\Delta\omega - \gamma H_1 I_x, \rho] \tag{52}$$

The magnetization is given by

$$\mathbf{M} = N \operatorname{Tr}\{\boldsymbol{\mu}\sigma\} = N\hbar\gamma \operatorname{Tr}\{\boldsymbol{I}\sigma\} \tag{53}$$

From (53) and (51) we find

$$M_z = \tfrac{1}{2}\hbar\gamma\,\Delta N \tag{54}$$

and

$$M_x + iM_y = (u + iv)e^{-i(\omega t - kz)} \tag{55}$$

where $u$, $v$ are defined by the equation

$$(u + iv) = N\hbar\gamma \langle - | \rho | + \rangle \tag{56}$$

Taking matrix elements of (52) we find for the time dependence of $u$, $v$, and $M_z$ after the addition of damping terms the equations

$$\begin{aligned} \frac{\partial u}{\partial t} + \Delta\omega v + \frac{u}{T_2} &= 0 \\ \frac{\partial v}{\partial t} - \Delta\omega u - \gamma H_1 M_z + \frac{v}{T_2} &= 0 \\ \frac{\partial M_z}{\partial t} + \gamma H_1 v + \frac{M_z - M_{0z}}{T_1} &= 0 \end{aligned} \tag{57}$$

If $H_1$ is independent of $z$, $\partial/\partial t \to d/dt$. This is the case in NMR spectroscopy.

We emphasize the comparison between the two problems because the magnetic resonance case has been treated extensively.[6] Much of the theory developed for NMR may be carried over directly to microwave, infrared, or visible spectroscopy. In addition, the NMR experiment has a simple classical interpretation. This classical picture may help lend motivation to the mathematical treatment, which the above analysis has shown is equivalent in either case. We now return to our discussion of (46).

So far we have assumed that the molecules are at rest. This is obviously not correct for a gas, so we now discuss the effects of translational motion. Consider a molecule moving with velocity $v$ in the $z$ direction. Let us transform from the laboratory rest frame $z$, $t$ to a frame $z'$, $t'$ moving with velocity $v$ in the $z$ direction where

$$z' = z - vt$$

and (58)

$$t' = t$$

With respect to the moving frame the molecule is at rest and, owing to the Doppler effect, experiences a field that from (2) and (58) is given by

$$E = 2\mathscr{E}(z' + vt', t') \cos(\omega' t' - k'z') \tag{59}$$

where

$$\omega' = \omega\left(1 - \frac{v}{c}\right)$$
$$k' = \frac{\omega'}{c} \tag{60}$$

In the moving frame the Hamiltonian is, from (1) and (58)

$$\mathscr{H}' = \mathscr{H}_0 - 2\mu\mathscr{E}(z' + vt', t') \cos(\omega' t' - k'z') \tag{61}$$

From the Schrödinger equation

$$i\hbar\frac{\partial\Psi'}{\partial t'} = \mathscr{H}'\Psi' \tag{62}$$

we derive the density matrix equation in the moving frame

$$i\hbar\frac{\partial\sigma'}{\partial t'} = [\mathscr{H}', \sigma'] \tag{63}$$

This leads to a set of equations exactly analogous to (46). For a weakly absorbing system $\mathscr{E}(z, t)$ in (46) is independent of $z$. Then (59) may be rewritten as

$$E = 2\mathscr{E}(t') \cos(\omega' t' - k'z') \tag{64}$$

From (46) and (64) it is apparent that in the moving frame $P_r'$, $P_i'$, and $\Delta N'$ depend on the velocity $v$ only through the term in $\Delta\omega'$ given by

$$\Delta\omega' = \omega_0 - \omega\left(1 - \frac{v}{c}\right) \tag{65}$$

In this case the polarization is given by, using (16),

$$P'(v, t) = [P_r'(v, t') + iP_i'(v, t')]e^{i(\omega' t' - k' z')} + \text{c.c.} \tag{66}$$

Transforming back to the laboratory rest frame we have

$$P(v, t) = [P_r'(v, t) + iP_i'(v, t)]e^{i(\omega t - kz)} + \text{c.c.} \tag{67}$$

$P_r'$ and $P_i'$ are the solutions to (46) with $\Delta\omega'$ given by (65). For simplicity we drop the primes on $P_r'$, $P_i'$, and $\Delta\omega'$ hereafter. Equation (67) by analogy with (40) leads to an expression for the absorption coefficient as a function of $v$, $t$ given by

$$\gamma(v, t) = -\frac{4\pi\omega}{c}\frac{P_i(v, t)}{\mathscr{E}} \tag{68}$$

For a gas the total absorption coefficient is obtained by convoluting $\gamma(v, t)$ with the normalized Maxwell distribution of velocities. We then have

$$\gamma(t) = \left(\frac{M}{2\pi kT}\right)^{1/2} \int_{-\infty}^{\infty} e^{-Mv^2/2kT}\gamma(v, t)\, dv \tag{69}$$

In (69) through (73), $k$ is Boltzmann's constant and $T$ is temperature. Writing

$$\gamma(v, t) \equiv \gamma\left(\omega_0 - \omega\left(1 - \frac{v}{c}\right), t\right) \tag{70}$$

and making the substitution

$$\frac{\omega v}{c} = \omega'$$

(69) becomes

$$\gamma(t) = \frac{q}{\pi^{1/2}} \int_{-\infty}^{\infty} e^{-q^2\omega'^2}\gamma(\omega_0 - \omega + \omega', t)\, d\omega' \tag{71}$$

where

$$q = \frac{\sqrt{\ln 2}}{\Delta\omega_D} \tag{72}$$

with the Doppler half-width $\Delta\omega_D$ given by

$$\Delta\omega_D = \frac{\omega}{c}\sqrt{\frac{2kT \ln 2}{M}} \tag{73}$$

We consider now one final restriction which may be lifted. In almost all cases of practical interest each of the two-level quantum systems is actually a group of degenerate levels. The above analysis then requires modification and in general it is no longer possible to derive a set of simple equations similar to those in (46). We now consider a particular degenerate two-level system, where by making some reasonable approximations, it is possible to derive a set of equations similar to those in (46).

Consider two rotational energy levels in a molecule interacting with the linearly polarized wave given by

$$\mathbf{E} = \hat{\mathbf{e}}2\mathscr{E} \cos(\omega t - kz) \tag{74}$$

where $\hat{\mathbf{e}}$ is the polarization direction of $\mathbf{E}$. We represent the lower and upper states by the eigenkets $|ajm\rangle$ and $|bj'm'\rangle$, respectively. The Hamiltonian is then given by

$$\mathscr{H} = \mathscr{H}_0 - 2\boldsymbol{\mu} \cdot \hat{\mathbf{e}}\mathscr{E} \cos(\omega t - kz) \tag{75}$$

where $\boldsymbol{\mu}$ is the dipole moment operator. We denote the space-fixed axes by $x, y, z$ and the molecular-fixed principal inertial axes by $X, Y, Z$. We also specify that $\boldsymbol{\mu}$ has only one component which we choose to lie along the $X$ molecular-fixed axis. Choosing the polarization direction to be along the $x$ space-fixed axis we have

$$\boldsymbol{\mu} \cdot \hat{\mathbf{e}} = \mu_x = \varphi_{Xx}\mu_X \tag{76}$$

where $\varphi_{Xx}$ is the direction cosine between the $X$ molecular-fixed and $x$ space-fixed axes. The matrix elements of $\varphi_{Xx}$ are diagonal in $m$, the quantum number specifying the projection of $\mathbf{j}$ on the space-fixed $x$ axis. We may write, therefore,

$$\varphi_{Xx}\mu_X = \mu \tag{77}$$

with

$$\langle ajm|\,\mu\,|bj'm'\rangle = \langle ajm|\,\mu\,|bj'm\rangle\delta_{mm'} \tag{78}$$

As far as the electric field interaction is concerned, (78) effectively decouples the degenerate $jm$ levels. Following the analysis of (3) through (28) we find for the polarization the equation

$$P = \sum_m [(P_{rm} + P_{im})e^{i(\omega t - kz)} + \text{c.c.}] \tag{79}$$

with $P_{rm}$, $P_{im}$, and $\Delta N_m$ satisfying the equations

$$\begin{gathered} \frac{\partial P_{rm}}{\partial t} + \Delta\omega P_{im} = 0 \\ \frac{\partial P_{im}}{\partial t} - \Delta\omega P_{rm} + \kappa_m{}^2\mathscr{E}\left(\frac{\hbar\,\Delta N_m}{4}\right) = 0 \\ \frac{\partial}{\partial t}\left(\frac{\hbar\,\Delta N_m}{4}\right) - \mathscr{E}P_{im} = 0 \end{gathered} \tag{80}$$

where

$$\kappa_m = \frac{2}{\hbar} |\langle ajm| \mu |bj'm\rangle| \tag{81}$$

In considering the effect of relaxation it is tempting to add damping terms of the form $P_{rm}/T_2$, $P_{im}/T_2$, and $(\hbar/4)(\Delta N_m - \Delta N_{0m})/T_1$ to the left-hand sides of (80). This would be justified only if transitions between the different $m$ levels were not very important. Even if these transitions are important we may proceed as follows. First we sum (80) over $m$ putting

$$P_r = \sum_m P_{rm}$$

and

$$P_i = \sum_m P_{im} \tag{82}$$

and replacing $\sum_m \kappa_m^2 \Delta N_m$ by $\kappa^2 \sum_m \Delta N_m = \kappa^2 \Delta N$ where $\kappa^2$ is some average of $\kappa_m^2$. It is now quite reasonable to add simple damping terms to give equations similar to those in (46).

## III. STEADY-STATE EXPERIMENTS

In this and the following sections we discuss the applications of the theory outlined in Section II to experimental situations. We assume always that we are dealing with a weakly absorbing medium. As discussed in Section II this considerably simplifies the problem.

### a. Power Saturation and Line Shape Analysis

First we derive the steady-state solutions to (46). In the steady state we apply

$$\frac{dP_r}{dt} = \frac{dP_i}{dt} = \frac{d}{dt}\left(\frac{\hbar \, \Delta N}{4}\right) = 0$$

to (46), giving

$$\begin{aligned} \Delta\omega P_i + \frac{P_r}{T_2} &= 0 \\ -\Delta\omega P_r + \kappa^2 \mathscr{E}\left(\frac{\hbar \, \Delta N}{4}\right) + \frac{P_i}{T_2} &= 0 \\ -\mathscr{E} P_i + \frac{\hbar}{4}\frac{(\Delta N - \Delta N_0)}{T_1} &= 0 \end{aligned} \tag{83}$$

which describe the system when it does not change in time. Solving these equations we find

$$\begin{aligned}
\frac{P_r}{\Delta N_0} &= \frac{\frac{1}{4}\hbar\kappa^2 \mathscr{E}\Delta\omega}{(1/T_2^2) + (\Delta\omega)^2 + (T_1/T_2)\kappa^2\mathscr{E}^2} \\
\frac{P_i}{\Delta N_0} &= \frac{-\frac{1}{4}\hbar\kappa^2\mathscr{E}(1/T_2)}{(1/T_2^2) + (\Delta\omega)^2 + (T_1/T_2)\kappa^2\mathscr{E}^2} \\
\frac{\Delta N}{\Delta N_0} &= \frac{(1/T_2^2) + (\Delta\omega)^2}{(1/T_2^2) + (\Delta\omega)^2 + (T_1/T_2)\kappa^2\mathscr{E}^2}
\end{aligned} \tag{84}$$

The absorption coefficient in the presence of Doppler broadening is, from (68), (71), and (84),

$$\gamma(\omega) = \frac{\pi^{1/2}\hbar\kappa^2\omega}{c}\,\Delta N_0 q \int_{-\infty}^{\infty} \frac{e^{-q^2\omega'^2}(1/T_2)\,d\omega'}{(1/T_2^2) + (T_1/T_2)\kappa^2\mathscr{E}^2 + (\omega_0 - \omega + \omega')^2} \tag{85}$$

This is the expression for the Doppler-Lorentz convoluted absorption coefficient in the presence of power saturation effects, which is used to interpret gas phase rotational line shapes.[2]

In the pressure broadened limit the Gaussian in (85) can be written as a delta function and the integral over $\omega'$ leads to a standard Lorentz line shape. Noting (69), (72), and (85), we write

$$\lim_{q\to 0}\{G(\omega')\} = \lim_{q\to 0}\{q\sqrt{\pi}\,e^{-q^2\omega'^2}\} = \delta(\omega') \tag{86}$$

where $\delta(\omega')$ is the Dirac delta function. Substituting into (85) gives

$$\begin{aligned}
\gamma(\omega) &= \frac{\hbar\kappa^2\omega\,\Delta N_0}{c}\int_{-\infty}^{+\infty}\frac{1}{T_2}\,\delta(\omega')\left[\frac{1}{(1/T_2^2) + (T_1/T_2)\kappa^2\mathscr{E}^2 + (\omega_0 - \omega - \omega')^2}\right] \\
&= \frac{\hbar\kappa^2\omega\,\Delta N_0}{c}\left[\frac{1/T_2}{(1/T_2^2) + (T_1/T_2)\kappa^2\mathscr{E}^2 + (\omega_0 - \omega)^2}\right]
\end{aligned} \tag{87}$$

in the pressure broadened limit.

It is evident from (87) that $T_2$ is readily measured in the pressure broadened low-power limit. The ratio $T_1/T_2$ can then be measured by increasing the power and measuring the new line shape when power saturation is evident. In principle, therefore, (85) or (87) can be used to measure $T_1$ and $T_2$ and no new information about molecular relaxation can be gained by experimental work in the time domain. Experimentally, however, there are problems involved in measuring the power in the steady state. It is difficult to measure absolute microwave and infrared powers and to achieve an even power distribution over the sample of molecules being irradiated. These difficulties may be the reason that so few measurements of $T_1$ and $T_2$ by steady-state line shape methods have been attempted. The results available in the microwave region, though rather limited in terms of number of molecules and transitions studied, indicate that $T_1$ is approximately equal to $T_2$.[7,8] Although $T_1$ and $T_2$ were introduced phenomenologically, it is possible to interpret them on a molecular level. Since $T_1$ measures the change in population difference, only molecular collisions which result in a change in $j$ will contribute to $T_1$. Such collisions will also contribute to $T_2$, since if the state of the molecule is changed the macroscopic polarization in general will be altered as well. However, $T_2$, the relaxation time for the polarization, may also reflect other processes. An elastic collision which leaves the molecule with the same energy (no change in $j$) may nevertheless alter the coherence of the polarization by changing the orientation of the molecule (change in $m$) or changing the phase of its oscillation. Much more than this cannot be said without a theory relating molecular relaxation to the details of collisional processes.

The above conclusions are based on a small amount of steady-state data. One of the purposes of this paper is to point out that there are advantages in measuring $T_1$ and $T_2$ in the time domain. In the time domain these relaxations can be measured directly without having a detailed knowledge of the radiation power or distribution. We describe these measurements in Sections IV through VII.

### B. Standing Wave Lamb Dip Experiments

Lamb dip experiments are important in the Doppler broadened limit where the line width is dominated by the velocity distribution in the gas. Lamb dip or saturation effects can lead to increased accuracy of measurement because the effects observed have widths, dependent on the relaxation times, that are considerably narrower than the Doppler width.

In order to describe the Lamb dip we now consider the absorption coefficient when the radiation is present as a standing wave of the form

$$E = 4\mathscr{E} \cos \omega t \cos kz \tag{88}$$

We may decompose this standing wave into two traveling waves $E_\pm$ propagating in the direction of $\mp z$. We have

$$E = E_+ + E_-$$

where

$$E_\pm = 2\mathscr{E} \cos (\omega t \pm kz)$$

If $E_+$ interacts with molecules moving with velocity $-v$ then $E_-$ interacts with molecules moving with velocity $+v$. From (68) we write the absorption coefficient as

$$\gamma(v) = -\frac{4\pi\omega}{c\mathscr{E}} [P_{i+}(-v) + P_{i-}(+v)] \tag{89}$$

Where $P_{i\pm}$ are the Fourier components of the polarization, $P_i$, corresponding to the fields $E_\pm$. $P_{i\pm}$ is determined from (84) with $\Delta N_0$ replaced by $\Delta N_\mp$. To see this one only has to consider the effect of switching off one of the $E_\pm$ fields. If $E_\pm$ is switched off $\Delta N$ decays not to $\Delta N_0$ but rather to $\Delta N_\mp$.

From (84) we have

$$P_{i\pm}(v) = \frac{-\frac{1}{4}\hbar\kappa^2\mathscr{E}(1/T_2)\,\Delta N_\mp}{(1/T_2^2) + (T_1/T_2)\kappa^2\mathscr{E}^2 + [\omega_0 - \omega \mp (\omega v/c)]} \tag{90}$$

and from (84)

$$\Delta N_\mp(v) = \frac{\Delta N_0\{(1/T_2^2) + [\omega_0 - \omega \pm (\omega v/c)]^2\}}{(1/T_2^2) + (T_1/T_2)\kappa^2\mathscr{E}^2 + [\omega_0 - \omega \pm (\omega v/c)]^2} \tag{91}$$

Combining (89), (90), and (91) and integrating over the velocity distribution function, we obtain

$$\gamma(\omega) = \frac{\pi^{1/2}\hbar\kappa^2\omega\,\Delta N_0}{cT_2}\,q\int_{-\infty}^{\infty} e^{-q^2\omega'^2} \times \left\{\frac{(2/T_2^2) + (\omega_0 - \omega - \omega')^2 + (\omega_0 - \omega + \omega')^2}{[(1/T_2^2) + (\omega_0 - \omega - \omega')^2 + (T_1/T_2)\kappa^2\mathscr{E}^2] \times [(1/T_2^2) + (\omega_0 - \omega + \omega')^2 + (T_1/T_2)\kappa^2\mathscr{E}^2]}\right\} d\omega' \tag{92}$$

Rearranging the terms in the integrand in (92) we have

$$\gamma(\omega) = \frac{2\pi^{1/2}\hbar\kappa^2\omega\,\Delta N_0}{cT_2}\,q\left\{\int_{-\infty}^{\infty} \frac{e^{-q^2\omega'^2}\,d\omega'}{[(1/T_2^2) + (\omega_0 - \omega + \omega')^2 + (T_1/T_2)\kappa^2\mathscr{E}^2]} - \int_{-\infty}^{\infty} \frac{e^{-q^2\omega'^2}(T_1/T_2)\kappa^2\mathscr{E}^2\,d\omega'}{[(1/T_2^2) + (\omega_0 - \omega - \omega')^2 + (T_1/T_2)\kappa^2\mathscr{E}^2] \times [(1/T_2^2) + (\omega_0 - \omega + \omega')^2 + (T_1/T_2)\kappa^2\mathscr{E}^2]}\right\} \tag{93}$$

Equation (93) describes the reflected wave Lamb dip experiment.[9] Comparison of this equation with (85) shows that the first term is just the sum of the values of the steady-state absorption coefficients for the two waves $E_+$ and $E_-$. The second integral in (93) has its maximum value at $\omega = \omega_0$. The effect of the second term is to give a dip at the line center. The half-width of the dip is of the order $[(1/T_2^2) + (T_1/T_2)\kappa^2\mathscr{E}^2]^{1/2}$.

Fig. 1 shows a plot of $\gamma$ versus frequency for three different values of the microwave power assuming $T_1 = T_2$. In the first graph the power ($\mathscr{E}^2$) is very low so that the second term in (93) is negligible. The curve is just the Doppler-Lorentz convoluted line shape of the transition where the Doppler width exceeds the Lorentzian relaxation width. In the second graph the onset of the Lamb dip is observed as one moves to higher power.

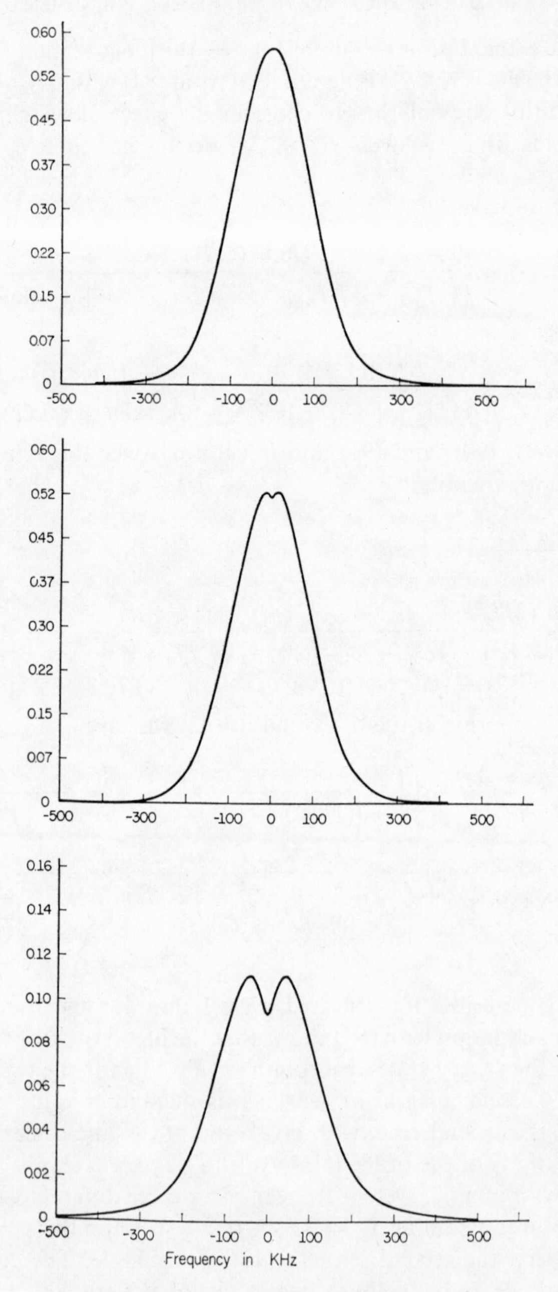


**Fig. 1.** The steady-state Lamb dip: Equation (93) is plotted for three values of the power. (*a*) The power is too low to saturate so no dip is seen. (*b*) Saturation begins. (*c*) As the power is increased the dip deepens and broadens.

The third graph shows the increase of both the depth and the width of the dip as the power is increased. Note also the decrease in intensity and increase in width of the curve as a whole.

These results are in agreement with experimental observations of the Lamb dip. The Lamb dip has been observed in microwave[10] and in infrared and optical spectroscopy.[11] The use of the Lamb dip to overcome Doppler broadening in high-resolution infrared spectroscopy is discussed by Brewer.[12]

### C. Steady-State Double Resonance Saturation Effects

The steady-state solutions of (84) may also be used to discuss steady-state double resonance effects when the power levels of the pump and signal sources are low enough that two photon effects are negligible.[13] Although we ignore two photon effects in this article, the density matrix formulation presented in Section II could easily be extended to include them. The results in that case cannot be represented as simple Bloch-type equations, but aside from more difficult mathematics, the basic principles involved are identical.[14]

Consider a three-level system designated in order of increasing energy as $a$, $b$, $c$. The transition $b \rightarrow c$ is pumped by a higher power source at frequency $\omega_p$. A weak signal source of frequency $\omega_s$ is used to probe the $a \rightarrow c$ transition. We use the subscripts $p$ and $s$ to denote variables involved in the pump and signal transitions, respectively. The Doppler averaged absorption coefficient for the signal transition is from (85) in the low-power limit

$$\gamma_s = \frac{\pi^{1/2}\hbar\kappa_s^2\omega_s}{c} q \int_{-\infty}^{\infty} \frac{e^{-q^2\omega'^2}\Delta N_s(1/T_{2s})\,d\omega'}{(1/T_{2s}^2) + (\omega_{0s} - \omega_s + \omega')^2} \tag{94}$$

In the presence of the pump power $\Delta N_s$ is given by

$$\Delta N_s = \Delta N_{0s} + \tfrac{1}{2}(\Delta N_p - \Delta N_{0p}) \tag{95}$$

which follows from the fact that every molecule that is taken from $b$ to $c$ by the pump increases the population difference $\Delta N_s$ by one. For molecules moving with velocity $v$, $\Delta N_p$ is, from (84),

$$\Delta N_p = \frac{\Delta N_{0p}\{(1/T_{2p}^2) + [\omega_{0p} - \omega_p + \sigma\omega_p(v/c)]^2\}}{(1/T_{2p}^2) + [\omega_{0p} - \omega_p + \sigma\omega_p(v/c)]^2 + (T_{1p}/T_{2p})\kappa_p^2\mathscr{E}_p^2} \tag{96}$$

The factor $\sigma$ is $\pm 1$ depending on whether the pump radiation is traveling in the same or the opposite direction as the signal radiation. From (96) we find for $\Delta N_p - \Delta N_{0p}$ the expression

$$\Delta N_p - \Delta N_{0p} = \frac{-\Delta N_{0p}(T_{1p}/T_{2p})\kappa_p^2\mathscr{E}_p^2}{(1/T_{2p}^2) + [\omega_{0p} - \omega_p + \sigma\omega_p(v/c)]^2 + (T_{1p}/T_{2p})\kappa_p^2\mathscr{E}_p^2} \tag{97}$$

Combining (95) and (97) and substituting $\omega_s v/c = \omega'$ we obtain $\Delta N_s$ which when substituted into (94) gives the following result:

$$\gamma_s = \frac{\pi^{1/2}\hbar\kappa_s{}^2\omega_s}{c} q \int e^{-q^2\omega'^2} \left\{ \frac{\Delta N_{0s}/T_{2s}}{(1/T_{2s}{}^2) + (\omega_{0s} - \omega_s + \omega')^2} - \frac{\frac{1}{2}(\Delta N_{0p}/T_{2s})\kappa_p{}^2\mathscr{E}_p{}^2 T_{1p}/T_{2p}}{[(1/T_{2s}{}^2] + (\omega_{0s} - \omega_s + \omega')^2] \times [(1/T_{2p}{}^2) + (\omega_{0p} - \omega_p + r\sigma\omega')^2 + (T_{1p}/T_{2p})\kappa^2\mathscr{E}_p{}^2]} \right\} d\omega' \tag{98}$$

where

$$r = \frac{\omega_p}{\omega_s} \tag{99}$$

From (98) we see that the absorption coefficient $\gamma_s$ is just the difference between two terms. The first term is the absorption coefficient in the absence of the pump radiation, and the second term has its maximum value when

$$\omega_{0s} - \omega_s = \frac{\omega_{0p} - \omega_p}{\sigma r} \tag{100}$$

When the width of the second term of (98) is dominated by power broadening, it is much broader than the first term, which is independent of pump power. The observed result is then simply an attenuation of the entire absorption curve. There have been several applications of this effect in microwave spectroscopy.[15] When (98) is not dominated by power broadening the second term is narrower than the first, resulting in the appearance of a dip centered at the frequency $\omega_s$ given by (100). For the important case when the pressure broadened line width $1/T_{2s}$ is much less than the Doppler broadened line width, we may simplify (98) by using the delta function for the Lorentzian since

$$\lim_{T_{2s}\to 0} \left\{ \frac{1/T_{2s}}{(1/T_{2s})^2 + (\omega_{0s} - \omega_s + \omega')^2} \right\} = \pi\delta(\omega_{0s} - \omega_s + \omega') \tag{101}$$

where $\delta(x)$ is again the Dirac delta function. Equation (98) then becomes

$$\gamma_s = \frac{\pi^{1/2}\hbar\kappa_s{}^2\omega_s}{c} q \exp[-q^2(\omega_s - \omega_{0s})^2] \times \left\{ \pi\Delta N_{0s} - \frac{(\pi/2)\,\Delta N_{0p}(T_{1p}/T_{2p})\kappa_p{}^2\mathscr{E}_p{}^2}{(1/T_{2p}{}^2) + [\omega_{0p} - \omega_p + \sigma r(\omega_s - \omega_{0s})]^2 + (T_{1p}/T_{2p})\kappa_p{}^2\mathscr{E}_p{}^2} \right\} \tag{102}$$

When $1/T_{2p}^2 + (T_{1p}/T_{2p})\kappa_p^2\mathscr{E}_p^2 \ll 1/q$ [see (72) for $q$] we see from the second term in (102) that the dip is a Lorentzian centered at $\omega_s$ where the depth relative to the absorption coefficient in the absence of the pump power is

$$\frac{1}{2}\frac{\Delta N_{0p}}{\Delta N_{0s}}\left\{\frac{(T_{1p}/T_{2p})\kappa_p^2\mathscr{E}_p^2}{(1/T_{2p}^2) + (T_{1p}/T_{2p})\kappa_p^2\mathscr{E}_p^2}\right\} \tag{103}$$

and where the half-width at half-height is

$$\left[\frac{1}{T_{2p}^2} + \frac{T_{1p}}{T_{2p}}\kappa_p^2\mathscr{E}_p^2\right]^{1/2} \tag{104}$$

Saturation dips of the kind described here have not yet been observed in the microwave region. This is due partly to the difficulty in this region of achieving the condition $1/T_{2s}$ much less than the Doppler width. In infrared and visible spectroscopy this condition is easily satisfied and saturation dips of the form predicted by (102) have been reported.[16] However, a more detailed description of the curves cannot be given without the addition of two photon effects.

## IV. TRANSIENT ABSORPTION

For the remainder of this article we discuss transient phenomena. It is here that the theory of Section II really proves its worth. Consider first the case where the two-level system is instantaneously brought into contact with the radiation. This is the electric field analogue of transient nutation in NMR. The time dependent solutions to (46) are discussed in Appendix A1, where it is shown that the general solution can be written in the form

$$F = Ae^{-at} + Be^{-bt}\cos\Omega t + \frac{Ce^{-bt}}{\Omega}\sin\Omega t + D \tag{105}$$

where $F$ stands for any of the three components $P_r$, $P_i$, or $\Delta N$. As shown in Appendix A1, explicit expressions for the constants in (105) may be given for two special cases, the on-resonance case and the case where $T_1 = T_2$. We now derive formulas for the absorption coefficient appropriate to these two cases.

### A. The On-Resonance Case, $\Delta\omega = 0$

From (A1.26) we have, for the initial conditions

$$P_r(0) = P_i(0) = 0\,, \qquad \Delta N(0) = m_0\,\Delta N_0$$

where $m_0$ reflects any deviation from the equilibrium population difference that may be present at the beginning of the experiment, the result

$$P_i = \frac{\kappa^2 \hbar \, \Delta N_0 \mathscr{E}}{4T_2} \left\{ \left[ \frac{e^{-t/T} \cos \Omega t - 1}{(1/T_2^2) + (T_1/T_2)\kappa^2 \mathscr{E}^2} \right] + \left[ \frac{\frac{1}{2}[(1/T_1) + (1/T_2)]}{(1/T_2^2) + (T_1/T_2)\kappa^2 \mathscr{E}^2} - m_0 T_2 \right] \frac{e^{-t/T} \sin \Omega t}{\Omega} \right\} \tag{106}$$

where

$$\Omega = \left[ \kappa^2 \mathscr{E}^2 - \frac{1}{4} \left( \frac{1}{T_2} - \frac{1}{T_1} \right)^2 \right]^{1/2} \tag{107}$$

and

$$T = \frac{2}{(1/T_1) + (1/T_2)} \tag{108}$$

Combining (40) and (106) we find for the absorption coefficient the expression

$$\gamma = \frac{\pi \hbar \kappa^2 \omega \, \Delta N_0}{c T_2} \left\{ \left[ \frac{1 - e^{-t/T} \cos \Omega t}{(1/T_2^2) + (T_1/T_2)\kappa^2 \mathscr{E}^2} \right] - \left[ \frac{\frac{1}{2}[(1/T_1) + (1/T_2)]}{(1/T_2^2) + (T_1/T_2)\kappa^2 \mathscr{E}^2} - m_0 T_2 \right] \frac{e^{-t/T} \sin \Omega t}{\Omega} \right\} \tag{109}$$

Note that from (107) when

$$\frac{1}{T_2} - \frac{1}{T_1} > 2\kappa \mathscr{E} \tag{110}$$

$\Omega$ is imaginary. In that case the absorption coefficient is given by the formula

$$\gamma = \frac{\pi \hbar \kappa^2 \omega \, \Delta N_0}{c T_2} \left\{ \left[ \frac{1 - e^{-t/T} \cosh \Omega t}{(1/T_2^2) + (T_1/T_2)\kappa^2 \mathscr{E}^2} \right] - \left[ \frac{\frac{1}{2}[(1/T_1) + (1/T_2)]}{(1/T_2^2) + (T_1/T_2)\kappa^2 \mathscr{E}^2} - m_0 T_2 \right] \frac{e^{-t/T} \sinh \Omega t}{\Omega} \right\} \tag{111}$$

where now

$$\Omega = \left[ \frac{1}{4} \left( \frac{1}{T_2} - \frac{1}{T_1} \right)^2 - \kappa^2 \mathscr{E}^2 \right]^{1/2} \tag{112}$$

For gases these expressions apply only at resonance in the pressure broadened limit. No closed form solution for this case can be worked out when Doppler broadening is dominant.

Equation (109) predicts a decay constant in the absorption coefficient proportional to a combination of $T_1$ and $T_2$. Thus the relation between $T_1$

and $T_2$ may be explored simply by keeping the saturating frequency on resonance.

### B. The $T_1 = T_2$ Case

From (A1.30) we have when

$$P_r(0) = P_i(0) = 0\,, \qquad \Delta N(0) = m_0\,\Delta N_0$$

$$P_i = \frac{\hbar\kappa^2\,\Delta N_0}{4}\,\mathscr{E}\left\{\frac{(e^{-t/T}\cos\Omega t - 1)(1/T)}{(1/T^2) + \Omega^2} + \left[\frac{1/T^2}{(1/T^2) + \Omega^2} - m_0\right]\frac{e^{-t/T}\sin\Omega t}{\Omega}\right\} \tag{113}$$

where

$$\Omega = [\kappa^2\mathscr{E}^2 + (\Delta\omega)^2]^{1/2} \tag{114}$$

and

$$T = T_1 = T_2$$

Combining (40) and (113) we find for the absorption coefficient the expression

$$\gamma = \frac{\pi\hbar\kappa^2\omega\,\Delta N_0}{c}\left\{\frac{(1 - e^{-t/T}\cos\Omega t)(1/T)}{(1/T^2) + \Omega^2} - \left[\frac{1/T^2}{(1/T^2) + \Omega^2} - m_0\right]\frac{e^{-t/T}\sin\Omega t}{\Omega}\right\} \tag{115}$$

For the normal condition where $m_0 = 1$, (115) becomes

$$\gamma = \frac{\pi\hbar\kappa^2\omega\,\Delta N_0}{c}\,\frac{(1/T)[1 - e^{-t/T}(\cos\Omega t - \Omega T\sin\Omega t)]}{[(1/T)^2 + \Omega^2]} \tag{116}$$

For gases the above equations hold in the pressure broadened limit. When Doppler broadening is important we may derive a formula for the absorption coefficient by integrating (115) over the velocity distribution function. Substituting $\Delta\omega = (\omega_0 - \omega + \omega')$ we have

$$\gamma = \frac{\pi^{1/2}\hbar\kappa^2\omega\,\Delta N_0}{c}\,q\int_{-\infty}^{\infty} e^{-q^2\omega'^2}\left\{\frac{(1 - e^{-t/T}\cos\Omega t)(1/T)}{(1/T^2) + (\omega_0 - \omega + \omega')^2 + \kappa^2\mathscr{E}^2} - \left[\frac{1/T^2}{(1/T^2) + (\omega_0 - \omega + \omega')^2 + \kappa^2\mathscr{E}^2} - m_0\right]\frac{e^{-t/T}\sin\Omega t}{\Omega}\right\}d\omega' \tag{117}$$

When $\Delta\omega_D \gg [(1/T^2) + \kappa^2\mathscr{E}^2]^{1/2}$ and $\kappa\mathscr{E} \ll 1/T$, that is, in the Doppler broadened low-power limit we may simplify the integral in (117) by making the substitutions

$$\frac{1/T}{(1/T^2) + (\omega_0 - \omega + \omega')^2 + \kappa^2\mathscr{E}^2} \approx \pi\delta(\omega_0 - \omega + \omega')$$

$$\frac{1/T^2}{(1/T^2) + (\omega_0 - \omega + \omega')^2 + \kappa^2\mathscr{E}^2} \approx 0 \tag{118}$$

$$\Omega \approx (\omega_0 - \omega + \omega')$$

Substituting (118) into (117) we have

$$\gamma = \frac{\pi^{1/2}\hbar\kappa^2\omega\,\Delta N_0}{c} q \int_{-\infty}^{\infty} e^{-q^2\omega'^2} \times \left\{[1 - e^{-t/T}\cos(\omega_0 - \omega + \omega')t]\pi\delta(\omega_0 - \omega + \omega') + m_0 e^{-t/T}\frac{\sin(\omega_0 - \omega + \omega')t}{(\omega_0 - \omega + \omega')}\right\} d\omega' \tag{119}$$

Integrating over the delta function, (119) becomes

$$\gamma = \frac{\pi^{1/2}\hbar\kappa^2\omega\,\Delta N_0}{c} q\{e^{-q^2(\omega_0-\omega)^2}\pi[1 - e^{-t/T}] + m_0 e^{-t/T} I\} \tag{120}$$

where the integral $I$ is given by

$$I = \int_{-\infty}^{\infty} e^{-q^2\omega'^2}\frac{\sin(\omega_0 - \omega + \omega')t}{(\omega_0 - \omega + \omega')} d\omega' \tag{121}$$

This integral is of the form

$$I = \int_{-\infty}^{\infty} e^{-q^2\omega'^2}\frac{\sin(\omega' + \varphi)t}{(\omega' + \varphi)} d\omega' \tag{122}$$

where

$$\varphi = (\omega_0 - \omega) \tag{123}$$

In Appendix A2 we show that this integral may be approximated for the following two cases by the expressions

*Case a* $\quad t \ll 2q$:

$$I = \frac{\sqrt{\pi}}{q\varphi}\sin\varphi t \tag{124}$$

*Case b* $\quad t \gg 2q$:

$$I = \pi e^{-\varphi^2 q^2} \tag{125}$$

Substituting (123), (124), and (125) into (120), we find for the absorption coefficient in the two limits the expressions

*Case a* $t \ll 2q$:

$$\gamma^{(a)} = \frac{\pi^{1/2}\hbar\kappa^2\omega\,\Delta N_0}{c}\, q\left\{e^{-q^2(\omega_0-\omega)^2}\pi[1 - e^{-t/T}] + \frac{m_0\sqrt{\pi}}{q(\omega_0 - \omega)}\, e^{-t/T} \sin(\omega_0 - \omega)t\right\} \quad (126)$$

*Case b* $t \gg 2q$:

$$\gamma^{(b)} = \frac{\pi^{1/2}\hbar\kappa^2\omega\,\Delta N_0}{c}\, q\{e^{-q^2(\omega_0-\omega)^2}\pi[1 - e^{-t/T}] + m_0\pi e^{-q^2(\omega-\omega_0)^2} e^{-t/T}\} \quad (127)$$

Equation (127) may be simplified to give the result

$$\gamma^{(b)} = \frac{\pi^{3/2}\hbar\kappa^2\omega\,\Delta N_0}{c}\, q e^{-q^2(\omega_0-\omega)^2}[e^{-t/T}(m_0 - 1) + 1] \quad (128)$$

Notice that for $m_0 = 1$ there is no time dependence in (128). The absorption coefficient is just the steady-state value for a Doppler broadened transition, which is what is expected for $t$ much greater than the characteristic Doppler half-width.

The available evidence seems to indicate that in the gas phase $T_1$ is approximately equal to $T_2$ in both the microwave and infrared regions. There are thus several experiments with which the predictions of this section may be compared. Hocker and Tang[17] and Brewer and Shoemaker[18] have observed optical nutation in the infrared region. Here Doppler broadening is dominant so (126) and (127) should apply. Although quantitative comparison is difficult, there seems to be at least general agreement.

In the microwave region the pressure broadened expression (115) or its simplification in (116) should apply. In the microwave experiment, nutation can be achieved by using a Stark field to bring a transition suddenly close in frequency to a powerful microwave source. The growth of the signal in time is then observed giving an oscillatory behavior described by (116).

In Fig. 2 we present plots of (116) for various values of the parameter $\Omega$. On resonance at low power the usual exponential growth of the signal is observed. As $\Omega$ is increased oscillations begin to develop in the approach to equilibrium. Notice that the applied power ($\kappa\mathscr{E}$) and the displacement from equilibrium ($\Delta\omega$) appear symmetrically in the equation for $\Omega$, so that equivalent oscillatory behavior in the absorption coefficient can be

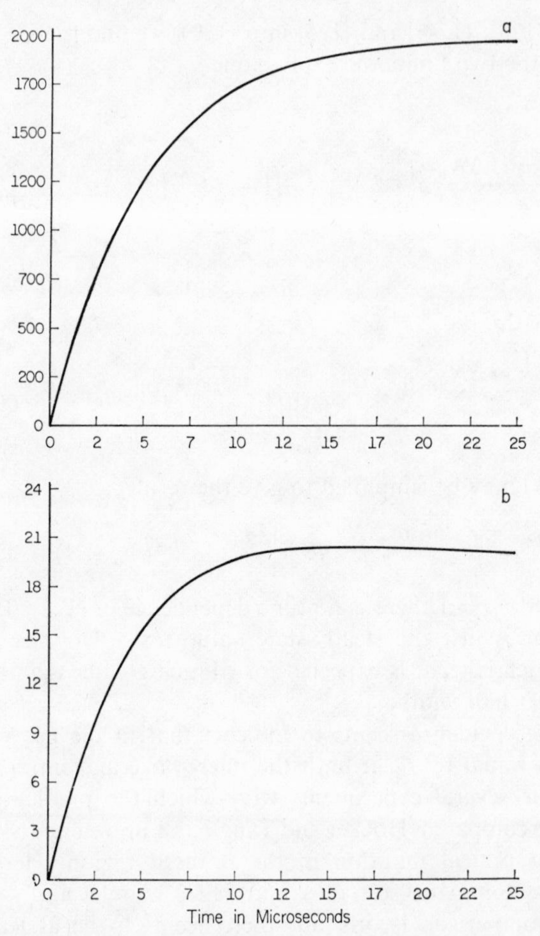


**Fig. 2.** Transient nutation: The absorption coefficient of (116) is presented as a function of $\Omega$. A relaxation time of 5 $\mu$sec is assumed. (*a*) $\Omega = 10^4$. The approach to equilibrium is a simple exponential. (*b*) $\Omega = 10^5$. $\Omega$ is still insignificant compared to $1/T$. (*c*) $\Omega = 2 \times 10^5$. The beginning of oscillatory behavior is observed. (*d*) $\Omega = 5 \times 10^5$. The oscillations become more obvious. (*e*) $\Omega = 10^6$. The sine term is now able to drive the absorption coefficient negative at some times. (*f*) $\Omega = 10^7$. The oscillatory behavior is now dominant. Notice the marked decrease in intensity as $\Omega$ is increased.

obtained either by increasing the power or by moving the applied frequency off-resonance. Both effects are reported experimentally. It should also be noted that the intensity of the absorption coefficient falls off as $1/\Omega$ for $\Omega > T$. Hence the absorption coefficient becomes quite small before many oscillations develop.

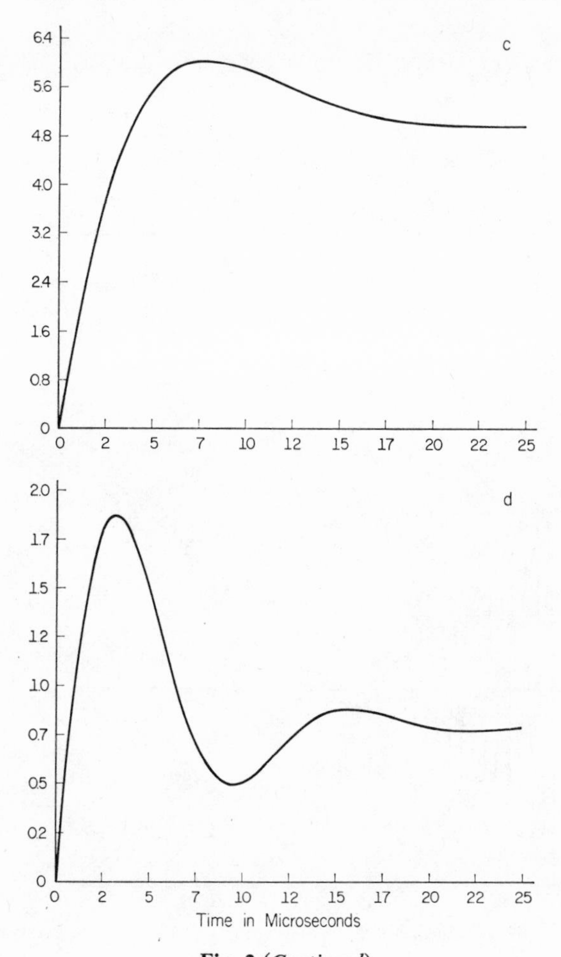


**Fig. 2** (*Continued*)

Equation (116) alone is somewhat misleading, however. The observed absorption is actually proportional to the applied power times the absorption coefficient as can be seen by expanding the exponential in (43). Therefore if $\Omega$ is increased by increasing the power, the $1/\Omega$ dependence of the absorption coefficient is cancelled, whereas if $\Omega$ is increased by moving off-resonance the intensity of absorption decreases, as one would expect. Thus it would seem that nutation is best observed by increasing the power. The practical difficulty of achieving such high microwave power may, however, necessitate off-resonance experiments with more sensitive detection of a weaker signal.

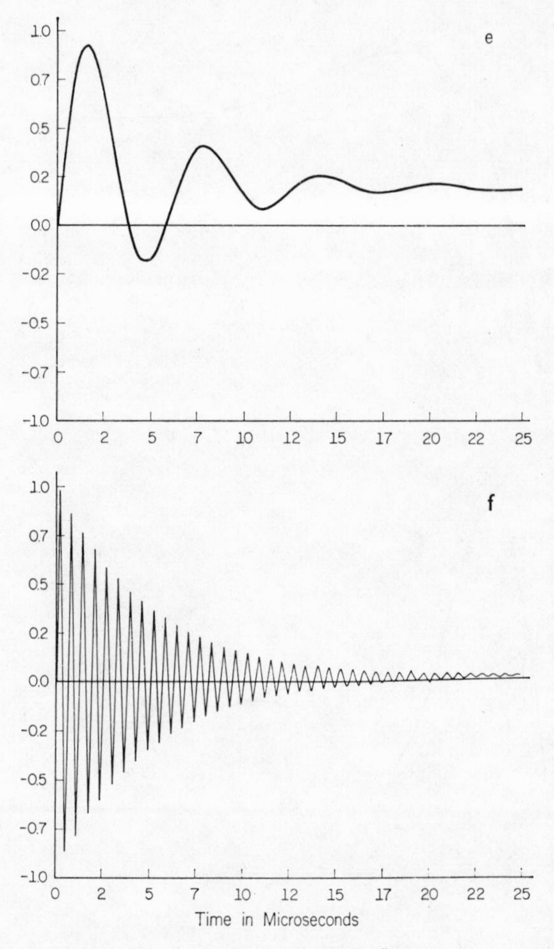


**Fig. 2** (*Continued*)

The observation of microwave transient nutation as described above has been reported several times.[19–24] Some authors have also developed the theoretical interpretation of these results as far as the expression given in (116),[25,26] including another more recent derivation of (116) using the analogy with the Bloch equations.[27] However, Levy et al.[20,27] interpret their fast passage experiment by formulas derived for transient nutation. This is only partially correct. The dominant effect is fast passage which is best described as a polarization of the molecules followed by emission. Brittain et al.[22] have observed transient nutation, emission, and the combination of these as observed in fast passage type experiments. However they have confused the physics involved in these different phenomena by incorrectly interpreting all their data with (116). This is only correct for the case of transient nutation. Amano and Shimizu[23] have also incorrectly interpreted experiments that are clearly dominated by transient emission phenomena in terms of transient absorption theory. It is easy to illustrate the differences in the phenomena of transient absorption and transient emission.[24] Figure 3 shows the signal observed when the microwave frequency is tuned to the OCS $J = 0 \rightarrow 1$ transition at $\nu_0 = 12{,}162$ MHz and a 50 kHz 200-V Stark field is applied. When the Stark field switches off, the transition comes into resonance with the radiation and transient absorption is observed as described by (109) with dependence on both $T_1$ and $T_2$. When the Stark field switches on, the molecular transition shifts to $\nu_0 + 1$ MHz out of resonance with the radiation. The molecules now emit coherent radiation at the Stark shifted frequency. This radiation mixes with the microwave radiation at the zero field frequency ($\nu_0$) in the detector to produce a beat frequency corresponding to the Stark shift of 1 MHz. According to the results in the next section, the amplitude of the emission decays with the relaxation time $T_2$.

In spite of some claims that new effects are being found, the basic time dependence in microwave nutation as described here is identical to the corresponding time dependence observed in the analogous NMR experiments which were explained 20 years earlier. Furthermore, we point out that transient absorption or emission experiments as described here and in the next sections enable only a measurement of $T_1$ and $T_2$. Thus no new information about molecular relaxation is won from transient experiments that cannot in principle be measured by steady-state line width experiments as described in Section III. Any information present in the time domain must also he present in the frequency domain. As pointed out earlier, however, there are reasons involving sensitivity and ease of measurement that may make transient experiments useful to pursue.

## V. TRANSIENT EMISSION

An initial polarization in a two-level system may be obtained by irradiating with a coherent source. If the radiation is applied for an appropriate length of time and then switched off, the molecules emit. This is the electric dipole analogue of free induction decay in NMR.[6] This coherent emission is sometimes called superradiance, though technically superradiance is a purely quantum phenomenon not predicted by the present semiclassical theory. It was first discussed by Dicke.[28]

Let the stimulating radiation of frequency $\omega$ be switched off at time $t = t_1$. At this instant the polarization of the molecules is given from (16) by

$$P(t_1) = [P_r(t_1) + iP_i(t_1)]e^{i(\omega t_1 - kz)} + \text{c.c.} \tag{129}$$

The values of $P_r(t_1)$ and $P_i(t_1)$ may be determined by solving (46) for $t \leqslant t_1$. We now determine the time development of $P_r(t)$ and $P_i(t)$ for $t > t_1$. When $\mathscr{E} = 0$ (46) takes the form

$$\begin{aligned} \frac{dP_r}{dt} + \Delta\omega P_i + \frac{P_r}{T_2} &= 0 \\ \frac{dP_i}{dt} - \Delta\omega P_r + \frac{P_i}{T_2} &= 0 \\ \frac{d\,\Delta N}{dt} + \frac{\Delta N - \Delta N_0}{T_1} &= 0 \end{aligned} \tag{130}$$

As shown in (A1.38), these equations have the general solutions

$$\begin{aligned} P_r &= e^{-t/T_2}(A \cos \Delta\omega t - B \sin \Delta\omega t) \\ P_i &= e^{-t/T_2}(B \cos \Delta\omega t + A \sin \Delta\omega t) \\ \Delta N &= \Delta N_0(1 + C)e^{-t/T_1} \end{aligned} \tag{131}$$

The constants $A$, $B$, and $C$ are determined by the boundary conditions at $t = t_1$. Using these boundary conditions gives

$$\begin{aligned} P_r &= e^{-(t-t_1)/T_2}\{P_r(t_1) \cos [\Delta\omega(t - t_1)] - P_i(t_1) \sin [\Delta\omega(t - t_1)]\} \\ P_i &= e^{-(t-t_1)/T_2}\{P_i(t_1) \cos [\Delta\omega(t - t_1)] + P_r(t_1) \sin [\Delta\omega(t - t_1)]\} \\ N &= \Delta N_0 + (\Delta N(t_1) - \Delta N_0)e^{-(t-t_1)/T_1} \end{aligned} \tag{132}$$

From (16) the polarization at time $t \geq t_1$ is

$$P(t) = 2[P_r \cos (\omega t - kz) - P_i \sin (\omega t - kz)] \tag{133}$$

where $P_r$ and $P_i$ are given by (132).

Let the emitted field have the form

$$E(t) = 2[\mathscr{E}_i \cos (\omega t - kz) + \mathscr{E}_r \sin (\omega t - kz)] \tag{134}$$

Following through an analysis similar to that in (29) to (32) we find, neglecting the terms $(1/c)(\partial\mathscr{E}_i/\partial t)$ and $(1/c)(\partial\mathscr{E}_r/\partial t)$,

$$\begin{aligned} \frac{\partial\mathscr{E}_i}{\partial z} &= \frac{2\pi\omega}{c} P_i \\ \frac{\partial\mathscr{E}_r}{\partial z} &= \frac{2\pi\omega}{c} P_r \end{aligned} \tag{135}$$

If $l$ is the length of the emitting sample, then integration of (135) from 0 to $l$ yields the result

$$\mathscr{E}_i = \frac{2\pi\omega l}{c} P_i$$
$$\mathscr{E}_r = \frac{2\pi\omega l}{c} P_r \tag{136}$$

Combining equations (132), (134), and (136) we find for the emitted field the result

$$E(t) = \frac{4\pi\omega l}{c} e^{-(t-t_1)/T_2}\{P_i(t_1)\cos[(\omega + \Delta\omega)(t - t_1) - kz + \omega t_1] + P_r(t_1)\sin[(\omega + \Delta\omega)(t - t_1) - kz + \omega t_1]\} \tag{137}$$

Substituting into this equation the velocity shifted $\Delta\omega$,

$$\Delta\omega = \omega_0 - \omega\left(1 - \frac{v}{c}\right)$$

we have for molecules moving with velocity $v$

$$E(v, t) = \frac{4\pi\omega l}{c} e^{-(t-t_1)/T_2}\left\{P_i(t_1)\cos\left[\left(\omega_0 + \frac{\omega v}{c}\right)(t - t_1) - kz + \omega t_1\right] + P_r(t_1)\sin\left[\left(\omega_0 + \frac{\omega v}{c}\right)(t - t_1) - kz + \omega t_1\right]\right\} \tag{138}$$

From this equation we see that the molecules emit coherent radiation at frequency $(\omega_0 - \omega v/c)$. The emission decays with the relaxation time $T_2$. The average power emitted, $W$, is proportional to the time average of $E^2$. From (138) we see that $W$ is proportional to $[P_i^2(t_1) + P_r^2(t_1)]$ and hence to the number density squared of molecules. This is in marked contrast to spontaneous emission where the emitted power is proportional to the first power of the number density of molecules present.

Thus far we have considered only the behavior of the system after the saturating power is turned off. We now discuss the preparation of the system prior to $t_1$. Let us first examine the case where prior to $t_1$ the molecules are in equilibrium with radiation of frequency $\omega$ and amplitude $2\mathscr{E}$. For simplicity we take $t_1 = 0$. The Fourier components of the polarization at $t = 0$, $P_r(0)$ and $P_i(0)$, are just the steady-state values given in

(84). Substituting these values for $P_r(0)$ and $P_i(0)$ into (138) and integrating over the velocity distribution function gives for the emission after $t = 0$

$$E(t) = -\frac{\pi^{1/2}\hbar\kappa\omega l\,\Delta N_0 q}{c}\, e^{-t/T_2}$$

$$\times \int_{-\infty}^{\infty} e^{-q^2\omega'^2}\left\{\frac{(\kappa\mathscr{E}/T_2)\cos[(\omega_0+\omega')t - kz]}{(1/T_2^2) + (\omega_0 - \omega + \omega')^2 + (T_1/T_2)\kappa^2\mathscr{E}^2}\right.$$

$$\left. - \frac{\kappa\mathscr{E}(\omega_0 - \omega + \omega')\sin[(\omega_0+\omega')t - kz]}{(1/T_2^2) + (\omega_0 - \omega + \omega')^2 + (T_1/T_2)\kappa^2\mathscr{E}^2}\right\} d\omega' \tag{139}$$

The integral in (139) may be approximated in closed form for two special cases. In the Doppler broadened limit where

$$\left[\frac{1}{T_2^{\,2}} + \frac{T_1}{T_2}\kappa^2\mathscr{E}^2\right]^{1/2} \ll \Delta\omega_D \tag{140}$$

the term $e^{-q^2\omega'^2}$ is slowly varying and so may be taken outside the integral to give

$$E(t) = -\frac{\pi^{1/2}\hbar\kappa\omega l\,\Delta N_0 q}{c}\, e^{-t/T_2} e^{-q^2(\omega-\omega_0)^2}$$

$$\times \int_{-\infty}^{\infty}\left\{\frac{(\kappa\mathscr{E}/T_2)\cos[(\omega_0+\omega')t - kz]}{(1/T_2^2) + (\omega_0 - \omega + \omega')^2 + (T_1/T_2)\kappa^2\mathscr{E}^2}\right.$$

$$\left. - \frac{\kappa\mathscr{E}(\omega_0 - \omega + \omega')\sin[(\omega_0+\omega')t - kz]}{(1/T_2^2) + (\omega_0 - \omega + \omega')^2 + (T_1/T_2)\kappa^2\mathscr{E}^2}\right\} d\omega \tag{141}$$

The integral in (141) may be evaluated by contour integration to give

$$E(t) = \frac{\pi^{3/2}\hbar\kappa^2\mathscr{E}\omega l\,\Delta N_0 q}{c}\, e^{-q^2(\omega-\omega_0)^2}[1 - (1 + T_1T_2\kappa^2\mathscr{E}^2)^{-1/2}]$$

$$\times e^{-t/T_2}\exp\left\{-\left(\frac{1}{T_2^{\,2}} + \frac{T_1}{T_2}\kappa^2\mathscr{E}^2\right)^{1/2} t\right\}\cos(\omega t - kz) \tag{142}$$

In the pressure-power broadened limit where

$$\left[\frac{1}{T_2^{\,2}} + \frac{T_1}{T_2}\kappa^2\mathscr{E}^2\right]^{1/2} \gg \Delta\omega_D \tag{143}$$

we can approximate (139) by

$$E(t) = -\frac{\pi^{1/2}\hbar\kappa\omega l\,\Delta N_0 q}{c}\,e^{-t/T_2}$$
$$\times\left\{\frac{(\kappa\mathscr{E}/T_2)}{(1/T_2^2) + (\omega_0-\omega)^2 + (T_1/T_2)\kappa^2\mathscr{E}^2}\right.$$
$$\times\int_{-\infty}^{\infty} e^{-q^2\omega'^2}\cos\,[(\omega_0+\omega')t - kz]\,d\omega'$$
$$-\frac{\kappa\mathscr{E}(\omega_0-\omega)}{(1/T_2^2) + (\omega_0-\omega)^2 + (T_1/T_2)\kappa^2\mathscr{E}^2}$$
$$\left.\times\int_{-\infty}^{\infty} e^{-q^2\omega'^2}\sin\,[(\omega_0+\omega')t - kz]\,d\omega'\right\} \quad (144)$$

The first integral in (144) is of a form evaluated in Appendix A2. From (A2.4) we have the result

$$\int_{-\infty}^{\infty} e^{-q^2\omega'^2}\cos\,[(\omega_0+\omega')t - kz]\,d\omega' = \frac{\pi^{1/2}}{q}\,e^{-t^2/4q^2}\cos\,(\omega_0 t - kz) \quad (145)$$

The second integral in (144) may be evaluated using similar techniques with the result

$$\int_{-\infty}^{\infty} e^{-q^2\omega'^2}\sin\,[(\omega_0+\omega')t - kz]\,d\omega' = \frac{\pi^{1/2}}{q}\,e^{-t^2/4q^2}\sin\,(\omega_0 t - kz) \quad (146)$$

Substituting (145) and (146) into (144) yields the result

$$E(t) = -\frac{\pi\hbar\kappa^2\mathscr{E}\omega l\,\Delta N_0}{c}\,e^{-t/T_2}e^{-t^2/4q^2}$$
$$\times\left[\frac{(1/T_2)\cos\,(\omega_0 t - kz)}{(1/T_2)^2 + (\omega_0-\omega)^2 + (T_1/T_2)\kappa^2\mathscr{E}^2}\right.$$
$$\left.-\frac{(\omega_0-\omega)\sin\,(\omega_0 t - kz)}{(1/T_2^2) + (\omega_0-\omega)^2 + (T_1/T_2)\kappa^2\mathscr{E}^2}\right] \quad (147)$$

From (142) we see that the coherent emission in the Doppler limit, (140), decays as

$$e^{-t/T_2}\exp\left\{-\left[\left(\frac{1}{T_2}\right)^2 + \frac{T_1}{T_2}\kappa^2\mathscr{E}^2\right]^{1/2} t\right\}$$

with a power dependent relaxation time. Equation (147) shows that in the pressure-power broadened limit, (143), the emission decays as

$$e^{-(t/T_2+t^2/4q^2)}.$$

In the pressure broadened limit $q$ is much greater than $T_2$ and the decay is determined by pure $T_2$. In the power broadened limit, $\kappa\mathscr{E} \gg \Delta\omega_D$, if $q \ll T_2$ the decay is determined by the Doppler width.

Equation (142) also contains a factor of $\cos(\omega t - kz)$ where $\omega$ is the frequency of the polarizing radiation, while (147) contains a factor of $\cos(\omega_0 t - kz)$ and $\sin(\omega_0 t - kz)$ where $\omega_0$ is the molecular transition frequency. However, the power emitted is proportional to the average of the square of $E(t)$ and hence to the average of $\cos^2(\omega t)$ or $\cos^2(\omega_0 t)$ and $\sin^2(\omega_0 t)$. Both $\omega$ and $\omega_0$ are very large numbers so that these factors are rapidly oscillating. They can therefore be replaced by the average value of $\frac{1}{2}$. The decay of the emission signal is then a pure exponential in either the pressure or the Doppler broadened limits. Notice that the emitted power decays with a decay constant twice that of the emitted field.

A more common experiment is to observe the beat of the pure emission signal with some other external frequency source. This produces oscillatory behavior similar in appearance to that obtained in the nutation experiment. We emphasize here that free induction decay oscillations are qualitatively different from the transient absorption experiments (nutation), resulting from nothing more than the interference between two frequencies; the emitted and reference frequencies mix in a nonlinear detector to produce the beat frequency. Using the emitted fields from the Doppler limit in (142), or the pressure broadened limit in (147) we write the intensity of the heterodyned signal between these emitted fields $E(t)$ and the field of a reference oscillator $E_{\text{ref}}(t) = 2E_{\text{ref}} \cos \omega_r t$ as

$$I(t) = \mathscr{B} E_{\text{ref}}(t) E(t) \tag{148}$$

where $\mathscr{B}$ is the conversion efficiency or gain of the detector. It is quite evident from this equation and from (142) and (147) that a nonlinear detector will give rise to the sum and difference (beat) frequencies between the emitter [(142) or (147)] and the reference.

One simple way of achieving a beat between the emitting radiation and a reference is to employ the Stark field to change the frequency of the emitted radiation. In this experiment the zero-field transition is saturated with high-power resonant radiation. Then the Stark field is turned on, shifting the energy levels suddenly so that the original radiation is now off-resonance. The molecules now emit their radiation at a resonant frequency dependent upon the Stark field and the emitted radiation mixes with the

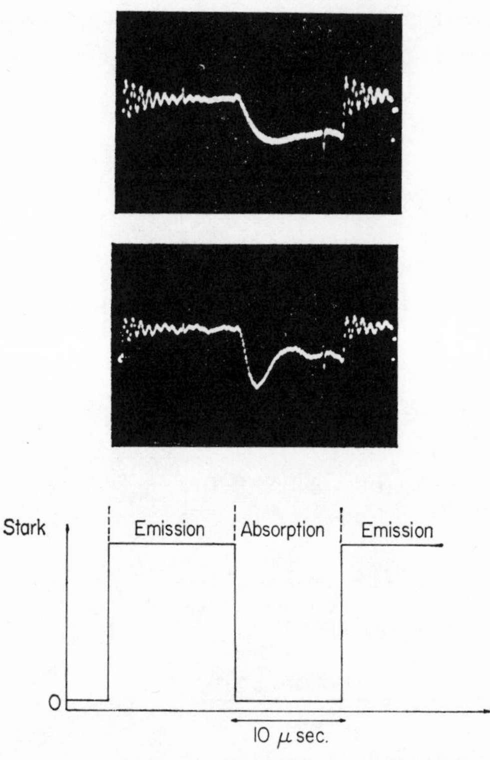


**Fig. 3.** Diagram and oscilloscope photographs showing transient absorption (field off) and transient emission (field on) as obtained from Stark switching. The lower photograph is taken at a higher incident power and the transient absorption is starting to demonstrate the oscillatory behavior which is so evident in Fig. 2.

original perturbing radiation that is still present. The beat frequency is directly proportional to the magnitude of the Stark effect. Figure 3 illustrates this discussion.

We now discuss the preparation of the system prior to $t_1$ by the application of pulses of radiation of duration short compared to the relaxation times $T_1$ and $T_2$. When $t \ll T_1$ or $T_2$, (46) takes the form

$$\begin{aligned} \frac{dP_r}{dt} + \Delta\omega P_i &= 0 \\ \frac{dP_i}{dt} - \Delta\omega P_r + \kappa^2 \mathscr{E} \frac{\hbar\,\Delta N}{4} &= 0 \\ \frac{d}{dt}\frac{\hbar\,\Delta N}{4} - \mathscr{E} P_i &= 0 \end{aligned} \tag{149}$$

The solution to these equations is derived in Appendix A1. From (A1.44) to (A1.47) we have

$$P_r = \left[\kappa^2\mathscr{E}^2 P_r(0) + \frac{\hbar}{4}\kappa^2\mathscr{E}\,\Delta\omega\,\Delta N(0)\right]\frac{1}{\Omega^2} + \left[(\Delta\omega)^2 P_r(0) - \frac{\hbar}{4}\kappa^2\mathscr{E}\,\Delta\omega\,\Delta N(0)\right]\frac{\cos\Omega t}{\Omega^2} - \Delta\omega P_i(0)\frac{\sin\Omega t}{\Omega}$$

$$P_i = P_i(0)\cos\Omega t + \left[\Delta\omega P_r(0) - \frac{\hbar\kappa^2\mathscr{E}}{4}\Delta N(0)\right]\frac{\sin\Omega t}{\Omega}$$

$$\frac{\hbar\kappa}{4}\Delta N = \left[\frac{\hbar\kappa}{4}(\Delta\omega)^2\,\Delta N(0) + \kappa\mathscr{E}\,\Delta\omega P_r(0)\right]\frac{1}{\Omega^2} + \left[\frac{\hbar}{4}\kappa^3\mathscr{E}^2\,\Delta N(0) - \kappa\mathscr{E}\,\Delta\omega P_r(0)\right]\frac{\cos\Omega t}{\Omega^2} + \kappa\mathscr{E}P_i(0)\frac{\sin\Omega t}{\Omega} \tag{150}$$

where

$$\Omega = [\kappa^2\mathscr{E}^2 + (\Delta\omega)^2]^{1/2}$$

When $P_r(0) = P_i(0) = 0$ we have from (150)

$$P_r(t) = \frac{\hbar\kappa^2\mathscr{E}\,\Delta\omega\,\Delta N(0)}{4\Omega^2}(1 - \cos\Omega t)$$

$$P_i(t) = -\frac{\hbar\kappa^2\mathscr{E}}{4}\Delta N(0)\frac{\sin\Omega t}{\Omega} \tag{151}$$

$$\Delta N(t) = \frac{\Delta N(0)}{\Omega^2}[(\Delta\omega)^2 + \kappa^2\mathscr{E}^2\cos\Omega t]$$

When Doppler broadening is unimportant we may obtain an expression for the field emitted following a pulse of duration $t_1$ by substituting (151) into (137). Writing $\omega + \Delta\omega = \omega_0$ we find

$$E(t) = -\frac{\pi\hbar\kappa^2\mathscr{E}\omega l\,\Delta N_0}{c\Omega}e^{-(t-t_1)/T_2}\left\{\sin\Omega t_1\cos[\omega_0(t - t_1) - kz + \omega t_1 - \frac{\Delta\omega}{\Omega}(1 - \cos\Omega t_1)\sin[\omega_0(t - t_1) - kz + \omega t_1]\right\} \tag{152}$$

From (152) when $\Delta\omega = 0$ the emitted field is a maximum following a $\pi/2$ pulse defined by

$$\Omega t = \frac{\pi}{2} \tag{153}$$

and is zero following a $\pi$ pulse defined by

$$\Omega t = \pi \tag{154}$$

In NMR $\pi/2$ and $\pi$ pulses are familiar concepts. From (151) when $\Delta\omega = 0$ we note that the population difference is zero following a $\pi/2$ pulse and is inverted following a $\pi$ pulse.

$$\begin{aligned} \Delta N\left(\Omega t = \frac{\pi}{2}\right) &= 0 \\ \Delta N(\Omega t = \pi) &= -\Delta N(0) \end{aligned} \tag{155}$$

The polarization following nonresonant pulses may be worked out provided that the power of the pulse is sufficiently high. We require

$$\kappa\mathscr{E} \gg \Delta\omega \tag{156}$$

where, when Doppler broadening is important, this carries with it the requirement

$$\kappa\mathscr{E} \gg \Delta\omega_D$$

Then we have

$$\Omega \approx \kappa\mathscr{E} \tag{157}$$

and (150) takes the form

$$\begin{aligned} P_r(t) &\approx P_r(0) \\ P_i(t) &\approx P_i(0) \cos \kappa\mathscr{E} t - \frac{\hbar\kappa}{4} \Delta N(0) \sin \kappa\mathscr{E} t \\ \Delta N(t) &\approx \Delta N(0) \cos \kappa\mathscr{E} t \end{aligned} \tag{158}$$

From (158) we have the following results for $\pi/2$ and $\pi$ pulses.
*$\pi/2$ pulse:*

$$\begin{aligned} t_{\pi/2} &= \frac{\pi}{2\kappa\mathscr{E}} \\ P_r(t_{\pi/2}) &\approx P_r(0) \\ P_i(t_{\pi/2}) &\approx -\frac{\hbar\kappa}{4} \Delta N(0) \\ \Delta N(t_{\pi/2}) &\approx 0 \end{aligned} \tag{159}$$

*$\pi$ pulse:*

$$\begin{aligned} t_\pi &= \frac{\pi}{\kappa \mathscr{E}} \\ P_r(t_\pi) &\approx P_r(0) \\ P_i(t_\pi) &\approx -P_i(0) \\ \Delta N(t_\pi) &\approx -\Delta N(0) \end{aligned} \tag{160}$$

Equations (159) and (160) together with (138) again show that when $P_r(0) = P_i(0) = 0$ the emission is a maximum following a $\pi/2$ pulse and zero following a $\pi$ pulse.

Free induction decay has been reported in the literature using pulse times both long and short relative to the relaxation time. In these experiments a Stark pulse or some other perturbation was used to shift a line suddenly out of resonance. Brewer and Shoemaker[29] have observed free induction decay in the infrared region (Doppler limit). Free induction decay has been observed in the microwave region (pressure broadening limit) using short pulses by Dicke and Romer[30] and by Hill et al.[31] The dependence of the free induction decay on pulse length has been illustrated[21] and discussed[32] in the work of Mache and Glorieux and by

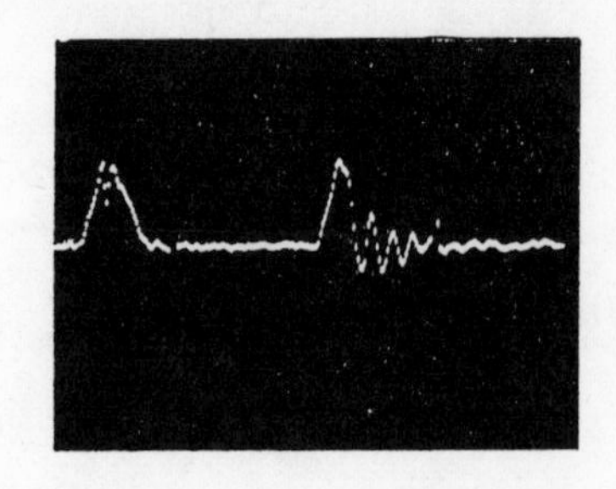

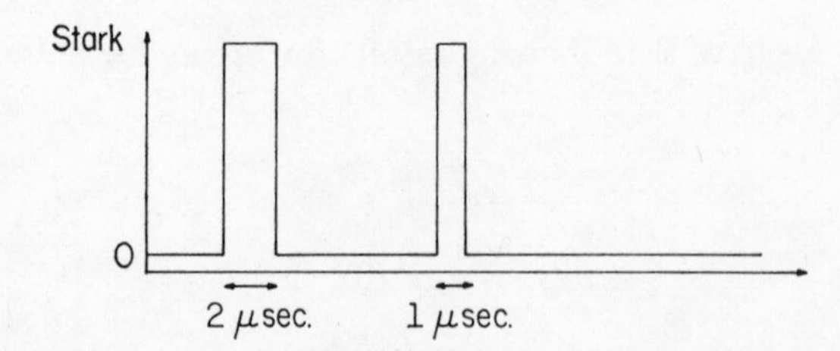


**Fig. 4.** The effects following resonant stimulation of the $J = 0 \rightarrow 1$ transition in OCS with $\pi$ (2 $\mu$sec) and $\pi/2$ (1 $\mu$sec) pulses. There is no coherent emission following the $\pi$ pulse. Coherent emission is readily observed following the $\pi/2$ pulse.

McGurk et al.[24] A good approximation to the $\pi/2$ and $\pi$ pulses may be obtained experimentally in a microwave absorption cell. Short pulses of radiation were applied by McGurk et al.[24] to the molecules by tuning the microwave frequency into resonance with the Stark shifted transition and then applying square wave Stark pulses of variable duration using the pulse generator. Figure 4 shows the emission following 2 and 1 $\mu$sec pulses exciting the $J = 0 \rightarrow 1$ transition in OCS at a pressure of 5 mtorr. The microwave power was adjusted to minimize the emission following the 2 $\mu$sec pulse. The 2 $\mu$sec pulse approximates a $\pi$ pulse with very weak emission, while the 1 $\mu$sec pulse approximates a $\pi/2$ pulse with strong emission.

## VI. MULTIPLE PULSE EXPERIMENTS

In this section we start by discussing pulse-echo experiments that can lead to rapid, accurate, and independent measurements of $T_1$ and $T_2$. Following this we develop the equations necessary to understand pulse-induced Fourier transform spectroscopy.

### A. $\pi,\pi/2$ and $\pi/2,\pi$ Pulse Trains and the Measurement of $T_1$ and $T_2$

We first investigate the emission following the application of a $\pi/2$ pulse at time $t = 0$ of duration $t_1$ followed at time $t = t_2$ by a $\pi$ pulse of duration $2t_1$. For molecules moving with velocity $v$ under the conditions of (156) we have from (132), (159), and (160) the following results:

$t = 0$

$$P_r(0) = P_i(0) = 0\,, \qquad \Delta N(0) = \Delta N_0 \tag{161}$$

$t = t_1$

$$P_i(t_1) = -\frac{\hbar\kappa}{4}\,\Delta N_0 \tag{162}$$

$t_1 \leqslant t \leqslant t_2$

$$P_i(\omega', t) = e^{-(t-t_1)/T_2} P_i(t_1) \cos\left[(\omega_0 - \omega + \omega')(t - t_1)\right] \tag{163}$$

$t = t_2 + 2t_1$

$$P_i(\omega', t_2 + 2t_1) = P_i(\omega', t_2) \tag{164}$$

where from now on we use the variable $\omega' = \omega v/c$. From (158) we see that $P_r(t)$ will remain zero for all $t \geqslant 0$, so we do not discuss it further. Substituting the value of $P_i(\omega', t_2)$ obtained from (163) into (164) we have

$$P_i(\omega', t_2 + 2t_1) = e^{-(t_2-t_1)/T_2} P_i(t_1) \cos\left[(\omega_0 - \omega + \omega')(t_2 - t_1)\right] \tag{165}$$

From (138) and (165) the emission is given by

$$E(\omega', t) = \frac{4\pi\omega l}{c} e^{-(t-t_1)/T_2} P_i(t_1) \cos [(\omega_0 + \omega')(t - t_1) - kz + \omega t_1] \tag{166}$$

for $t_1 \leqslant t \leqslant t_2$ and for $t \geqslant t_2 + 2t_1$

$$E(\omega', t) = \frac{4\pi\omega l}{c} e^{-(t-3t_1)/T_2} P_i(t_1) \cos [(\omega_0 - \omega + \omega')(t_2 - t_1)] \times \cos [(\omega_0 + \omega')(t - t_2 - 2t_1) - kz + \omega(t_2 + 2t_1)] \tag{167}$$

Integrating (166) over the velocity distribution function, we obtain

$$E(t) = \frac{4\pi^{1/2}\omega l q}{c} e^{-(t-t_1)/T_2} P_i(t_1) I \tag{168}$$

where the integral $I$ is of the form

$$I = \int_{-\infty}^{\infty} e^{-q^2\omega'^2} \cos [(\omega_0 + \omega')\tau + \varphi] \, d\omega' \tag{169}$$

where

$$\tau = t - t_1 \tag{170}$$

and

$$\varphi = \omega t_1 - kz \tag{171}$$

This integral is of a form evaluated in Appendix A2. We have from (A2.4)

$$I = \frac{\pi^{1/2}}{q} e^{-\tau^2/4q^2} \cos (\omega_0 \tau + \varphi) \tag{172}$$

Combining (168), (169), (170), (171), and (172) we have for the emitted radiation

$$E(t) = \frac{4\pi\omega l}{c} e^{-(t-t_1)/T_2} e^{-(t-t_1)^2/4q^2} P_i(t_1) \cos [\omega_0(t - t_1) - kz + \omega t_1] \tag{173}$$

On substitution of $P_i(t_1)$ from (162) we have the result for $t_1 \leqslant t \leqslant t_2$

$$E(t) = -\frac{\pi\hbar\kappa\omega l}{c} \Delta N_0 e^{-(t-t_1)/T_2} e^{-(t-t_1)^2/4q^2} \cos [\omega_0(t - t_1) - kz + \omega t_1] \tag{174}$$

Using the identity

$$2 \cos A \cos B = \cos (A + B) + \cos (A - B) \tag{175}$$

we can rewrite (167) in the form

$$E(\omega', t) = \frac{2\pi\omega l}{c} e^{-(t-3t_1)/T_2} P_i(t_1)$$
$$\times \{\cos [\omega'(t - 3t_1) + \omega_0(t - 3t_1) + 3\omega t_1 - kz]$$
$$+ \cos [\omega'(t - 2t_2 - t_1) + \omega_0(t - 2t_2 - t_1) + \omega(2t_2 + t_1) - kz]\} \quad (176)$$

Integrating this equation over the velocity distribution function we have

$$E(t) = \frac{2\pi^{1/2}\omega l q}{c} e^{-(t-3t_1)/T_2} P_i(t_1)[I_1 + I_2] \quad (177)$$

with the integrals $I_1$ and $I_2$ given by the equations

$$I_1 = \int_{-\infty}^{\infty} e^{-q^2\omega'^2} \cos [\omega'(t - 3t_1) + \varphi_1] \quad (178)$$

where

$$\varphi_1 = [\omega_0(t - 3t_1) + 3\omega t_1 - kz] \quad (179)$$

and

$$I_2 = \int_{-\infty}^{\infty} e^{-q^2\omega'^2} \cos [\omega'(t - 2t_2 - t_1) + \varphi_2] \quad (180)$$

where

$$\varphi_2 = [\omega_0(t - 2t_2 - t_1) + \omega(2t_2 + t_1) - kz] \quad (181)$$

The integrals $I_1$ and $I_2$ are of a form that is evaluated in Appendix A2. Comparison with (A2.4) yields the results

$$I_1 = \frac{\pi^{1/2}}{q} e^{-(t-3t_1)^2/4q^2} \cos \varphi_1 \quad (182)$$

and

$$I_2 = \frac{\pi^{1/2}}{q} e^{-(t-t_1-2t_2)^2/4q^2} \cos \varphi_2 \quad (183)$$

Substituting (162), (179), (181), (182), and (183) into (177) yields the final result for $t \geqslant t_2 + 2t_1$:

$$E(t) = -\frac{\pi\hbar\kappa\omega l \,\Delta N_0}{2c} e^{-(t-3t_1)/T_2}$$
$$\times \{e^{-(t-3t_1)^2/4q^2} \cos [\omega_0(t - 3t_1) + 3\omega t_1 - kz]$$
$$+ e^{-(t-t_1-2t_2)^2/4q^2} \cos [\omega_0(t - 2t_2 - t_1) + \omega(2t_2 + t_1) - kz]\} \quad (184)$$

If Doppler broadening is dominant, $E(t)$ at time $t = 2t_2 + t_1$ has the approximate form

$$E(t) = -\frac{\pi\hbar\kappa\omega l\,\Delta N_0}{2c}\, e^{-2(t_2-t_1)/T_2} \cos\,[\omega(2t_2 + t_1) - kz] \tag{185}$$

since the first term in (184) has become negligibly small with respect to the second if $q$ is sufficiently small.

Consider now the situation where at time $t = 0$ a $\pi$ pulse is applied to the system of duration $2t_1$ followed at time $t_2$ by a $\pi/2$ pulse. For molecules moving with velocity $v$ under the conditions of (156) we have from (132), (150), (159), and (160) the following results:

$t = 0$

$$P_r(0) = P_i(0) = 0\,, \qquad \Delta N(0) = \Delta N_0 \tag{186}$$

$t = 2t_1$

$$P_i(2t_1) = 0 \tag{187}$$

$$\Delta N(2t_1) = -\Delta N_0 \tag{188}$$

$2t_1 \leqslant t \leqslant t_2$

$$\frac{\Delta N(t)}{\Delta N_0} = 1 - 2e^{-(t-2t_1)/T_1} \tag{189}$$

$t = t_2$

$$P_i(t_2) = 0 \tag{190}$$

$$\frac{\Delta N(t_2)}{\Delta N_0} = 1 - 2e^{-(t_2-2t_1)/T_1} \tag{191}$$

$t = t_2 + t_1$

$$P_i(t_2 + t_1) = -\frac{\hbar\kappa\,\Delta N(t_2)}{4} \tag{192}$$

$P_r(t)$ again, of course, remains zero. Substituting (191) into (192) we have

$$P_i(t_2 + t_1) = -\frac{\hbar\kappa}{4}\,\Delta N_0[1 - 2e^{-(t_2-2t_1)/T_1}] \tag{193}$$

From (138), (187), and (193) the emission is given by

$2t_1 \leqslant t \leqslant t_2$

$$E(t) = 0 \tag{194}$$

$t \geqslant t_2 + t_1$

$$E(\omega', t) = -\frac{\pi\hbar\kappa\omega l\,\Delta N_0}{c}\, e^{-(t-t_2-t_1)/T_2} \cos\,[(\omega_0 + \omega')(t - t_2 - t_1) - kz + \omega(t_2 + t_1)]\{1 - 2e^{-(t_2-2t_1)/T_1}\} \tag{195}$$

where $\omega' = \omega v/c$.

Integrating over the velocity distribution function we have

$$E(t) = -\frac{\pi^{1/2}\hbar\kappa\omega l\,\Delta N_0 q}{c}\, e^{-(t-t_2-t_1)/T_2}[1 - 2e^{-(t_2-2t_1)/T_1}]I \qquad (196)$$

where $I$ is the integral defined by

$$I = \int_{-\infty}^{\infty} e^{-q^2\omega'^2} \cos\,[\omega'(t - t_2 - t_1) + \varphi]\, d\omega' \qquad (197)$$

where

$$\varphi = [\omega_0(t - t_2 - t_1) - kz + \omega(t_2 + t_1)] \qquad (198)$$

The integral in (197) is again of a form evaluated in Appendix A2. From (A2.4) we have

$$I = \frac{\pi^{1/2}}{q} \cos \varphi e^{-(t-t_2-t_1)^2/4q^2} \qquad (199)$$

Substituting (198) and (199) into (196) yields the result

$$E(t) = -\frac{\pi\hbar\kappa\omega l\,\Delta N_0}{c}\, e^{-(t-t_2-t_1)/T_2} e^{-(t-t_2-t_1)^2/4q^2}[1 - 2e^{-(t_2-2t_1)/T_1}]$$

$$\times \cos\,[\omega_0(t - t_2 - t_1) - kz + \omega(t_2 + t_1)] \qquad (200)$$

From (200) the field amplitude $\mathscr{E}$ at time $t = t_2 + t_1$ has the form

$$\mathscr{E}(t_2 + t_1) = \frac{\pi\hbar\kappa\omega l\,\Delta N_0}{c}\,[1 - 2e^{-(t_2-2t_1)/T_1}] \qquad (201)$$

The equations for the consecutive $\pi/2$, $\pi$ or $\pi$, $\pi/2$ pulses were derived for the case where

$$\kappa\mathscr{E} \gg \Delta\omega_D \qquad (202)$$

Although this is quite feasible in the microwave spectral region it is unrealistic for transitions in the infrared. When condition (202) is not satisfied it is not difficult to show that a reasonable approximation is to replace the term in the equations of the form

$$qe^{-t^2/4q^2} \cos\,(\omega_0 t + \varphi) \qquad (203)$$

by a term of the form

$$\frac{q\kappa\mathscr{E}}{(q^2\kappa^2\varepsilon^2 + 1)^{1/2}}\, e^{-[\kappa^2\mathscr{E}^2t^2/4(q^2\kappa^2\mathscr{E}^2+1)]} e^{-[q^2(\omega-\omega_0)^2/(q^2\kappa^2\mathscr{E}^2+1)]}$$

$$\times \cos\left[\left(\omega_0 + \frac{\omega - \omega_0}{q^2\kappa^2\mathscr{E}^2 + 1}\right)t + \varphi\right] \qquad (204)$$

To see this let us consider the emission following a $\pi/2$ pulse of duration $t_1$. For

$$P_r(0) = P_i(0) = 0\,, \qquad \Delta N(0) = \Delta N_0$$

we have from (150) for molecules moving with velocity $v$

$$P_i(t_1) = -\frac{\hbar\kappa}{4}\,\Delta N_0 \kappa \mathscr{E}\,\frac{\sin \Omega t_1}{\Omega} \tag{205a}$$

$$P_r(t_1) = -\frac{\hbar\kappa}{4}\,\Delta N_0 \kappa \mathscr{E}(\omega_0 - \omega + \omega') \cos \Omega t_1 \tag{205b}$$

where

$$\Omega = [\kappa^2 \mathscr{E}^2 + (\omega_0 - \omega + \omega')^2]^{1/2} \tag{206}$$

and for a $\pi/2$ pulse

$$\kappa \mathscr{E} t_1 = \frac{\pi}{2} \tag{207}$$

Again we have used the variable $\omega' = \omega v/c$. Let us now examine the function $\kappa\mathscr{E}(\sin \Omega t_1/\Omega)$. We have from (206) and (207)

*when* $\omega' = \omega - \omega_0$

$$\frac{\kappa \mathscr{E} \sin \Omega t_1}{\Omega} = 1 \tag{209}$$

*when* $\omega' \gg \omega - \omega_0 + \kappa\mathscr{E}$

$$\frac{\kappa \mathscr{E} \sin \Omega t_1}{\Omega} \approx \frac{\kappa \mathscr{E} \sin (\omega_0 - \omega + \omega')t_1}{(\omega_0 - \omega + \omega')} \tag{210}$$

In the integral over the velocity distribution function it is apparent that the function $\kappa\mathscr{E} \sin \Omega t_1/\Omega$ will behave in the same way as the function $e^{-(\omega_0-\omega+\omega')/\kappa^2\mathscr{E}^2}$. For we have

*when* $\omega' = \omega - \omega_0$

$$e^{-(\omega_0-\omega+\omega')^2/\kappa^2\mathscr{E}^2} = 1 \tag{211}$$

*and when* $\omega' \gg \omega - \omega_0 + \kappa\mathscr{E}$

$$e^{-(\omega_0-\omega+\omega')^2/\kappa^2\mathscr{E}^2} \approx 0 \tag{212}$$

As far as the integral is concerned the right-hand sides of (210) and (212) are equivalent since the rapidly oscillating sine function in (210) makes only a vanishingly small contribution to the integral, so we replace (205a) by

$$P_i(t_1) = -\frac{\hbar\kappa}{4}\,\Delta N_0 e^{-(\omega_0-\omega+\omega')^2/\kappa^2\mathscr{E}^2} \tag{213}$$

It may be shown using similar techniques that to the same approximation (205b) may be replaced by

$$P_r(t_1) = 0 \tag{214}$$

Substituting (213) and (214) into (138) and integrating over the velocity distribution function yields the result for the emission following the $\pi/2$ pulse

$$E(t) = -\frac{\pi^{1/2}\hbar\kappa\omega l\,\Delta N_0}{c}\, e^{-(t-t_1)/T_2} qI \tag{215}$$

where

$$I = \int_{-\infty}^{\infty} e^{-[q^2\omega'^2 + (\omega_0-\omega+\omega')^2/\kappa^2\mathscr{E}^2]} \cos\left[(\omega_0+\omega')(t-t_1)+\varphi\right] d\omega' \tag{216}$$

and

$$\varphi = (\omega t_1 - kz)$$

Completing the square in the exponent in (216) we have

$$I = \exp\left\{-\left[\frac{q^2(\omega-\omega_0)^2}{q^2\kappa^2\mathscr{E}^2+1}\right]\right\}\int_{-\infty}^{\infty}\exp\left\{-\left[\left(\frac{q^2\kappa^2\mathscr{E}^2+1}{\kappa^2\mathscr{E}^2}\right)\left(\omega' + \frac{(\omega_0-\omega)}{q^2\kappa^2\mathscr{E}^2+1}\right)^2\right]\right\} \times \cos\left[(\omega_0+\omega')(t-t_1)+\varphi\right] d\omega' \tag{217}$$

Making the substitution

$$\omega' + \frac{(\omega_0-\omega)}{q^2\kappa^2\mathscr{E}^2+1} = \bar{\omega} \tag{218}$$

gives

$$I = \exp\left\{-\left[\frac{q^2(\omega-\omega_0)^2}{q^2\kappa^2\mathscr{E}^2+1}\right]\right\}\int_{-\infty}^{\infty}\exp\left\{-\left[\left(\frac{q^2\kappa^2\mathscr{E}^2+1}{\kappa^2\mathscr{E}^2}\right)\bar{\omega}^2\right]\right\} \times \cos\left[\left(\omega_0 + \frac{\omega-\omega_0}{q^2\kappa^2\mathscr{E}^2+1} + \bar{\omega}\right)(t-t_1)+\varphi\right] d\bar{\omega} \tag{219}$$

The integral in (219) is of a form evaluated in (A2.1). From (A2.4) we have the result

$$I = \exp\left\{-\left[\frac{q^2(\omega-\omega_0)^2}{q^2\kappa^2\mathscr{E}^2+1}\right]\right\}\exp\left\{-\left[\frac{\kappa^2\mathscr{E}^2(t-t_1)^2}{4(q^2\kappa^2\mathscr{E}^2+1)}\right]\right\}\frac{\pi^{1/2}\kappa\mathscr{E}}{(q^2\kappa^2\mathscr{E}^2+1)^{1/2}} \times \cos\left[\left(\omega_0 + \frac{\omega-\omega_0}{q^2\kappa^2\mathscr{E}^2+1}\right)(t-t_1)+\varphi\right] \tag{220}$$

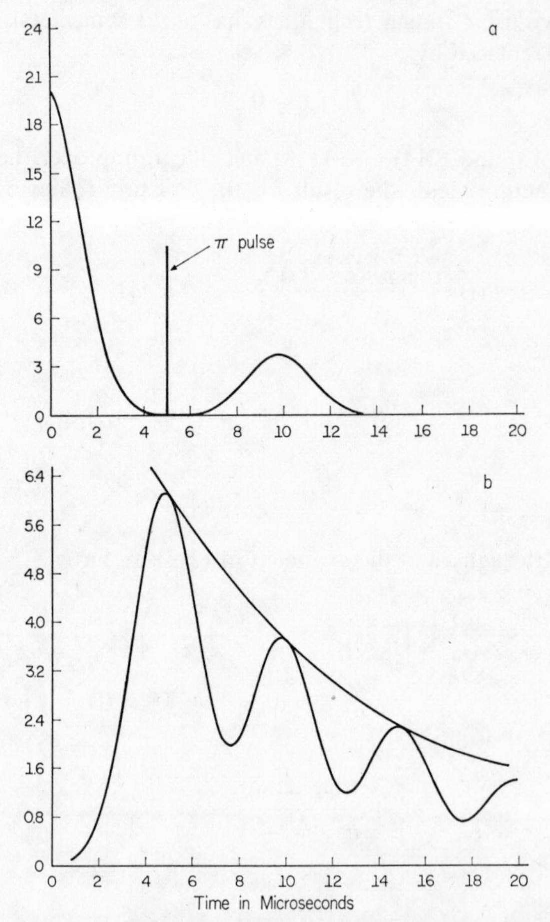


**Fig. 5.** Pulse echo: (*a*) Equations (174) and (184) are presented for a single pulse echo experiment. A $\pi/2$ pulse is applied at $t = 0$. At $t_2 = 5$ μsec (vertical line) a $\pi$ pulse is applied and an echo is observed at time $2t_2$. (*b*) The results of several such experiments ($t_2 = 2.5$, 5.0, 7.5, and 10.0 μsec) are plotted on the same axis. The decay envelope is $e^{-t/T_2}$.

Substituting (220) into (215) yields the final result

$$E(t) = -\frac{\pi\hbar\kappa\omega l\,\Delta N_0}{c} e^{-(t-t_1)/T_2}$$

$$\times \exp\left\{-\left[\frac{\kappa^2\mathscr{E}^2(t-t_1)^2}{4(q^2\kappa^2\mathscr{E}^2+1)}\right]\right\}\frac{\pi^{1/2}q\kappa\mathscr{E}}{(q^2\kappa^2\mathscr{E}^2+1)^{1/2}}\exp\left\{-\left[\frac{q^2(\omega-\omega_0)^2}{q^2\kappa^2\mathscr{E}^2+1}\right]\right\}$$

$$\times \cos\left[\left(\omega_0 + \frac{\omega-\omega_0}{q^2\kappa^2\mathscr{E}^2+1}\right)(t-t_1) - kz + \omega t_1\right] \quad (221)$$

Equation (221) is the same as would be obtained from (174) by making the substitution given in (203) and (204). Note that the decay time of the emission increases as $\kappa\mathscr{E}$ increases. An analysis similar to that given above shows again that there is no emission following a $\pi$ pulse. Hence the results for the $\pi/2$ and $\pi$ pulse trains remain valid.

We have now derived the equations for the electric resonance analogue of the NMR spin-echo experiments. They predict subsequent rises of the emission signal after periods of no emission. To our knowledge, no attempt has been made to observe this effect in microwave spectroscopy where (174) and (184) or (194) and (200) would be valid. It has, however, been observed in the infrared region.[35,18] Here one is saturating only a narrow band under the Doppler curve with a $\pi/2$ followed by a $\pi$ pulse. The appropriate theoretical description is then (174) and (184) as modified by the procedure demonstrated in (202) through (221).

The above equations suggest a very elegant method of observing $T_1$ and $T_2$ independently. Equation (185) shows that the intensity of the echo pulse in the $\pi/2$, $\pi$ case is a function of $t_2$, the time of the $\pi$ pulse. The intensity shows a decay constant of pure $T_2$. Therefore by repeating the echo experiment at several values of $t_2$ and plotting the decay envelope one may obtain an unambiguous measure of $T_2$.

Similarly (201) shows that the emission intensity in the $\pi$, $\pi/2$ experiment is a function of $t_2$. Its decay constant, however, is pure $T_1$. The two experiments together, then, provide a direct comparison of $T_1$ and $T_2$ in the same system.

Fig. 5*a* presents a graph of the expected results for the $\pi/2$, $\pi$ experiment. Fig. 5*b* shows the echoes from several values of $t_2$ on the same axis system with the decay envelope also drawn in.

### B. Pulse-Induced Fourier Transform Spectroscopy

At this point we find it instructive to return to (138). For simplicity we put $t_1 = 0$. Under conditions where $P_r(0) = 0$ and $P_i(0)$ is constant over the velocity distribution function the coherent emission for $t > 0$ is given by

$$E(t) = \frac{4\pi^{1/2}\omega lq}{c} P_i(0)e^{-t/T_2}\int_{-\infty}^{\infty} e^{-q^2\omega'^2} \cos\left[(\omega_0 + \omega')t - kz\right] d\omega' \quad (222)$$

Define the frequency spectrum $E(\omega_1)$ as the cosine Fourier transform of $E(t)$.

$$E(\omega_1) = \int_{-\infty}^{\infty} E(t) \cos \omega_1 t \, dt \quad (223)$$

Since $E(t) = 0$ for $t < 0$ we have on substituting (222) into (223)

$$E(\omega_1) = \frac{4\pi^{1/2}\omega lq}{c} P_i(0)\int_0^{\infty}\int_{-\infty}^{\infty} e^{-q^2\omega'^2} e^{-t/T_2} \times \cos\,[(\omega_0 + \omega')t + kz] \cos \omega_1 t \, d\omega' \, dt \quad (224)$$

Using the relation

$$2 \cos\,[(\omega_0 + \omega')t - kz] \cos \omega_1 t = \cos\,[(\omega_0 + \omega' + \omega_1)t - kz] + \cos\,[(\omega_0 + \omega' - \omega_1)t - kz] \quad (225)$$

and interchanging the order of integration in (224) we have

$$E(\omega_1) = \frac{2\pi^{1/2}\omega lq}{c} P_i(0)\int_{-\infty}^{\infty} e^{-q^2\omega'^2} \int_0^{\infty} e^{-t/T_2} \times \{\cos\,[(\omega_0 + \omega' + \omega_1)t - kz] + \cos\,[(\omega_0 + \omega' - \omega_1)t - kz]\} \, dt \, d\omega' \quad (226)$$

Rewriting (226) in the form

$$E(\omega_1) = \frac{\pi^{1/2}\omega lq}{c} P_i(0)\int_{-\infty}^{\infty} e^{-q^2\omega'^2}\left\{\int_0^{\infty} e^{-ikz}[e^{it[(\omega_0+\omega'+\omega_1)+i/T_2]} + e^{it[(\omega_0+\omega'-\omega_1)+i/T_2]}] \, dt + cc\right\} d\omega' \quad (227)$$

we have on carrying out the integral over $t$

$$E(\omega_1) = \frac{\pi^{1/2}\omega lq}{c} P_i(0)\int_{-\infty}^{\infty} e^{-q^2\omega'^2}\left\{e^{-ikz}\left[\frac{(1/T_2) + i(\omega_0 + \omega' + \omega_1)}{(1/T_2^2) + (\omega_0 + \omega' + \omega_1)^2} + \frac{(1/T_2) + i(\omega_0 + \omega' - \omega_1)}{1/T_2^2 + (\omega_0 + \omega' - \omega_1)^2}\right] + cc\right\} d\omega' \quad (228)$$

The integral in (228) is only large when $\omega_1 \approx \omega_0$. Then the first term in the integrand is negligible compared to the second and we can write

$$E(\omega_1) = \frac{2\pi^{1/2}\omega lq}{c} P_i(0)\left\{\cos kz \int_{-\infty}^{\infty} e^{-q^2\omega'^2} \frac{(1/T_2)\, d\omega'}{(1/T_2^2) + (\omega_0 - \omega_1 + \omega')^2} + \sin kz \int_{-\infty}^{\infty} e^{-q^2\omega'^2} \frac{(\omega_0 - \omega_1 + \omega')\, d\omega'}{(1/T_2^2) + (\omega_0 - \omega_1 + \omega')^2}\right\} \quad (229)$$

If we detect the emission at a distance $z$ such that

$$kz = n\pi \qquad n = 0, 1, 2, \cdots \tag{230}$$

then

$$E(\omega_1) = (-1)^n \frac{2\pi^{1/2}\omega lq}{c} P_i(0) \int_{-\infty}^{\infty} e^{-q^2\omega'^2} \frac{(1/T_2)}{(1/T_2)^2 + (\omega_0 - \omega_1 + \omega')^2} d\omega' \tag{231}$$

We thus have the important result that when $P_r(0) = 0$ and $P_i(0)$ is constant over the Doppler curve, the Fourier cosine transform of the coherent emission under condition (230) is proportional to the Doppler broadened absorption line shape function in the low power limit. If the emission is detected at a distance $z$ such that

$$kz = (n + \tfrac{1}{2})\pi \qquad n = 0, 1, 2, \ldots \tag{232}$$

then the Fourier cosine transform gives the dispersive line shape function

$$E(\omega_1) = (-1)^n \frac{2\pi^{1/2}\omega lq}{c} P_i(0) \int_{-\infty}^{\infty} e^{-q^2\omega'^2} \frac{(\omega_0 - \omega_1 + \omega')}{(1/T_2^2) + (\omega_0 - \omega_1 + \omega')^2} d\omega' \tag{233}$$

When more than one line is present, the coherent emission from all of them can be detected as a beat against a single reference frequency. From (222) the emitted field from the $j$th line is

$$E(t) = \frac{4\pi^{1/2}\omega lq_j}{c} [P_i(0)]_j e^{-t/T_{2_j}} \int_{-\infty}^{\infty} e^{-q^2\omega'^2} \cos [(\omega_j + \omega')t - k_j z] \, d\omega' \tag{234}$$

where $\omega_j$ is the $j$th molecular transition frequency. Let the field of the reference oscillator be

$$E_{\text{ref}}(t) = 2\mathscr{E}_{\text{ref}} \cos \omega_{\text{ref}} t \tag{235}$$

Then from (148) the total signal from a nonlinear detector when only the difference frequencies are retained is

$$I(t) = \frac{4\pi^{1/2}\omega l\beta\mathscr{E}_{\text{ref}}}{c} \sum_j q_j [P_i(0)]_j e^{-t/T_{2_j}} \int_{-\infty}^{\infty} e^{-q_j^2\omega'^2} \times \cos [(\omega_j - \omega_{\text{ref}} + \omega')t - k_j z] \, d\omega' \tag{236}$$

If the frequency spread of the transitions is not too large then condition

(232) can be satisfied for all $j$ and the cosine Fourier transform of (236) gives the frequency spectrum

$$I(\omega_1) = (-1)^n \frac{2\pi^{1/2}\omega l\beta\mathscr{E}_{\text{ref}}}{c} \sum_j q_j[P_i(0)]_j$$

$$\times \int e^{-q_j^2\omega'^2} \frac{(1/T_{2j})}{(1/T_{2j})^2 + (\omega_j - \omega_{\text{ref}} - \omega_1 + \omega')^2} d\omega' \quad (237)$$

which is just the sum of the Doppler broadened line shape functions for all $j$ transitions centered at the frequencies $(\omega_j - \omega_{\text{ref}})$.

The preceding result holds as long as the power is sufficiently high that $\kappa\mathscr{E} \gg \omega_j - \omega_p$ (where $\omega_p$ is the frequency of the polarizing radiation) for all the lines involved. If it is not, $P_i(t_1)$ will not be constant for each transition. The resulting spectrum will be distorted, with the distortion increasing as $\omega_p$ moves away from $\omega_j$.

This result also suggests an interesting experiment. It is well known in NMR Fourier transform spectroscopy that gathering data in the time domain can lead to an increase in signal to noise ratio in a spectral band which is proportional to the square root of the spectral band to line width ratio.[6,34] McGurk et al. have shown that the same thing is possible in microwave spectroscopy.[35] By saturating a multiplet and then taking the Fourier transform of the beat pattern of the decaying emission from the various lines, one reproduces the spectrum. The maximum emission is obtainable by saturating with a $\pi/2$ pulse. This method may prove valuable in resolving closely spaced microwave lines or in observing very weak lines.

In Fig. 6*a* we present an idealized two-line spectrum. Fig. 6*b* shows the decaying interference pattern expected when the emission from such a spectrum is mixed with a reference frequency with the important relationships indicated. This is a simple, but typical, example of the emission pattern in a Fourier transform experiment in microwave spectroscopy.

An alternative technique that would yield the same information without the power limitation is to perform the microwave analogue of the NMR fast passage experiment.[36] If the saturating frequency is not pulsed, but instead is swept through resonance in a time short compared to the relaxation time, there will again be an induced polarization. Experiments which are in fact microwave fast passage have been reported several times but incorrectly interpreted in terms of transient absorption theory.[22,27] McGurk et al. have recently developed microwave fast passage theory and the experimental verification as applied to Fourier transform spectroscopy.[36]

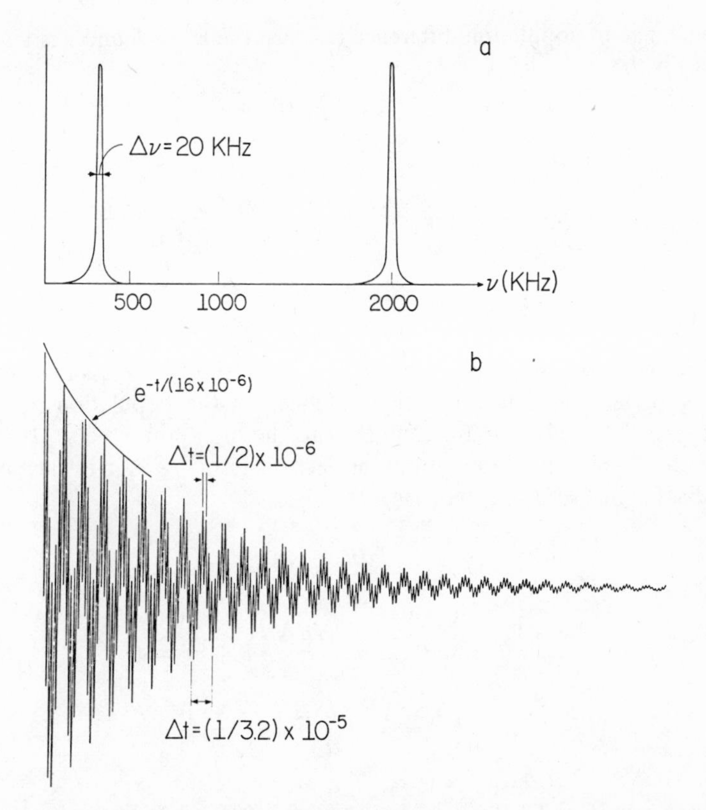


**Fig. 6.** Fourier transform microwave spectroscopy: (*a*) A two-line spectrum is given as a displacement from a reference frequency at zero. One transition is centred at 320 kHz and the second at 2000 kHz. (*b*) The emission signal when both lines are excited with a short pulse of radiation shows two superimposed oscillations with periods corresponding to the frequencies of the two lines relative to the reference. The damping is a function of the relaxation time. (*a*) and (*b*) are Fourier transforms of each other.

## VII. TRANSIENT DOUBLE RESONANCE

We now proceed to discuss some transient double resonance effects. Consider a three-level system designated in order of increasing energy by $a$, $b$, and $c$. We suppose that at time $t = 0$ a high-power pulse of radiation is applied to the transition $b \to c$. The effect of this pulse is monitored by a low-power signal source, constant in time, applied to the transition $a \to b$. We shall distinguish the variables involved in the pulse transition by the subscript $p$. For a square pulse of amplitude $2\mathscr{E}_p$ and duration $t_p$ such that

$$t_p \ll T_{1p}, T_{2p}$$

the change in population difference between the levels $b$ and $c$ is given by (150) with

$$\Delta N_p(0) = \Delta N_{0p}\,, \qquad P_{rp}(0) = P_{ip}(0) = 0$$

We have then

$$\frac{\Delta N_p}{\Delta N_{0p}} = \frac{(\Delta\omega_p)^2}{\kappa_p{}^2\mathscr{E}_p{}^2 + (\Delta\omega_p)^2} + \frac{\kappa_p{}^2\mathscr{E}_p{}^2 \cos \Omega t_p}{\kappa_p{}^2\mathscr{E}_p{}^2 + (\Delta\omega_p)^2} \tag{238}$$

where

$$\Omega = [\kappa_p{}^2\mathscr{E}_p{}^2 + (\Delta\omega_p)^2]^{1/2} \tag{239}$$

We must now solve (46) for the signal transition. We assume that the weak signal power has a negligible effect on the population difference between the levels $b$ and $c$ compared to the high-power pulse. It should then be a good approximation to neglect the term $-\mathscr{E}P_i$ in the last of (46) and solve instead the equations

$$\begin{aligned} \frac{dP_r}{dt} + \Delta\omega P_i + \frac{P_r}{T_2} &= 0 \\ \frac{dP_i}{dt} - \Delta\omega P_r + \kappa^2\mathscr{E}\left(\frac{\hbar\,\Delta N}{4}\right) + \frac{P_i}{T_2} &= 0 \\ \frac{d}{dt}\left(\frac{\hbar\,\Delta N}{4}\right) + \frac{\hbar}{4}\frac{(\Delta N - \Delta N_0)}{T_1} &= 0 \end{aligned} \tag{240}$$

It is shown in (A1.53) that the solution for $P_i$ is given by

$$\begin{aligned} \frac{P_i}{\Delta N_0} = &-\frac{\hbar\kappa^2\mathscr{E}(m_0 - 1)}{4} \\ &\times \left\{\frac{(1/T_2) - (1/T_1)}{[(1/T_2) - (1/T_1)]^2 + (\Delta\omega)^2}\,(e^{-t/T_1} - e^{-t/T_2}\cos \Delta\omega t) \right. \\ &\left. + \left[1 - \frac{[(1/T_2) - (1/T_1)]^2}{[(1/T_2) - (1/T_1)]^2 + (\Delta\omega)^2}\right] e^{-t/T_2}\frac{\sin \Delta\omega t}{\Delta\omega}\right\} \\ &- \frac{\hbar\kappa^2\mathscr{E}}{4}\,\frac{1/T_2}{(1/T_2^2) + (\Delta\omega)^2} \end{aligned} \tag{241}$$

where $m_0$ is the value of $\Delta N(0)/\Delta N_0$. It is easy to show, using (95), that

$$(m_0 - 1) = \frac{\Delta N_{0p}}{2\,\Delta N_0}\left(1 - \frac{\Delta N_p}{\Delta N_{0p}}\right) \tag{242}$$

since the three-level system is considered closed, which from (238) may be rewritten in the form

$$(m_0 - 1) = \frac{\Delta N_{0p}}{2\,\Delta N_0}\left[\frac{\kappa_p^2 \mathscr{E}_p^2}{\kappa_p^2 \mathscr{E}_p^2 + (\Delta\omega_p)^2}\right](1 - \cos \Omega t_p) \tag{243}$$

We shall at this point specialize to two different cases. In the first case we suppose that

$$\kappa_p \mathscr{E}_p \gg \Delta\omega_{Dp} \tag{244}$$

and

$$\Omega t_p \approx \kappa_p \mathscr{E}_p t_p = \pi$$

This case would be appropriate for microwave-microwave double resonance when very high power is being used to saturate a narrow Doppler line.

From (243) we have then

$$(m_0 - 1) = \frac{\Delta N_{0p}}{\Delta N_0} \tag{245}$$

For molecules moving with velocity $v$ the absorption coefficient is from (68)

$$\gamma(v) = -\frac{4\pi\omega P_i(v)}{c\mathscr{E}} \tag{246}$$

Before integrating $\gamma(v)$ over the velocity distribution function we make some simplifying assumptions. We assume

$$\left(\frac{1}{T_2} - \frac{1}{T_1}\right) \ll \Delta\omega_D \tag{247}$$

Then it should be a good approximation to put

$$\frac{(1/T_2) - (1/T_1)}{[(1/T_2) - (1/T_1)]^2 + (\omega_0 - \omega + \omega')^2} \approx \pi\,\delta(\omega_0 - \omega + \omega') \tag{248}$$

and

$$\frac{[(1/T_2) - (1/T_1)]^2}{[(1/T_2) - (1/T_1)]^2 + (\omega_0 - \omega + \omega')^2} \approx 0 \tag{249}$$

We neglect the steady-state term in $\gamma$ and compute only the transient term $\gamma_t$. Combining (241), (245), (246), (248), and (249) we have, integrating over the velocity distribution function,

$$\gamma_t = \frac{\pi^{1/2}\hbar\kappa^2\,\Delta N_{0p}}{c}\,q[e^{-t/T_1}I_1 - e^{-t/T_2}I_2 + I_3 e^{-t/T_2}] \tag{250}$$

where

$$I_1 = \pi \int_{-\infty}^{\infty} e^{-q^2\omega'^2} \delta(\omega_0 - \omega + \omega')\, d\omega' \tag{251}$$

$$I_2 = \pi \int_{-\infty}^{\infty} e^{-q^2\omega'^2} \delta(\omega_0 - \omega + \omega') \cos(\omega_0 - \omega + \omega')t\, d\omega' \tag{252}$$

$$I_3 = \int_{-\infty}^{\infty} e^{-q^2\omega'^2} \frac{\sin(\omega_0 - \omega + \omega')t}{(\omega_0 - \omega + \omega')}\, d\omega' \tag{253}$$

The integrals $I_1$ and $I_2$ may be evaluated immediately with the result

$$I_1 = I_2 = \pi e^{-q^2(\omega_0-\omega)^2} \tag{254}$$

In Appendix A2 we show that an integral of the form $I_3$ may be approximated in two important cases:

*Case* 1*a* $t \ll 2q$

$$I^{(a)} = \frac{\pi^{1/2}}{q(\omega - \omega_0)} \sin(\omega - \omega_0)t \tag{255}$$

*Case* 1*b* $t \gg 2q$

$$I^{(b)} = \pi e^{-(\omega-\omega_0)^2 q^2} \tag{256}$$

Substituting (254), (255), and (256) into (250) we have for the absorption coefficient in the two cases:

*Case* 1*a* $t \ll 2q$

$$\gamma_t^{(a)} = \frac{\pi^{3/2}\hbar\kappa^2 \Delta N_{0p} q}{c} \left\{ e^{-q^2(\omega-\omega_0)^2}[e^{-t/T_1} - e^{-t/T_2}] + e^{-t/T_2} \frac{\pi^{-1/2}}{q(\omega - \omega_0)} \sin(\omega - \omega_0)t \right\} \tag{257}$$

*Case* 1*b* $t \gg 2q$

$$\gamma_t^{(b)} = \frac{\pi^{3/2}\hbar\kappa^2 \Delta N_{0p} q}{c} e^{-q^2(\omega-\omega_0)^2} e^{-t/T_1} \tag{258}$$

For the second case we suppose that

$$\kappa_p \mathscr{E}_p \sim \Delta\omega_{Dp}$$

and that the pulse duration is such that $\cos \Omega t_p$ makes several cycles of oscillation over the Doppler half-width, $\Delta\omega_{Dp}$. When we integrate over the velocity distribution function the rapidly oscillating term in $\cos \Omega t_p$ then makes only a very small contribution to the integrals and so it may be neglected. This case is appropriate to infrared-microwave double resonance, where the width of the exciting laser output is approximately equal to the Doppler width of the line being pumped, or to low-power microwave-microwave double resonance.

For molecules moving with velocity $v$, $(m_0 - 1)$ is, from (243)

$$(m_0 - 1) = \left(\frac{\Delta N_{0p}}{2\,\Delta N_0}\right)\frac{\kappa_p^2\mathscr{E}_p^2}{\kappa_p^2\mathscr{E}_p^2 + (\omega_{0p} - \omega_p + r\omega')^2} \tag{259}$$

where

$$\omega' = \frac{\omega v}{c}$$
$$r = \frac{\omega_p}{\omega} \tag{260}$$

In order to evaluate the integrals in closed form we approximate the Lorentzian function in (259) by a Gaussian of equal height and half-width. Then we have

$$(m_0 - 1) = \frac{\Delta N_{0p}}{2\,\Delta N_0}\exp\left[-\frac{\ln 2}{\kappa_p^2\mathscr{E}_p^2}(\omega_{0p} - \omega_p + r\omega')^2\right] \tag{261}$$

Again we assume

$$\left(\frac{1}{T_2} - \frac{1}{T_1}\right) \ll \Delta\omega_D \tag{262}$$

and make the approximations given in (248) and (249). Combining (241), (246), (248), (249), and (261) and integrating over the velocity distribution function yields the result

$$\gamma_t = \frac{\pi^{1/2}\hbar\kappa^2\,\Delta N_{0p}q}{2c}\left[e^{-t/T_1}I_1 - e^{-t/T_2}I_2 + I_3e^{-t/T_2}\right] \tag{263}$$

where

$$I_1 = \pi\int_{-\infty}^{\infty} e^{-q^2\omega'^2}e^{-Q^2(\omega_{0p}-\omega_p+r\omega')^2}\,\delta(\omega_0 - \omega + \omega')\,d\omega' \tag{264}$$

$$I_2 = \pi\int_{-\infty}^{\infty} e^{-q^2\omega'^2}e^{-Q^2(\omega_{0p}-\omega_p+r\omega')^2}\cos(\omega_0 - \omega + \omega')t\;\delta(\omega - \omega_0 + \omega')\,d\omega' \tag{265}$$

$$I_3 = \int_{-\infty}^{\infty} e^{-q^2\omega'^2}e^{-Q^2(\omega_{0p}-\omega_p+r\omega')^2}\,\frac{\sin(\omega_0 - \omega + \omega')t}{(\omega_0 - \omega + \omega')}\,d\omega' \tag{266}$$

and

$$Q = \frac{\ln 2}{\kappa_p^2\mathscr{E}_p^2} \tag{267}$$

The first two integrals may be evaluated immediately to give

$$I_1 = I_2 = \pi\exp\{-[q^2(\omega - \omega_0)^2 + Q^2(\omega_{0p} - \omega_p + r(\omega - \omega_0))^2]\} \tag{268}$$

$I_3$ may be evaluated by first completing the square of the exponent. This results in an integral of the form given in (A2.5). It is shown there that the integral may be approximated for two cases. Then we have

*Case 2a* $\quad t^2 \ll 4(q^2 + r^2Q^2)$:

$$I_3^{(a)} = \frac{\pi^{1/2}}{(q^2 + r^2Q^2)^{1/2}} \exp\left\{-\left[\frac{q^2Q^2(\omega_{0p} - \omega_p)^2}{q^2 + r^2Q^2}\right]\frac{\sin \Theta t}{\Theta}\right\} \tag{269}$$

where

$$\Theta = (\omega - \omega_0) + \frac{rQ^2(\omega_{0p} - \omega_p)}{(q^2 + r^2Q^2)} \tag{270}$$

*Case 2b* $\quad t^2 \gg 4(q^2 + r^2Q^2)$:

$$I_3^{(b)} = \pi \exp\{-[q^2(\omega - \omega_0)^2 + Q^2[(\omega_{0p} - \omega_p) + r(\omega - \omega_0)]^2]\} \tag{271}$$

Combining (261), (268), (269), and (271) we have for the absorption coefficient $\gamma_t$ in the two special cases the following results.

*Case 2a* $\quad t^2 \ll 4(q^2 + r^2Q^2)$:

$$\gamma_t = \frac{\pi^{1/2}\hbar\kappa^2\,\Delta N_{0p}q}{2c} \times \left(\pi \exp\{-[q^2(\omega - \omega_0)^2 + Q^2(\omega_{0p} - \omega_p + r(\omega - \omega_0))^2]\} \times (e^{-t/T_1} - e^{-t/T_2}) + e^{-t/T_2}\frac{\pi^{1/2}}{(q^2 + r^2Q^2)^{1/2}} \times \exp\left\{-\left[\frac{q^2Q^2(\omega_{0p} - \omega_p)^2}{q^2 + r^2Q^2}\right]\frac{\sin \Theta t}{\Theta}\right\}\right) \tag{272}$$

with $\Theta$ given by (270).

*Case 2b* $\quad t^2 \gg 4(q^2 + r^2Q^2)$:

$$\gamma_t = \frac{\pi^{3/2}\hbar\kappa^2\,\Delta N_{0p}q}{2c} e^{-t/T_1} \times \exp\{-[q^2(\omega - \omega_0)^2 + Q^2(\omega_{0p} - \omega_p + r(\omega - \omega_0))^2]\} \tag{273}$$

The idea behind these double resonance experiments is really quite simple. A nonequilibrium population difference between two levels is created by pumping a connected transition. The decay of this nonequilibrium distribution when the pumping is stopped is then a measure of molecular relaxation. Although the above analysis is for an increased population difference leading to a stronger absorption, the same treatment can be given to experiments in which the pumping leads to a smaller population difference and decreased absorption.

Time resolved low-power microwave-microwave double resonance has been reported by Unland and Flygare.[38] Shimizu and Oka[39] have observed infrared-microwave double resonance, but they made no effort to follow their experiment in time. Time resolved infrared-microwave double resonance has been investigated by Shoemaker and Flygare,[40] by Levy et al.,[20] by Frenkel et al.,[41] and by Jetter et al.[42] Infrared-infrared double resonance has been reported by Burak et al.[43]

There has been some controversy over the correct interpretation of the results of these experiments. Equations (257), (258), (272), and (273) should provide clarification. Infrared-microwave double resonance at short times has sometimes been interpreted by assuming that $T_1$ is about equal to $T_2$. Notice that (272), which describes this experiment, makes such an assumption, though probably valid, unnecessary. For observations several Doppler widths off-resonance the coefficient of the term in $(e^{-t/T_1} - e^{-t/T_2})$ is very small no matter what the values of $T_1$ and $T_2$.

Equations (257) and (272) predict a short-time damping proportional to $T_2$ only, at either high or moderate powers, whenever the observing frequency is far off-resonance. For the moderate-power case there are oscillations in the approach to equilibrium if the pumping frequency is not exactly on-resonance. In both cases there are oscillations if the probe frequency is not on-resonance.

At longer times the equations become simpler. Equations (258) and (273) predict a transient decay proportional only to $T_1$ for both high and moderate powers. There are no oscillations in either case. This is in agreement with the work of Jetter et al.,[42] in which they observed a pure exponential decay that was a function of pressure only. These equations also correctly predict the falloff in intensity as $\omega$ is moved away from $\omega_0$ and, in the moderate-power case, as $\omega_p$ is moved away from $\omega_{0p}$. Therefore, both $T_1$ and $T_2$ can be obtained from double resonance measurements, first by looking off-resonance at short times for $T_2$, then by looking on-resonance at long times for $T_1$.

## APPENDIX A1

We derive here the transient solutions to (46) that were used in Sections IV through VII. First let us rewrite (46) in a form more convenient for solution:

$$\frac{dP_r}{d\tau} + \beta P_r + \delta P_i = 0$$
$$\frac{dP_i}{d\tau} + \beta P_i - \delta P_r + M = 0 \qquad \text{(A1.1)}$$
$$\frac{dM}{d\tau} + \alpha M - P_i = \alpha M_0$$

where

$$\begin{aligned}
\tau &= \kappa \mathscr{E} t \\
\alpha &= \frac{1}{\kappa \mathscr{E} T_1} \\
\beta &= \frac{1}{\kappa \mathscr{E} T_2} \\
\delta &= \frac{\Delta\omega}{\kappa \mathscr{E}} \\
M &= \frac{\kappa \hbar \, \Delta N}{4} \\
M_0 &= \frac{\kappa \hbar \, \Delta N_0}{4}
\end{aligned} \tag{A1.2}$$

Torrey[44] has discussed the solutions to (A1.1) appropriate to NMR experiments. As a convenience to the reader we now reproduce his solutions for the general case and for the two special cases $\Delta\omega = 0$ and $T_1 = T_2$. The equations are solved by the Laplace transform technique. The Laplace transform of $f(\tau)$ is defined by

$$\bar{f}(u) = \int_0^\infty f(\tau) e^{-u\tau} \, d\tau \tag{A1.3}$$

Taking Laplace transforms of (A1.1) we have

$$\begin{aligned}
(u + \beta)\bar{P}_r + \delta \bar{P}_i &= r_0 M_0 \\
-\delta \bar{P}_r + (u + \beta)\bar{P}_i + \bar{M} &= i_0 M_0 \\
-\bar{P}_i + (u + \alpha)\bar{M} &= \frac{\alpha M_0}{u} + m_0 M_0
\end{aligned} \tag{A1.4}$$

where $\bar{P}_r$, $\bar{P}_i$, and $\bar{M}$ are the transforms and $r_0 M_0$, $i_0 M_0$, $m_0 M_0$ the initial values of $P_r$, $P_i$, and $M$, respectively. The solutions to (A1.4) are

$$\begin{aligned}
\frac{u \, \Delta(u) \bar{P}_r}{M_0} &= r_0 u[1 + (u + \alpha)(u + \beta)] + \delta\alpha + \delta u m_0 - i_0 \delta u(u + \alpha) \\
\frac{u \, \Delta(u) \bar{P}_i}{M_0} &= r_0 \delta u(u + \alpha) + i_0 u(u + \alpha)(u + \beta) - (\alpha + m_0 u)(u + \beta) \\
\frac{u \, \Delta(u) \bar{M}}{M_0} &= r_0 \delta u + i_0 u(u + \beta) + (\alpha + m_0 u)[(u + \beta)^2 + \delta^2]
\end{aligned} \tag{A1.5}$$

where $\Delta(u)$ is the determinant of the coefficients in (A1.4). We have

$$\Delta(u) = (u + \alpha)(u + \beta)^2 + u + \beta + \delta^2(u + \alpha) \tag{A1.6}$$

Let $f(\tau)$ stand for any one of the components $P_r/M_0$, $P_i/M_0$, or $M/M_0$. We have then from (A1.5)

$$\bar{f}(u) = \frac{g(u)}{u\,\Delta(u)} \tag{A1.7}$$

where $g(u)$ is a cubic form in $u$ which is different for the three components $\bar{P}_r$, $\bar{P}_i$, and $\bar{M}$. The equation $\Delta(x) = 0$ has at least one real negative root. If this is $-a$, then $\Delta(u)$ can be factored into

$$\Delta(u) = (u + a)[(u + b)^2 + s^2] \tag{A1.8}$$

Equation (A1.7) may now be expanded in partial fractions to give

$$\bar{f}(u) = \frac{A}{u + a} + \frac{B(u + b) + C}{(u + b)^2 + s^2} + \frac{D}{u} \tag{A1.9}$$

The inverse transform of this expression is

$$f = Ae^{-a\tau} + Be^{-b\tau} \cos s\tau + \frac{C}{s} e^{-b\tau} \sin s\tau + D \tag{A1.10}$$

The first three terms give the transient effects and the last term the steady state. The coefficients $A$ and $D$ are given by

$$D = \lim_{u \to 0} [u\bar{f}(u)] \tag{A1.11}$$

$$A = \lim_{u \to -a} [(u + a)\bar{f}(u)] \tag{A1.12}$$

The other coefficients are most conveniently found by applying the boundary conditions at $t = 0$. We find

$$D = \frac{g(0)}{a(b^2 + s^2)} \tag{A1.13}$$

$$A = \frac{-g(-a)}{a[(b - a)^2 + s^2]} \tag{A1.14}$$

*For* $f = P_r/M_0$

$$B = -(A + D) + r_0 \tag{A1.15}$$

$$C = aA + bB - \beta r_0 - \delta i_0 \tag{A1.16}$$

*For* $f = P_i/M_0$

$$B = -(A + D) + i_0 \tag{A1.17}$$

$$C = aA + bB - m_0 + \delta r_0 - \beta i_0 \tag{A1.18}$$

*For* $f = M/M_0$

$$B = -(A + D) + m_0 \tag{A1.19}$$

$$C = aA + bB + \alpha(1 - m_0) + i_0 \tag{A1.20}$$

The time constants $a$, $b$, and $s$ may be found by expanding (A1.6) and (A1.8) in powers of $u$ and equating coefficients. We find

$$\begin{aligned} 2b + a &= 2\beta + \alpha \\ b^2 + s^2 + 2ab &= 2\alpha\beta + \beta^2 + \delta^2 + 1 \\ a(b^2 + s^2) &= \alpha(\beta^2 + \delta^2) + \beta \end{aligned} \tag{A1.21}$$

Substituting the value for $a(b^2 + s^2)$ obtained in the last of (A1.21) into (A1.13) we have

$$D = \frac{g(0)}{[\alpha(\beta^2 + \delta^2) + \beta]} \tag{A1.22}$$

Substituting $g(0)$ from (A1.5) into (A1.22) yields the following results.

*For* $f = P_r/M_0$

$$D = \frac{\alpha\delta}{[\alpha(\beta^2 + \delta^2) + \beta]} \tag{A1.23a}$$

*For* $f = P_i/M_0$

$$D = \frac{-\alpha\beta}{\alpha(\beta^2 + \delta^2) + \beta} \tag{A1.23b}$$

*For* $f = M/M_0$

$$D = \frac{\alpha(\beta^2 + \delta^2)}{\alpha(\beta^2 + \delta^2) + \beta} \tag{A1.23c}$$

These solutions are identical to those given in (84). There are two special cases where the equation $\Delta(x) = 0$ has simple roots:

1. If $\delta = 0$, a real root of $\Delta(x) = 0$ is $x = -\beta$; then

$$a = \beta \tag{A1.24a}$$

and from (A1.21)

$$\begin{aligned} b &= \tfrac{1}{2}(\alpha + \beta) \\ s^2 &= 1 - \tfrac{1}{4}(\beta - \alpha)^2 \end{aligned} \tag{A1.24b}$$

The coefficients $A$, $B$, and $C$ for the three components of $f$ are, from (A1.14) through (A1.20) together with the conditions $r_0 = i_0 = 0$, as follows.

*For* $f = P_r/M_0$

$$\begin{aligned} A &= 0 \\ B &= 0 \\ C &= 0 \end{aligned} \tag{A1.25}$$

*For* $f = P_i/M_0$

$$A = 0$$

$$B = \frac{\alpha}{1 + \alpha\beta} \tag{A1.26}$$

$$C = \frac{\alpha(\alpha + \beta)}{2(1 + \alpha\beta)} - m_0$$

*For* $f = M/M_0$

$$A = 0$$

$$B = m_0 - \frac{\alpha\beta}{1 + \alpha\beta} \tag{A1.27}$$

$$C = (\beta - \alpha)\frac{m_0}{2} + \alpha\left[1 - \frac{\beta(\alpha + \beta)}{2(1 + \alpha\beta)}\right]$$

Substituting the values of $A$, $B$, $C$ given in (A1.26) and the value of $D$ for $\delta = 0$ obtained from (A1.23b) into (A1.10) gives, changing the variables using (A1.2), (106).

2. If $\alpha = \beta$, $x = -\beta$ is again a root of $\Delta(x) = 0$, then

$$a = \beta \tag{A1.28a}$$

and from (A1.21)

$$b = \beta \tag{A1.28b}$$

$$s^2 = 1 + \delta^2 \tag{A1.28c}$$

The coefficients $A$, $B$, $C$, under the conditions $r_0 = i_0 = 0$, are, from (A1.14) through (A1.20), as follows.

*For* $f = P_r/M_0$

$$A = -\frac{\delta(1 - m_0)}{1 + \delta^2}$$

$$B = \frac{\delta}{1 + \delta^2}\left(\frac{\beta^2}{1 + \beta^2 + \delta^2} - m_0\right) \tag{A1.29}$$

$$C = -\frac{\delta\beta}{1 + \beta^2 + \delta^2}$$

*For* $f = P_i/M_0$

$$A = 0$$

$$B = \frac{\beta}{1 + \beta^2 + \delta^2} \tag{A1.30}$$

$$C = -m_0 + \frac{\beta^2}{1 + \beta^2 + \delta^2}$$

*For* $f = M/M_0$

$$A = \frac{(m_0 - 1)\delta^2}{1 + \delta^2}$$

$$B = -\frac{(\beta^2 + \delta^2)}{1 + \beta^2 + \delta^2} - \frac{(m_0 - 1)\delta^2}{1 + \delta^2} + m_0 \tag{A1.31}$$

$$C = \frac{\beta}{1 + \beta^2 + \delta^2}$$

Substituting the values of $A$, $B$, $C$ given in (A1.30) and the value of $D$ for $\alpha = \beta$ obtained from (A1.23b) into (A1.10) gives, after changing variables using (A1.2), (113).

We now consider the solutions to (130), that is,

$$\begin{aligned} \frac{dP_r}{dt} + \Delta\omega P_i + \frac{P_r}{T_2} &= 0 \\ \frac{dP_i}{dt} - \Delta\omega P_r + \frac{P_i}{T_2} &= 0 \\ \frac{d\,\Delta N}{dt} + \frac{\Delta N - \Delta N_0}{T_1} &= 0 \end{aligned} \tag{A1.32}$$

We substitute

$$\begin{aligned} P_r &= e^{-t/T_2} f(t) \\ P_i &= e^{-t/T_2} g(t) \end{aligned} \tag{A1.33}$$

where $f(t)$, $g(t)$ are arbitrary functions of $t$ into (A1.32). We then have

$$\begin{aligned} \frac{df}{dt} + \Delta\omega g &= 0 \\ \frac{dg}{dt} - \Delta\omega f &= 0 \end{aligned} \tag{A1.34}$$

Differentiating the first of these equations and substituting $dg/dt$ obtained from the last equation, we have

$$\frac{d^2 f}{dt^2} + (\Delta\omega)^2 f = 0 \tag{A1.35}$$

Integrating this equation twice we have

$$f = A \cos \Delta\omega t - B \sin \Delta\omega t \tag{A1.36}$$

Differentiating this equation and substituting into the first of (A1.34), we have

$$g = B \cos \Delta\omega t + A \sin \Delta\omega t \tag{A1.37}$$

Combining (A1.33), (A1.36), and (A1.37) yields the result

$$P_r = e^{-t/T_2}(A \cos \Delta\omega t - B \sin \Delta\omega t) \tag{A1.38a}$$

$$P_i = e^{-t/T_2}(B \cos \Delta\omega t + A \sin \Delta\omega t) \tag{A1.38b}$$

The last of (A1.32) may be integrated directly with the result

$$\Delta N = \Delta N_0(1 + C)e^{-t/T_1} \tag{A1.38c}$$

Equations (A1.38) correspond to those given in (131).

To solve (149) we first transform the variables using (A1.2) to give

$$\begin{aligned} \frac{dP_r}{d\tau} + \delta P_i &= 0 \\ \frac{dP_i}{d\tau} - \delta P_r + M &= 0 \\ \frac{dM}{d\tau} - P_i &= 0 \end{aligned} \tag{A1.39}$$

Taking the Laplace transforms of these equations gives

$$\begin{aligned} u\bar{P}_r + \delta\bar{P}_i &= r_0 M_0 \\ u\bar{P}_i - \delta\bar{P}_r + \bar{M} &= i_0 M_0 \\ u\bar{M} - \bar{P}_i &= m_0 M_0 \end{aligned} \tag{A1.40}$$

where again $r_0M_0$, $i_0M_0$, and $m_0M_0$ are the initial values of $P_r$, $P_i$, and $M$, respectively. The solutions to (A1.40) are

$$\begin{aligned} \Delta(u)\frac{\bar{P}_r}{M_0} &= r_0(u^2 + 1) - i_0\delta u + \delta m_0 \\ \Delta(u)\frac{\bar{P}_i}{M_0} &= i_0 u^2 - m_0 u + r_0\delta u \\ \Delta(u)\frac{\bar{M}}{M_0} &= m_0 u^2 + i_0 u + m_0\delta^2 + r_0\delta \end{aligned} \tag{A1.41}$$

where $\Delta(u)$ is the determinant of the coefficients in (A1.40)

$$\Delta(u) = u[u^2 + 1 + \delta^2] \tag{A1.42}$$

If $f$ stands for any one of the three components, $P_r/M_0$, $P_i/M_0$, $M/M_0$, then expanding (A1.41) in partial fractions we may write

$$\bar{f} = \frac{A}{u} + \frac{Bu + C}{u^2 + 1 + \delta^2} \tag{A1.43}$$

This has the inverse transform

$$f = A + B\cos(1 + \delta^2)^{1/2}\tau + \frac{C}{(1 + \delta^2)^{1/2}}\sin(1 + \delta^2)^{1/2}\tau \qquad \text{(A1.44)}$$

The constants $A$, $B$, and $C$ are obtained most easily by comparing (A1.41) and (A1.43). We then have the following,

*For* $f = P_r/M_0$

$$A = \frac{r_0 + \delta m_0}{1 + \delta^2}$$
$$B = \frac{r_0\delta^2 - \delta m_0}{1 + \delta^2} \qquad \text{(A1.45)}$$
$$C = -i_0\delta$$

*For* $f = P_i/M_0$

$$A = 0$$
$$B = i_0 \qquad \text{(A1.46)}$$
$$C = r_0\delta - m_0$$

*For* $f = M/M_0$

$$A = \frac{m_0\delta^2 + r_0\delta}{1 + \delta^2}$$
$$B = \frac{m_0 - r_0\delta}{1 + \delta^2} \qquad \text{(A1.47)}$$
$$C = i_0$$

From (A1.44), (A1.45), (A1.46), and (A1.47) we obtain after transforming variables using (A1.2) the solutions given in (150).

Let us now consider the solution of (240) for $P_i$. First we transform variables using (A1.2) to give

$$\frac{dP_r}{d\tau} + \beta P_r + \delta P_i = 0$$
$$\frac{dP_i}{dt} + \beta P_i - \delta P_r + M = 0 \qquad \text{(A1.48)}$$
$$\frac{dM}{d\tau} + \alpha M = \alpha M_0$$

Taking Laplace transforms we have

$$(u + \beta)\bar{P}_r + \delta\bar{P}_i = r_0 M_0$$
$$-\delta\bar{P}_r + (u + \beta)\bar{P}_i + \bar{M} = i_0 M_0 \qquad \text{(A1.49)}$$
$$+(u + \alpha)\bar{M} = \frac{\alpha M_0}{u} + m_0 M_0$$

where again $r_0M_0$, $i_0M_0$, $m_0M_0$ are the initial values of $P_r$, $P_i$, and $M$. Solving (A1.49) for $\bar{P}_i$ we have

$$\frac{u\,\Delta(u)\bar{P}_i}{M_0} = (u+\beta)u(u+\alpha)i_0 + \delta r_0 u(u+\alpha) - (\alpha + m_0 u)(u+\beta) \quad \text{(A1.50)}$$

where $\Delta(u)$ is the determinant of the coefficients in (A1.49)

$$\Delta(u) = (u+\alpha)[(u+\beta)^2 + \delta^2] \quad \text{(A1.51)}$$

Comparing (A1.8) and (A1.51) we have from (A1.10) the inverse transform

$$\frac{P_i}{M_0} = Ae^{-\alpha\tau} + Be^{-\beta\tau}\cos\delta\tau + \frac{C}{\delta}e^{-\beta\tau}\sin\delta\tau + D \quad \text{(A1.52)}$$

From (A1.13), (A1.14), (A1.17), and (A1.18) the coefficients $A$, $B$, $C$, and $D$ are given by

$$\begin{aligned} A = -B &= -(m_0 - 1)\frac{(\beta-\alpha)}{(\beta-\alpha)^2+\delta^2} \\ C &= -(m_0-1) + \frac{(m_0-1)(\beta-\alpha)^2}{(\beta-\alpha)^2+\delta^2} \\ D &= -\frac{\beta}{\beta^2+\delta^2} \end{aligned} \quad \text{(A1.53)}$$

In obtaining (A1.53) we used the value of $r_0$ given by (A1.23a) in the low-power limit

$$r_0 = \frac{P_r(0)}{M_0} = \frac{\delta}{\beta^2+\delta^2} \quad \text{(A1.54)}$$

## APPENDIX A2

To evaluate the integral

$$I = \int_{-\infty}^{\infty} e^{-q^2\omega'^2}\cos(\omega' t + \varphi)\,d\omega' \quad \text{(A2.1)}$$

we first use the identity

$$\cos(\omega' t + \varphi) = \cos\omega' t\cos\varphi - \sin\omega' t\sin\varphi \quad \text{(A2.2)}$$

Substituting this into (A2.1) we have

$$I = \cos\varphi\int_{-\infty}^{\infty} e^{-q^2\omega'^2}\cos\omega' t\,d\omega' \quad \text{(A2.3)}$$

where we have used the fact that $\sin\omega' t$ is an odd function in $\omega'$ and so makes no contribution to the integral. The integral in (A2.3) is now a standard integral. We have

$$I = \cos\varphi\,\frac{\pi^{1/2}}{q}\,e^{-t^2/4q^2} \quad \text{(A2.4)}$$

We now discuss the integral defined by

$$I = \int_{-\infty}^{\infty} e^{-q^2\omega'^2} \frac{\sin(\omega' + \varphi)t}{\omega' + \varphi} d\omega' \tag{A2.5}$$

First we differentiate with respect to $t$

$$\frac{\partial I}{\partial t} = \int_{-\infty}^{\infty} e^{-q^2\omega'^2} \cos(\omega' + \varphi)t \, d\omega' \tag{A2.6}$$

Comparison with (A2.1) and (A2.4) yields the result

$$\frac{\partial I}{\partial t} = \cos \varphi t \frac{\pi^{1/2}}{q} e^{-t^2/4q^2} \tag{A2.7}$$

Since $I = 0$ when $t = 0$ we have, on integrating (A2.7),

$$I = \frac{\pi^{1/2}}{q} \int_0^t \cos \varphi t e^{-t^2/4q^2} \, dt \tag{A2.8}$$

*Case a* $\quad t \ll 2q$

We make in this case the approximation

$$e^{-t^2/4q^2} \approx 1 \tag{A2.9}$$

Then

$$I = \frac{\pi^{1/2}}{q} \int_0^t \cos \varphi t \, dt \tag{A2.10}$$

that is,

$$I = \frac{\pi^{1/2}}{q\varphi} \sin \varphi t \tag{A2.11}$$

*Case b* $\quad t \gg 2q$

In this case we have from (A2.8)

$$I \approx \frac{\pi^{1/2}}{q} \int_0^{\infty} \cos \varphi t e^{-t^2/4q^2} \, dt \tag{A2.12}$$

The integral in (A2.12) is standard and yields the result

$$I = \pi e^{-\varphi^2 q^2} \tag{A2.13}$$

## References


1. L. L. McCall and E. L. Hahn, *Phys. Rev.*, **183**, 457 (1969).
2. R. Karplus and J. Schwinger, *Phys. Rev.*, **73**, 1020 (1948).
3. M. V. Klein, *Optics*, Wiley, New York, 1970.
4. F. Bloch, *Phys. Rev.*, **70**, 460 (1946).
5. The formal equivalence of the two problems has been discussed by R. P. Feynman, R. W. Hellwarth, and F. L. Vernon, Jr., *J. Appl. Phys.*, **28**, 49 (1957).
6. A. Abragam, *The Principles of Nuclear Magnetism*, Oxford University Press, London and New York, 1961; C. P. Slichter, *Principles of Magnetic Resonance*, Harper, New York, 1963.

7. C. H. Townes and A. L. Schawlow, *Microwave Spectroscopy*, McGraw-Hill, New York, 1955.
8. H. A. Dijkerman, Ph.D. Thesis, University of Utrecht, Holland, 1966.
9. W. E. Lamb, *Phys. Rev.*, **134,** 1429 (1964).
10. C. C. Costain, *Can. J. Phys.*, **47,** 2431 (1969).
11. A. Syoke and A. Javan, *Phys. Rev. Lett.*, **10,** 521 (1963); R. S. Brewer, M. J. Kelley, and A. Javan, *Phys. Rev. Lett.*, **23,** 559 (1969).
12. R. G. Brewer, *Science*, **178,** 247 (1972).
13. G. W. Flynn, *J. Mol. Spectr.*, **28,** 1 (1968); F. G. Wodarcyyk and E. B. Wilson, *J. Mol. Spectr.*, **37,** 445 (1971).
14. M. S. Feld and A. Javan, *Phys. Rev.*, **177,** 540 (1968).
15. M. L. Unland, V. W. Weiss, and W. H. Flygare, *J. Chem. Phys.*, **42,** 2138 (1965); A. P. Cox, G. W. Flynn, and E. B. Wilson, *J. Chem. Phys.*, **42,** 3094 (1965); T. Oka, *J. Chem. Phys.*, **49,** 3135 (1968).
16. R. H. Cardover, P. A. Bonczyk, and A. Javan, *Phys. Rev. Lett.*, **18,** 730 (1967); W. G. Schweitzer, Jr., M. M. Birdy, and J. A. White, *J. Opt. Soc. Am.*, **97,** 1226 (1967).
17. B. Hocker and C. L. Tang, *Phys. Rev.*, **184,** 356 (1969).
18. R. G. Brewer and R. J. Shoemaker, *Phys. Rev. Lett.*, **27,** 631 (1971).
19. H. Harrington, Symposium on Molecular Structure and Spectroscopy, Columbus, Ohio, 1968.
20. J. M. Levy, J. H. S. Wang, S. G. Kukolich, and J. I. Steinfeld, *Phys. Rev. Lett.*, **29,** 395 (1972).
21. B. Mache and P. Glorieux, *Chem. Phys. Lett.*, **14,** 85 (1972).
22. A. H. Brittain, P. J. Manor, and R. H. Schwendeman, *J. Chem. Phys.*, **58,** 5735 (1973).
23. T. Amano and T. Shimizu, *J. Phys. Soc. Jap.* **35,** 237 (1973).
24. J. C. McGurk, T. R. Hofmann, and W. H. Flygare, *J. Chem. Phys.*, **60,** in press (1974).
25. L. L. Brown, *J. Chem. Phys.*, **56,** 1000 (1972).
26. R. Schwendeman and H. Pickett, *J. Chem. Phys.*, **57,** 3511 (1972).
27. J. H. S. Wang, J. M. Levy, S. G. Kukolich, and J. I. Steinfeld, *Chem. Phys.*, **1,** 141 (1973).
28. R. H. Dicke, *Phys. Rev.*, **93,** 99 (1954).
29. R. G. Brewer and R. L. Shoemaker, *Phys. Rev. A*, **6,** 2001 (1972).
30. R. H. Dicke and R. H. Romer, *Rev. Sci. Instr.*, **26,** 915 (1955).
31. R. M. Hill, D. E. Kaplan, G. F. Hermann, and S. K. Ichiki, *Phys. Rev. Lett.*, **18,** 105 (1967).
32. B. Mache and P. Glorieux, *Chem. Phys. Lett.*, **18,** 91 (1973).
33. J. P. Gordon, C. H. Wang, C. K. N. Patel, R. E. Slusher, and W. J. Tomlinson, *Phys. Rev.*, **179,** 294 (1969).
34. R. R. Ernst and W. A. Anderson, *Rev. Sci. Instr.*, **37,** 93 (1966).
35. J. C. McGurk, H. Mäder, R. T. Hofmann, and W. H. Flygare, *J. Chem. Phys.*, **60,** in press (1974).
36. R. R. Ernst, *Advances in Magnetic Resonance*, Vol. 2, J. S. Waugh, Ed., Academic Press, New York, 1966, p. 1.
37. J. C. McGurk, T. G. Schmalz, and W. H. Flygare, *J. Chem. Phys.*, **60,** in press (1974).
38. M. L. Unland and W. H. Flygare, *J. Chem. Phys.*, **45,** 2421 (1966).
39. T. Shimizu and T. Oka, *Phys. Rev. A*, **2,** 1177 (1970).

40. R. L. Shoemaker, Ph.D. Thesis, University of Illinois, 1971.
41. L. Frenkel, H. Marantz, and T. Sullivan, *Phys. Rev A*, **3,** 1640 (1971).
42. H. Jetter, E. F. Pearson, C. L. Norris, J. C. McGurk, and W. H. Flygare, *J. Chem. Phys.*, **59,** 1796 (1973).
43. J. Burak, A. W. Nowak, J. I. Steinfeld, and D. G. Sutton, *J. Chem. Phys.*, **51,** 2275 (1969).
44. H. C. Torrey, *Phys. Rev.*, **76,** 1059 (1949).

# CLASSICAL-LIMIT QUANTUM MECHANICS AND THE THEORY OF MOLECULAR COLLISIONS

WILLIAM H. MILLER

*Department of Chemistry, University of California, Berkeley, California*

## CONTENTS



## I. INTRODUCTION

An important contribution to atomic and molecular collision theory was made in 1959 when Ford and Wheeler[1] showed how elastic atom-atom scattering could be described semiclassically. The basic physical idea in Ford and Wheeler's treatment is that a quantum mechanical *description* of the scattering process is used—that is, scattering amplitudes are the primary objects of interest—but the dynamical parameters in the quantum mechanical formulas are evaluated by the appropriate use of classical mechanics; for elastic scattering these "dynamical parameters" are simply the scattering phase shifts, and the classical-limit approximations to them are the well-known WKB phase shifts. The use of scattering, or probability, *amplitudes* means that the quantum mechanical principle of superposition is built into the theory, and this is the only vestige of quantum mechanics that is retained.

The intervening years have seen numerous applications and extensions of the Ford and Wheeler analysis, and it is now clear that this combination of "classical dynamics plus quantum superposition" provides a quantitative description of essentially all the quantum effects in elastic scattering. Even such quantumlike phenomena as resonances in elastic scattering are accurately described if one analytically continues classical mechanics so as to include classically forbidden (i.e., tunneling) processes. Bernstein's 1966 review[2] is still one of the best summaries of the semiclassical theory of elastic scattering, and the more recent review by Berry and Mount[3] is recommended for the more recent developments as well as more mathematical aspects of the theory.

The semiclassical description of elastic scattering is important in two respects—it helps one to think and also to compute. The conceptual picture offered by the theory is extremely illuminating, providing simple physical interpretations of the quantum effects in elastic scattering that are not only interesting but also particularly useful in extracting quantitative information about the interatomic potential. The semiclassical theory also leads to useful methods for calculating cross sections, although this is no longer so important for elastic scattering, since it is usually feasible to carry out fully quantum mechanical calculations for the phase shifts and cross sections.

In view of the success in treating elastic scattering semiclassically, it is natural to attempt a similar description of more complex molecular collisions, such as that between an atom $A$ and diatomic molecule $BC$, which can undergo inelastic and reactive processes. Most approaches to a semiclassical theory of complex collisions have begun with the standard quantum mechanical formulation of the problem, namely, a coupled channel expansion,[4] and semiclassical approximations are then introduced in attempting to solve the coupled radial equations. The rotational/vibrational degrees of freedom are thus treated completely quantum mechanically via the expansion in their quantum states, whereas the semiclassical approximation has to do only with the translational degree of freedom. Although many useful semiclassical developments have been based on this formulation, the work of Cross[5] and his co-workers having pursued this approach most fully for the case of molecular systems, there are some inherent drawbacks. The essential difficulty is that for a molecular system the number of coupled channels, that is, the number of rotational/vibrational states, is much too large for any simple treatment to be possible. There is thus no way to solve the radial equations semiclassically without introducing explicit *dynamical* approximations, these most often being some variation of the sudden approximation,[5] the Eikonal approximation,[6] the Glauber approximation,[7] etc.

Another class of semiclassical approximations that have been developed recently[8–10]—and which are the primary topic of this review—circumvents the problem of the large number of rotational/vibrational states by not expanding in any quantum states; rather, all degrees of freedom, rotational/vibrational as well as translational, are treated classically, with quantum mechanics being incorporated only through the principle of superposition; quantization of the internal degrees of freedom is achieved through the boundary conditions of the classical equations of motion. For the $A + BC$ collision system, for example, quantum mechanical expressions for the cross-sections are used, but all dynamical parameters, the $S$-matrix elements, are evaluated by the appropriate use of classical mechanics. This approach is thus similar in spirit to the semiclassical description of elastic scattering,[1,2] the most awkward feature being that classical trajectories for the three-body problem, $A + BC$, cannot be obtained analytically. (For elastic scattering the dynamical problem reduces to a one-dimensional one after separation of angular momentum, and the classical trajectories can thus be obtained in closed form.) As is well known, however, they can be computed quite readily by numerical integration of the classical equations of motion; the use of classical trajectory calculations within a strictly classical framework, usually with Monte Carlo averaging procedures, has in fact become an extremely useful tool itself for studying the dynamics of inelastic and reactive collisions.[11,12]

The net result of this semiclassical theory is thus a prescription for using classical trajectories for an $A + BC$ collision system, say, to construct the classical-limit approximation to $S$-matrix elements which describe transitions between individual quantum states of the collision partners. The power of this "classical $S$-matrix" theory is that it allows one to incorporate the complete dynamics of all degrees of freedom exactly, albeit classically, and thus borrow from the considerable technology that has evolved for computing classical trajectories.[11] Furthermore, since quantum effects enter solely through the principle of superposition there is a clear distinction between *dynamical* effects, which appear in a classical or quantum description, and *quantum* effects, which lie outside the realm of a purely classical theory. The fact that quantum mechanics enters classical $S$-matrix theory only through the principle of superposition might at first seem to severely limit the quantum effects that the model can describe. As will be seen, however, essentially *all* quantum effects—interference, tunneling, quantization, and selection rules—are a direct result of superposition and are thus qualitatively, and often quantitatively, contained within the theory.

In addition to the conceptual insight regarding the nature of quantum effects in complex collisions, the classical $S$-matrix approach makes it possible to carry out reliable calculations for inelastic and reactive processes in molecular collisions. This computational facility is of greater practical significance for these complex collisions than it is for elastic scattering, for here it is often not possible to carry out quantum mechanical calculations for physically realistic models. As is discussed in Section IV, the feature of the theory that actually appears to be most important for applications is the ability to analytically continue classical mechanics within the framework of classical $S$-matrix theory in such a way as to describe *classically forbidden* processes, that is, those that cannot occur via ordinary classical dynamics.

The purpose of this review is to summarize the structure of classical-limit quantum mechanics in as general a framework as possible, with the principal applications being to molecular collision phenomena; Section VI discusses several applications other than to scattering. The primary emphasis is on theoretical aspects of the semiclassical approach, although some specific applications are discussed to illustrate and clarify the theoretical notions. A discussion of the methodology and various applications are the subject of a later review.[13]

The applications that are discussed have all incorporated exact (i.e., numerical) classical mechanics, the reason being to focus primary attention on the validity of the semiclassical model itself, that is, "classical mechanics plus quantum superposition." Once the nature and importance of the intrinsically quantum effects are fully established, however, it is clear that one may often wish to apply the semiclassical theory with approximate dynamics. Many times, in fact, it is the degree of applicability of simple dynamical models that is more interesting than numerically computed trajectories.

## II. GENERAL CORRESPONDENCE RELATIONS

As Dirac emphasizes in the preface to his book,[14] quantum mechanics is a mechanics of transformation relations, and the correspondence between classical mechanics and quantum mechanics accordingly has its most fundamental expression in the relation between canonical transformations of classical mechanics and unitary transformations of quantum mechanics. Therefore Section II.A first summarizes some of the basic properties of canonical transformations in classical mechanics.

### A. Canonical Transformations in Classical Mechanics

The ability to change from one set of coordinates and momenta to another canonical set, still preserving the Hamiltonian form of the equations of motion, provides great flexibility in classical mechanics. Standard texts on classical mechanics discuss the topic of canonical transformations[15] at length, and the reader is referred to these for more background.

Let $q$ and $p$ be a set of canonically conjugate coordinates and momenta for the system; $q$ and $p$ are $N$-dimensional vectors for a system of $N$ degrees of freedom, for example, $q = (q_1, q_2, \ldots, q_N)$ and $p = (p_1, p_2, \ldots, p_N)$, but multidimensional notation is not used explicitly unless it is required for clarity; in most cases the explicitly multidimensional expressions are obvious. One considers changing to a new set of coordinates and momenta $Q$ and $P$; these new variables are functions of the old ones, $P(p, q)$ and $Q(p, q)$. Equivalently, one may take the new variables $P$ and $Q$ as the independent ones and consider the old variables as a function of them, $p(P, Q)$ and $q(P, Q)$.

Not all functions $P(p, q)$ and $Q(p, q)$, however, are admissible as *canonical* variables. By "canonical" one simply means that the time dependence of the coordinates and momenta is given by Hamilton's equations. The old variables $p$ and $q$ are assumed to be canonical, so that

$$\dot{q} = \frac{\partial H(p, q)}{\partial p} \tag{2.1a}$$

$$\dot{p} = -\frac{\partial H(p, q)}{\partial q} \tag{2.1b}$$

where $\dot{q} \equiv (d/dt)q(t)$, etc., and $H(p, q)$ is the Hamiltonian function. The new variables $P$ and $Q$ are thus canonical if, and only if, their time derivatives are given by

$$\dot{Q} = \frac{\partial H(P, Q)}{\partial P} \tag{2.2a}$$

$$\dot{P} = -\frac{\partial H(P, Q)}{\partial Q} \tag{2.2b}$$

where $H(P, Q)$ is the Hamiltonian expressed in terms of the new variables.

To discover the condition on the functions $P(p, q)$ and $Q(p, q)$ for the $(p, q) \leftrightarrow (P, Q)$ transformation to be canonical, note the identity

$$H(p, q) = H(P(p, q), Q(p, q)) \tag{2.3}$$

Differentiating (2.3) with respect to $p$, holding $q$ constant, gives

$$\left(\frac{\partial H}{\partial p}\right)_q = \left(\frac{\partial H}{\partial P}\right)_Q \left(\frac{\partial P}{\partial p}\right)_q + \left(\frac{\partial H}{\partial Q}\right)_P \left(\frac{\partial Q}{\partial p}\right)_q \tag{2.4}$$

With (2.1) and (2.2) (i.e., assuming the transformation to be canonical), this becomes

$$\dot{q} = \dot{Q}\left(\frac{\partial P}{\partial p}\right)_q - \dot{P}\left(\frac{\partial Q}{\partial p}\right)_q \tag{2.5}$$

But

$$\dot{Q} \equiv \frac{d}{dt} Q(p, q) = \left(\frac{\partial Q}{\partial p}\right)_q \dot{p} + \left(\frac{\partial Q}{\partial q}\right)_p \dot{q} \tag{2.6a}$$

and similarly

$$\dot{P} = \left(\frac{\partial P}{\partial p}\right)_q \dot{p} + \left(\frac{\partial P}{\partial q}\right)_p \dot{q} \tag{2.6b}$$

so that (2.5) becomes

$$\dot{q} = \left[\left(\frac{\partial Q}{\partial p}\right)_q \dot{p} + \left(\frac{\partial Q}{\partial q}\right)_p \dot{q}\right]\left(\frac{\partial P}{\partial p}\right)_q - \left[\left(\frac{\partial P}{\partial p}\right)_q \dot{p} + \left(\frac{\partial P}{\partial q}\right)_p \dot{q}\right]\left(\frac{\partial Q}{\partial p}\right)_q \tag{2.7}$$

The first and third terms cancel, and others combine to imply

$$1 = \left(\frac{\partial P}{\partial p}\right)_q \left(\frac{\partial Q}{\partial q}\right)_p - \left(\frac{\partial P}{\partial q}\right)_p \left(\frac{\partial Q}{\partial p}\right)_q \tag{2.8}$$

The RHS of this is recognized as a Jacobian determinant, so that it becomes

$$1 = \left|\frac{\partial(P, Q)}{\partial(p, q)}\right| \equiv \det \begin{bmatrix} \left(\frac{\partial P}{\partial p}\right)_q & \left(\frac{\partial P}{\partial q}\right)_p \\ \left(\frac{\partial Q}{\partial p}\right)_q & \left(\frac{\partial Q}{\partial q}\right)_p \end{bmatrix} \tag{2.9}$$

(This same condition is also obtained if one differentiates (2.3) with respect to $q$, holding $p$ constant.) The $(p, q) \leftrightarrow (P, Q)$ transformation is canonical, therefore, if the Jacobian relating the new and old variables is unity.

The above criterion (2.9) is a useful way to *test* any proposed set of functions $P(p, q)$ and $Q(p, q)$ to see if they form a canonical set, but it does not provide a useful way to construct transformations that are guaranteed to be canonical. To do this one needs to consider not $p$ and $q$, or $P$ and $Q$, as the independent variables, but rather one must choose one old variable and one new variable as the independent variables, the other two then being considered as functions of these two. For example, one may choose $q$ and $P$ as the independent variables, and then $Q(q, P)$ and $p(q, P)$ are considered as functions of them. It is clear that there are four ways of choosing "one old variable and one new variable," $q$ and $P$, $q$ and $Q$, $p$ and $Q$, or $p$ and $P$.

Suppose, for example, one chooses the old coordinate $q$ and the new coordinate $Q$ as the independent variables; the momenta are then considered as functions of them, $p(q, Q)$ and $P(q, Q)$. Noting the identity

$$H(p(q, Q), q) = H(P(q, Q), Q) \tag{2.10}$$

one differentiates with respect to $q$, holding $Q$ constant:

$$\left(\frac{\partial H}{\partial p}\right)_q \left(\frac{\partial p}{\partial q}\right)_Q + \left(\frac{\partial H}{\partial q}\right)_p = \left(\frac{\partial H}{\partial P}\right)_Q \left(\frac{\partial P}{\partial q}\right)_Q \tag{2.11}$$

With (2.1) and (2.2) (i.e., assuming the transformation to be canonical), this becomes

$$\dot{q}\left(\frac{\partial p}{\partial q}\right)_Q - \dot{p} = \dot{Q}\left(\frac{\partial P}{\partial q}\right)_Q \tag{2.12}$$

But

$$\dot{p} = \frac{d}{dt}\, p(q, Q) = \left(\frac{\partial p}{\partial q}\right)_Q \dot{q} + \left(\frac{\partial p}{\partial Q}\right)_q \dot{Q}$$

so that (2.12) becomes

$$-\left(\frac{\partial p}{\partial Q}\right)_q \dot{Q} = \dot{Q}\left(\frac{\partial P}{\partial q}\right)_Q$$

That is, the condition that the $(p, q) \leftrightarrow (P, Q)$ transformation be canonical is that the functions $p(q, Q)$ and $P(q, Q)$ satisfy the relation

$$-\frac{\partial p(q, Q)}{\partial Q} = \frac{\partial P(q, Q)}{\partial q} \tag{2.13}$$

The most general way of satisfying (2.13) is to introduce an auxiliary function of the independent variables, the so-called generating function, or generator, $F(q, Q)$. If the functions $p(q, Q)$ and $P(q, Q)$ are given in terms of $F(q, Q)$ by

$$p(q, Q) = \frac{\partial F(q, Q)}{\partial q} \tag{2.14a}$$

$$P(q, Q) = -\frac{\partial F(q, Q)}{\partial Q} \tag{2.14b}$$

then it is immediately obvious that (2.13) is satisfied and the transformation is thus canonical.

Canonical transformations carried out by aid of a generator determine $P$ and $Q$ in terms of $p$ and $q$ only implicitly. The explicit expressions for $P$ and $Q$ in terms of $p$ and $q$ must be obtained by solving the basic equations of (2.14). Given the generating function $F(q, Q)$, for example, one first solves (2.14a),

$$p = \frac{\partial F(q, Q)}{\partial q}$$

for $Q$ in terms of $p$ and $q$; the resulting expression for $Q(p, q)$ may then be substituted into (2.14b) to obtain $P$ explicitly in terms of $p$ and $q$. The significance of this implicit nature of the $(p, q) \leftrightarrow (P, Q)$ canonical transformation is great (as will be seen later) and results from the fact that the solution of (2.14a) for $Q(p, q)$ may in general be *multivalued*.

The above generating function is the one denoted $F_1(q, Q)$ by Goldstein.[15] There are three other related generators corresponding to the three other ways of choosing "one old variable and one new variable" as the two independent variables. They are denoted by Goldstein[15] as $F_2(q, P)$, $F_3(p, Q)$, and $F_4(p, P)$ and are related to $F_1(q, Q)$ by

$$F_2(q, P) = F_1(q, Q) + PQ \tag{2.15a}$$

$$F_3(p, Q) = F_1(q, Q) - pq \tag{2.15b}$$

$$F_4(p, P) = F_1(q, Q) + PQ - pq \tag{2.15c}$$

where on the RHS of (2.15) the dependent variables are assumed to be expressed in terms of the independent ones, for example, in (2.15a) $Q = Q(q, P)$. Corresponding to each of these other three generators there are a pair of differential equations analogous to (2.14) which define the transformation; these are

$$p(q, P) = \frac{\partial F_2(q, P)}{\partial q} \tag{2.16a}$$

$$Q(q, P) = \frac{\partial F_2(q, P)}{\partial P} \tag{2.16b}$$

$$q(p, Q) = -\frac{\partial F_3(p, Q)}{\partial p} \tag{2.17a}$$

$$P(p, Q) = -\frac{\partial F_3(p, Q)}{\partial Q} \tag{2.17b}$$

$$q(p, P) = -\frac{\partial F_4(p, P)}{\partial p} \tag{2.18a}$$

$$Q(p, P) = \frac{\partial F_4(p, P)}{\partial P} \tag{2.18b}$$

A canonical transformation is specified, therefore, by choosing "one old variable and one new variable" as the independent variables, specifying the appropriate generating function (of the $F_1$, $F_2$, $F_3$, or $F_4$ type), and invoking the corresponding pair of differential equations [(2.14), (2.16), (2.17), or (2.18)]. In specific applications, however, certain ones of the four possible choices of independent variables may not be possible. The simplest of all transformations, the identity transformation, for example, requires that one use an $F_2$- or $F_3$-type generator. Thus with

$$F_2(q, P) = qP \tag{2.19}$$

it is easy to see that the defining equations (2.14) become

$$p = P$$

$$Q = q$$

That is, the new variables are identically the old ones. With this result established it is now possible to compute the value of the $F_1$-type generator, for example; from (2.15a) one has

$$F_1(q, Q) = F_2(q, P) - PQ$$

but with $F_2$ from (2.19) (and since $q = Q$) one has

$$F_1(q, Q) \equiv 0 \tag{2.20}$$

One similarly finds $F_4(p, P) \equiv 0$ (but $F_3(p, Q) = -pQ$), so it is now clear why one could not carry out this transformation with an $F_1$- or $F_4$-type generator.

Another slightly more general transformation is specified by an $F_2$-type generator of the form

$$F_2(q, P) = f(q)P \tag{2.21}$$

where $f(q)$ is any function of coordinates alone. The defining equations (2.14) give

$$p = f'(q)P$$
$$Q = f(q)$$

The explicit expression for the new variables in terms of the old is

$$Q = f(q) \tag{2.22a}$$

$$P = \frac{p}{f'(q)} \tag{2.22b}$$

This type of transformation is called a *point* coordinate transformation since the new coordinates are a function only of the old coordinates, and not of the old momenta, as is generally the case; the new momenta involve both the old coordinates and old momenta. Making use again of (2.15a) one can compute the value of the $F_1$-type generator for this transformation; with the aid of (2.21) and (2.22a) it is easy to see that

$$F_1(q, Q) \equiv 0 \tag{2.20}$$

That is, *the $F_1$-type generator for a point coordinate transformation is identically zero.*

As an interesting exercise the reader can show that the transformation specified by an $F_3$-type generator of the form

$$F_3(p, Q) = -f(p)Q$$

is a point momentum transformation—that is, the new momenta are functions only of the old momenta and not the old coordinates—and that the $F_4$-type generator for such a transformation is identically zero. [Equations (2.17) are the appropriate defining equations for this case.]

## B. Derivation of the Correspondence Relations

In the previous section the generating functions enter only as an intermediary to ensure that a $(p, q) \leftrightarrow (P, Q)$ transformation is canonical; they appear to have no particular meaning in themselves. The full significance of these generating functions of classical mechanics is revealed in the classical-limit of quantum mechanics.

The primary objects of interest in quantum mechanics are the elements of unitary transformations, that is, probability amplitudes; the square modulus of a particular transformation element has physical meaning as a transition probability between the two states, or more precisely, as a *conditional probability*. The ordinary wave function for a particle in one-dimensional motion, for example, is expressed in Dirac notation as

$$\psi_E(x) \equiv \langle x \mid E \rangle$$

and is thus the transformation element connecting the coordinate states $|x\rangle$ with energy states $|E\rangle$. The square modulus of this quantity

$$|\langle x \mid E \rangle|^2$$

is interpreted as the probability distribution in $x$ for a given value of $E$, or equivalently, as the probability distribution in $E$ given a particular value of $x$. The position $x$ is obviously the coordinate of the ordinary Cartesian coordinate and momentum variables, and as is discussed in Section II.C, the energy $E$ may be considered as the momentum of another set of coordinates and momentum (the coordinate in this set of variables is the time). The statement of unitarity of the transformation is

$$\int dx \langle E' \mid x \rangle \langle x \mid E \rangle = \delta(E' - E)$$

$$\int dE \langle x' \mid E \rangle \langle E \mid x \rangle = \delta(x' - x)$$

which in this case is simply the orthogonality and completeness relations for the wave functions.

Thus consider two different sets of canonical variables, $(p, q)$ and $(P, Q)$, and the transformation elements connecting them. There are four ways to express the matrix elements between these two different "representations," depending on whether one uses coordinate or momentum states in either case; that is, the various transformation elements of interest are

$$\langle q \mid Q \rangle, \qquad \langle q \mid P \rangle, \qquad \langle p \mid Q \rangle, \qquad \langle p \mid P \rangle \tag{2.23}$$

and as has been noted above, all objects of interest in quantum mechanics are quantities of this type. The transformation elements in (2.23) are complex numbers and can thus be written in the form of an amplitude times a phase factor

$$\langle q \mid Q \rangle = A_1(q, Q) \exp \left[ \frac{if_1(q, Q)}{\hbar} \right] \tag{2.24a}$$

$$\langle q \mid P \rangle = A_2(q, P) \exp \left[ \frac{if_2(q, P)}{\hbar} \right] \tag{2.24b}$$

$$\langle p \mid Q \rangle = A_3(p, Q) \exp \left[ \frac{if_3(p, Q)}{\hbar} \right] \tag{2.24c}$$

$$\langle p \mid P \rangle = A_4(p, P) \exp \left[ \frac{if_4(p, P)}{\hbar} \right] \tag{2.24d}$$

where the amplitude and phase functions $\{A_i\}$ and $\{f_i\}$ are to be determined.

The only fact that must be assumed is that regarding transformation elements between canonically conjugate variables, that is,

$$\langle q \mid p \rangle = (2\pi i\hbar)^{-1/2} \exp\left(\frac{iqp}{\hbar}\right) \tag{2.25a}$$

and the adjoint relation follows

$$\langle p \mid q \rangle = (-2\pi i\hbar)^{-1/2} \exp\left(\frac{-ipq}{\hbar}\right) \tag{2.25b}$$

As discussed by Dirac,[14] assuming the transformation relations in (2.25) is tantamount to assuming the commutation relation between the $q$ and $p$ operators

$$[q_{op}, p_{op}] = i\hbar$$

and is thus equivalent to assuming the uncertainty principle. The uncertainty principle is thus explicitly built into the semiclassical theory that is being constructed, so it is not surprising that difficulties regarding its "violation" are never encountered. Equations (2.25) are true for any canonically conjugate set, and thus for the other set of variables $(P, Q)$:

$$\langle Q \mid P \rangle = (2\pi i\hbar)^{-1/2} \exp\left(\frac{iQP}{\hbar}\right) \tag{2.26a}$$

$$\langle P \mid Q \rangle = (-2\pi i\hbar)^{-1/2} \exp\left(\frac{-iPQ}{\hbar}\right) \tag{2.26b}$$

To determine the classical-limit approximation to the amplitude and phase functions $\{A_i\}$ and $\{f_i\}$ in (2.24) one carries out matrix multiplication of various transformation elements and evaluates the necessary integrals within the stationary phase approximation. The stationary phase approximation,[16] that is,

$$\int dx\, g(x) \exp\left[\frac{if(x)}{\hbar}\right] \simeq g(x_0)\left[\frac{2\pi i\hbar}{f''(x_0)}\right]^{1/2} \exp\left[\frac{if(x_0)}{\hbar}\right] \tag{2.27a}$$

where $x_0$ is the point of stationary phase,

$$f'(x_0) = 0 \tag{2.27b}$$

becomes exact as $\hbar \to 0$ and is, in fact, the only approximation that need be introduced. In a very precise sense one may thus say that classical mechanics is the stationary phase approximation to quantum mechanics.

Consider now the quantum mechanical identities

$$\langle q \mid Q \rangle = \int dp \langle q \mid p \rangle \langle p \mid Q \rangle \tag{2.28a}$$

$$\langle q \mid Q \rangle = \int dP \langle q \mid P \rangle \langle P \mid Q \rangle \tag{2.28b}$$

Substituting the appropriate quantities from (2.24), (2.25), and (2.26) leads to

$$A_1(q, Q) \exp\left[\frac{if_1(q, Q)}{\hbar}\right] = \int dp \frac{A_3(p, Q)}{(2\pi i\hbar)^{1/2}} \exp\left\{\frac{i}{\hbar}[f_3(p, Q) + pq]\right\} \tag{2.29a}$$

$$A_1(q, Q) \exp\left[\frac{if_1(q, Q)}{\hbar}\right] = \int dP \frac{A_2(q, P)}{(-2\pi i\hbar)^{1/2}} \exp\left\{\frac{i}{\hbar}[f_2(q, P) - PQ]\right\} \tag{2.29b}$$

Applying the stationary phase approximation [(2.27)] to (2.29a) and equating the amplitude and phase of each side gives

$$f_1(q, Q) = f_3(p, Q) + pq \tag{2.30a}$$

$$A_1(q, Q) = A_3(p, Q)\left[\frac{\partial^2 f_3(p, Q)}{\partial p^2}\right]^{-1/2} \tag{2.30b}$$

with $p$ determined by the stationary phase relation

$$\frac{\partial f_3(p, Q)}{\partial p} + q = 0 \tag{2.30c}$$

Proceeding similarly with (2.29b) gives

$$f_1(q, Q) = f_2(q, P) - PQ \tag{2.31a}$$

$$A_1(q, Q) = A_2(q, P)\left[-\frac{\partial^2 f_2(q, P)}{\partial P^2}\right]^{-1/2} \tag{2.31b}$$

with $P$ determined by the stationary phase relation

$$\frac{\partial f_2(q, P)}{\partial P} - Q = 0 \tag{2.31c}$$

Analogous to (2.28) one has the following six identities (two each for $\langle q \mid p \rangle$, $\langle p \mid Q \rangle$, and $\langle p \mid P \rangle$):

$$\langle q \mid P \rangle = \int dQ \langle q \mid Q \rangle \langle Q \mid P \rangle \tag{2.32a}$$

$$\langle q \mid P \rangle = \int dp \langle q \mid p \rangle \langle p \mid P \rangle \tag{2.32b}$$

$$\langle p \mid Q \rangle = \int dq \langle p \mid q \rangle \langle q \mid Q \rangle \tag{2.33a}$$

$$\langle p \mid Q \rangle = \int dP \langle p \mid P \rangle \langle P \mid Q \rangle \tag{2.33b}$$

$$\langle p \mid P \rangle = \int dq \langle p \mid q \rangle \langle q \mid P \rangle \tag{2.34a}$$

$$\langle p \mid P \rangle = \int dQ \langle p \mid Q \rangle \langle Q \mid P \rangle \tag{2.34b}$$

Stationary phase evaluation of all these integrals leads to the six relations between the phase functions,

$$f_2(q, P) = f_1(q, Q) + QP \tag{2.35a}$$
$$f_2(q, P) = f_4(p, P) + qp \tag{2.35b}$$
$$f_3(p, Q) = f_1(q, Q) - pq \tag{2.36a}$$
$$f_3(p, Q) = f_4(p, P) - PQ \tag{2.36b}$$
$$f_4(p, P) = f_2(q, P) - pq \tag{2.37a}$$
$$f_4(p, P) = f_3(p, Q) + PQ \tag{2.37b}$$

the six amplitude relations,

$$A_2(q, P) = A_1(q, Q)\left[\frac{\partial^2 f_1(q, Q)}{\partial Q^2}\right]^{-1/2} \tag{2.38a}$$

$$A_2(q, P) = A_4(p, P)\left[\frac{\partial^2 f_4(p, P)}{\partial p^2}\right]^{-1/2} \tag{2.38b}$$

$$A_3(p, Q) = A_1(q, Q)\left[-\frac{\partial^2 f_1(q, Q)}{\partial q^2}\right]^{-1/2} \tag{2.39a}$$

$$A_3(p, Q) = A_4(p, P)\left[-\frac{\partial^2 f_4(p, P)}{\partial P^2}\right]^{-1/2} \tag{2.39b}$$

$$A_4(p, P) = A_2(q, P)\left[-\frac{\partial^2 f_2(q, P)}{\partial q^2}\right]^{-1/2} \tag{2.40a}$$

$$A_4(p, P) = A_3(p, Q)\left[\frac{\partial^2 f_3(p, Q)}{\partial Q^2}\right]^{-1/2} \tag{2.40b}$$

and the six corresponding stationary phase relationships,

$$\frac{\partial f_1(q, Q)}{\partial Q} + P = 0 \tag{2.41a}$$

$$\frac{\partial f_4(p, P)}{\partial p} + q = 0 \tag{2.41b}$$

$$\frac{\partial f_1(q, Q)}{\partial q} - p = 0 \tag{2.42a}$$

$$\frac{\partial f_4(p, P)}{\partial P} - Q = 0 \tag{2.42b}$$

$$\frac{\partial f_2(q, P)}{\partial q} - p = 0 \tag{2.43a}$$

$$\frac{\partial f_3(p, Q)}{\partial Q} + P = 0 \tag{2.43b}$$

respectively.

Note first the eight stationary phase relations, (2.30c), (2.31c), and (2.41) to (2.43); they are precisely the eight equations [(2.14) and (2.16) to (2.18)] that define the classical generators $F_1$, $F_2$, $F_3$, and $F_4$. Thus the solution for *the phase functions $f_i$ are the classical generating functions $F_i$* discussed in Section II.A. Furthermore, the eight phase equations [(2.30a), (2.31a), and (2.35) to (2.37)] are now seen to follow identically from the relations (2.15) between the various generators. Finally, the eight amplitude equations [(2.30b), (2.31b), and (2.38) to (2.40)] can also be shown to follow identically once the $f$'s are chosen to be the $F$'s.

At this stage, therefore, the phase functions $f_i$ in (2.24) have been determined, but the amplitudes $A_i$ are still unknown and cannot be determined from any of the relations established thus far. To determine them consider identities of the type

$$\langle p \mid q \rangle = \int dQ \langle p \mid Q \rangle \langle Q \mid q \rangle \tag{2.44}$$

Substituting in the appropriate expressions gives

$$(-2\pi i\hbar)^{-1/2} \exp\left(\frac{-iqp}{\hbar}\right)$$
$$= \int dQ A_1(q, Q)^* A_3(p, Q) \exp\left\{\frac{i}{\hbar}[f_3(p, Q) - f_1(q, Q)]\right\}$$

and equating amplitudes and phases after stationary phase evaluation of the integral gives

$$-qp = f_3(p, Q) - f_1(q, Q) \tag{2.45a}$$

$$(-2\pi i\hbar)^{-1/2} = (2\pi i\hbar)^{1/2} A_1(q, Q)^* A_3(p, Q) \left[\frac{\partial^2 f_3(p, Q)}{\partial Q^2} - \frac{\partial^2 f_1(q, Q)}{\partial Q^2}\right]^{-1/2} \tag{2.45b}$$

with the stationary phase relation

$$\frac{\partial f_3(p, Q)}{\partial Q} - \frac{\partial f_1(q, Q)}{\partial Q} = 0 \tag{2.45c}$$

Equations (2.45a) and (2.45c) contain nothing new, but (2.45b) does; eliminating $A_3$ between (2.45b) and (2.39a) gives

$$|A_1(q, Q)|^2 = (2\pi\hbar)^{-1}\left[-\frac{\partial^2 f_1(q, Q)}{\partial q^2}\right]^{1/2}\left[\frac{\partial^2 f_3(p, Q)}{\partial Q^2} - \frac{\partial^2 f_1(q, Q)}{\partial Q^2}\right]^{1/2} \tag{2.46}$$

$f_3$ can be eliminated from (2.46) by using the relation

$$f_3(p, Q) = f_1(q, Q) - pq$$

or more explicitly,

$$f_3(p, Q) = f_1[q(p, Q), Q] - pq(p, Q)$$

Differentiation twice with respect to $Q$, holding $p$ constant, gives

$$\frac{\partial^2 f_3(p, Q)}{\partial Q^2} = \frac{\partial^2 f_1(q, Q)}{\partial Q^2} + \frac{\partial^2 f_1(q, Q)}{\partial q\, \partial Q}\frac{\partial q(p, Q)}{\partial Q} \tag{2.47}$$

The factor $\partial q(p, Q)/\partial Q$ can be simplified by differentiating the identity

$$p = \frac{\partial f_1[q(p, Q), Q]}{\partial q}$$

with respect to $Q$, holding $p$ constant:

$$0 = \frac{\partial^2 f_1(q, Q)}{\partial q^2}\frac{\partial q(p, Q)}{\partial Q} + \frac{\partial^2 f_1(q, Q)}{\partial q\, \partial Q}$$

or

$$\frac{\partial q(p, Q)}{\partial Q} = -\frac{\partial^2 f_1(q, Q)/\partial q\, \partial Q}{\partial^2 f_1(q, Q)/\partial q^2}$$

Substituting this result into (2.47) gives

$$\frac{\partial^2 f_3(p, Q)}{\partial Q^2} = \frac{\partial^2 f_1(q, Q)}{\partial Q^2} - \frac{[\partial^2 f_1(q, Q)/\partial q\, \partial Q]^2}{\partial^2 f_1(q, Q)/\partial q^2}$$

and substituting this back into (2.46) gives the final result:

$$|A_1(q, Q)|^2 = \frac{\partial^2 f_1(q, Q)/\partial q\, \partial Q}{2\pi\hbar} \tag{2.48a}$$

In a similar manner one can show that the other amplitude functions are given similarly:

$$|A_2(q, P)|^2 = \frac{\partial^2 f_2(q, P)/\partial q\, \partial P}{2\pi\hbar} \tag{2.48b}$$

$$|A_3(p, Q)|^2 = \frac{\partial^2 f_3(p, Q)/\partial p\, \partial Q}{2\pi\hbar} \tag{2.48c}$$

$$|A_4(p, P)|^2 = \frac{\partial^2 f_4(p, P)/\partial p\, \partial P}{2\pi\hbar} \tag{2.48d}$$

As has been shown previously,[8a] it is also possible to arrive at the results in (2.48) by invoking directly the unitary of the transformation elements. Thus if $U(x, y)$ is any unitary "matrix" with continuous indices $x$ and $y$ of the form

$$U(x, y) = A(x, y) \exp\left[\frac{i\phi(x, y)}{\hbar}\right]$$

then it has been shown that in the limit $\hbar \to 0$ the amplitude function is related to the phase by

$$|A(x, y)|^2 = \frac{\partial^2 \phi(x, y)/\partial x\, \partial y}{2\pi\hbar}$$

The above conditions leave the amplitude functions undetermined with respect to an overall constant phase. Although constant overall phase factors have no physical significance (and we are often careless in getting them correct) it is possible to arrive at a unique assignment. In order for (2.24b) and (2.24c) to reduce to the known expressions in (2.25) for the case of the identity transformation $[F_2(q, P) = qP$ or $F_3(p, Q) = -pQ]$, one must choose

$$A_2(q, P) = \left[\frac{\partial^2 f_2(q, P)/\partial q\, \partial P}{2\pi i\hbar}\right]^{1/2} \tag{2.49a}$$

and

$$A_3(p, Q) = \left[\frac{\partial^2 F_3(p, Q)/\partial p\, \partial Q}{2\pi i\hbar}\right]^{1/2} \tag{2.49b}$$

From (2.38a) and (2.40b), for example, one can then determine that

$$A_1(q, Q) = \left[\frac{\partial^2 F_1(q, Q)/\partial q\, \partial Q}{(-2\pi i\hbar)}\right]^{1/2} \tag{2.49c}$$

$$A_4(p, P) = \left[\frac{\partial^2 F_4(p, P)/\partial p\, \partial P}{(-2\pi i\hbar)}\right]^{1/2} \tag{2.49d}$$

In summary, the classical limit of the unitary transformation elements connecting the $(p, q)$ and $(P, Q)$ representations are

$$\langle q \mid Q\rangle = \left[-\frac{\partial^2 F_1(q, Q)/\partial q\, \partial Q}{2\pi i\hbar}\right]^{1/2} \exp\left[\frac{iF_1(q, Q)}{\hbar}\right] \tag{2.50a}$$

$$\langle q \mid P\rangle = \left[\frac{\partial^2 F_2(q, P)/\partial q\, \partial P}{2\pi i\hbar}\right]^{1/2} \exp\left[\frac{iF_2(q, P)}{\hbar}\right] \tag{2.50b}$$

$$\langle p \mid Q\rangle = \left[\frac{\partial^2 F_3(p, Q)/\partial p\, \partial Q}{2\pi i\hbar}\right]^{1/2} \exp\left[\frac{iF_3(p, Q)}{\hbar}\right] \tag{2.50c}$$

$$\langle p \mid P\rangle = \left[-\frac{\partial^2 F_4(p, P)/\partial p\, \partial P}{2\pi i\hbar}\right]^{1/2} \exp\left[\frac{iF_4(p, P)}{\hbar}\right] \tag{2.50d}$$

where the $F$'s are the classical generating functions for the $(p, q) \leftrightarrow (P, Q)$ canonical transformation. It may also be worthwhile to note here the explicitly multidimensional form of (2.50): if $q$ and $Q$ are $N$-dimensional coordinate vectors, then (2.50) is modified by the replacement

$$\frac{\partial F_1(q, Q)/\partial q\, \partial Q}{(-2\pi i\hbar)} = \frac{\det |\partial F_1(\mathbf{q}, \mathbf{Q})/\partial q_i\, \partial Q_j|}{(-2\pi i\hbar)^N} \tag{2.51}$$

and similarly for the other cases. Equations (2.50) express classical-limit quantum mechanics in its most general form. Since the only approximation in the theory is the use of the stationary phase approximation to evaluate integrals over oscillatory integrands, so long as all such integrations are evaluated within this approximation one has a complete, internally consistent "semiclassical algebra"; this is because successive applications of the stationary phase approximation commute with one another. For example, if classical-limit expressions are used for the quantities in the integrand of

$$\langle p \mid Q\rangle = \int d\tilde{p} \langle p \mid \tilde{p}\rangle \langle \tilde{p} \mid Q\rangle$$

where $(\tilde{p}, \tilde{q})$ are a canonical set different from $(p, q)$ and $(P, Q)$, and if the integration over $\tilde{p}$ is evaluated by stationary phase, the result is the same as though the classical-limit approximation were used directly for the quantity $\langle p \mid Q \rangle$.

One complication that has been glossed over thus far is that "one old variable and one new variable" do not necessarily determine the other two variables uniquely; this point was mentioned in Section II.A in noting that the defining equations could give rise to multiple roots in constructing explicit expressions for $P$ and $Q$ in terms of $p$ and $q$. In terms of the development in this section consider (2.28a), for example,

$$\langle q \mid Q \rangle = \int dp \langle q \mid p \rangle \langle p \mid Q \rangle$$

Quantum mechanically all values of $p$ contribute to the $\langle q \mid Q \rangle$ transformation element; semiclassically, however, the integral over $p$ is evaluated by stationary phase and this singles out a particular value of $p$, the one that is determined classically by the specific values $q$ and $Q$. As noted, though, $p(q, Q)$ may be a multivalued function, meaning that there may be more than one point of stationary phase in the above integral over $p$. The modification of the stationary phase formula (2.27) for the case of several points of stationary phase is

$$\int dx\, g(x) \exp\left[\frac{if(x)}{\hbar}\right] \simeq \sum_k g(x_k) \left[\frac{2\pi i\hbar}{f''(x_k)}\right]^{1/2} \exp\left[\frac{if(x_k)}{\hbar}\right] \tag{2.52a}$$

$$f'(x_k) = 0 \tag{2.52b}$$

that is, simply a sum of similar terms from each point of stationary phase. Equations (2.50) are thus a sum of such terms over all multiple roots.

The amplitude, or preexponential factors in (2.50), have simple physical interpretations as classical probability factors. Consider, for example, the square modulus of (2.50a)

$$|\langle q \mid Q \rangle|^2 = \frac{|\partial^2 F_1(q, Q)/\partial q\, \partial Q|}{2\pi\hbar} \tag{2.53}$$

Taking account of the derivative relations for $F_1$, (2.14), this can be written as

$$|\langle q \mid Q \rangle|^2 = \frac{|\partial P(q, Q)/\partial q|}{2\pi\hbar} = \left[2\pi\hbar \left|\left(\frac{\partial q}{\partial P}\right)_Q\right|\right]^{-1} \tag{2.54a}$$

and also as

$$|\langle q \mid Q \rangle|^2 = \frac{|\partial p(q, Q)/\partial Q|}{2\pi\hbar} = \left[2\pi\hbar \left|\left(\frac{\partial Q}{\partial p}\right)_q\right|\right]^{-1} \tag{2.54b}$$

Thinking of $Q$ as being held constant in (2.54a), the LHS is the probability distribution in $q$; with $Q$ held constant, however, $P$ and $q$ are functionally related by the canonical transformation, so that their probability distributions are related:

$$\text{Prob}\,(q)\,dq = \text{Prob}\,(P)\,dP \tag{2.55}$$

With $Q$ held constant, though, the probability distribution in $P$ is constant, that is, any value of $P$ is equally likely (see the uncertainty principle), so that (2.55) leads to

$$\text{Prob}\,(q) = \frac{\text{constant}}{|(\partial q/\partial P)_Q|}$$

which is the same as (2.54a). A similar interpretation of (2.54b) follows if one imagines holding $q$ constant and interpreting $|\langle q \mid Q\rangle|^2$ as the probability distribution in $Q$.

Equations (2.54) can also be interpreted in terms of classical phase space distributions. The *a priori* distribution in $p$–$q$ phase space is constant,

$$\rho(p, q) = h^{-1} = (2\pi\hbar)^{-1} \tag{2.56}$$

so that the probability of $q$ and $Q$ taking on certain particular values is given by

$$|\langle q \mid Q\rangle|^2 \equiv \int dp' \int dq' \rho(p', q')\, \delta(q - q')\, \delta[Q - Q(p', q')] \tag{2.57}$$

where $Q(p', q')$ is the function determined by the canonical transformation $(p, q) \leftrightarrow (P, Q)$. With (2.56) this becomes

$$\begin{aligned} |\langle q \mid Q\rangle|^2 &= \int dp' \int dq' (2\pi\hbar)^{-1}\, \delta(q - q')\, \delta[Q - Q(p', q')] \\ &= (2\pi\hbar)^{-1} \int dp'\, \delta[Q - Q(p', q)] \\ &= \left[2\pi\hbar \left|\left(\frac{\partial Q}{\partial p}\right)_q\right|\right]^{-1} \end{aligned}$$

which is (2.54b). Equation (2.54a) results from a similar argument if one considers the phase space distribution in $P$–$Q$ phase space [which is also given by (2.56)].

In general, therefore, the preexponential Jacobian factors in (2.50) express purely classical probability relations. The phases of the transformation elements are the classical generators of the canonical transformation and thus enter into classical-limit quantum mechanics in a more direct manner than they appear in classical mechanics itself.

### C. Example 1: One-Dimensional Systems

To illustrate the use of the correspondence relations in (2.50) consider first the simple case of a one-dimensional dynamical system with Hamiltonian of the form

$$H(p, q) = \frac{p^2}{2m} + V(q) \tag{2.58}$$

$p$ and $q$ being ordinary Cartesian variables. The quantum mechanical object of interest is the coordinate wave function $\langle q \mid E \rangle$, $E$ being the total energy.

To apply the results of the previous section one thus needs to carry out a canonical transformation from $(p, q)$ to $(P, Q)$ where the new momentum $P$ is the energy $E$; that is, one defines $P(p, q)$ by

$$P(p, q) = \frac{p^2}{2m} + V(q) \tag{2.59}$$

What is needed is the generator for this transformation, and we search for one of the $F_2$-type. Equation (2.16a)

$$p(q, P) = \frac{\partial F_2(q, P)}{\partial q}$$

may be thought of as a differential equation for $F_2$; if $p(q, P)$ is known one can obtain $F_2$ by integration

$$F_2(q, P) = \int dq p(q, P) + C(P) \tag{2.60}$$

where $C(P)$ is an undetermined (and irrelevant) function of $P$. From (2.59), however, it is possible to solve for $p(q, P)$,

$$p(q, P) = \pm\{2m[P - V(q)]\}^{1/2} \tag{2.61}$$

and the result is seen to be multivalued. Substitution of (2.61) into (2.60) gives the desired generator:

$$F_2(q, P) = \pm \int dq\{2m[P - V(q)]\}^{1/2} + C(P) \tag{2.62}$$

Having found $F_2$ it is now possible to invoke the other derivative relation for $F_2$,

$$Q = \frac{\partial F_2(q, P)}{\partial P}$$

to see what the new coordinate is

$$Q = \pm \int dq \left\{ \frac{2}{m} [P - V(q)] \right\}^{-1/2} + C'(P) \tag{2.63}$$

which to within an additive constant is recognized as the time.

The preexponential factor in (2.50) requires the mixed second derivative of $F_2$, and it is easy to show that

$$\frac{\partial^2 F_2(q, P)}{\partial q\, \partial P} = \pm \left\{ \frac{2}{m} [P - V(q)] \right\}^{-1/2} \tag{2.64}$$

With (2.62) and (2.64) the general correspondence relation (2.50b) gives the wave function, normalized on the energy scale, as

$$\begin{aligned} \langle q \mid E \rangle = (2\pi\hbar)^{-1/2} &\left\{ \frac{2}{m} [E - V(q)] \right\}^{-1/4} \\ &\times \left\{ \exp\left( -\frac{i\pi}{4} + \int dq \{2m[E - V(q)]\}^{1/2} \right) \right. \\ &\left. + \exp\left( \frac{i\pi}{4} - \int dq \{2m[E - V(q)]\}^{1/2} \right) \right\} \end{aligned} \tag{2.65}$$

where $P$ has been replaced by the more familiar symbol $E$ and the overall constant phase $\exp[iC(P)/\hbar]$ has been dropped. There are two terms in (2.65) because of the double-valuedness of the relation in (2.61).

Equation (2.65) is the standard WKB approximation[17] for the wave function which is usually obtained by solving Schrödinger's equation asymptotic in $\hbar$. It arises here as a special case of the classical-limit transformation relations in (2.50). One interesting note is that, having kept proper track of the powers of $i$, the ubiquitous $\pi/4$ of WKB theory[17] emerges automatically from the general expressions without recourse to any "connection formulas."

The origin of the two terms in the wave function (2.65) was noted to be the double-valued nature of (2.61). This, of course, has a simple physical interpretation: there are two ways for the particle to be at position $q$ with energy $E$, going to the right (positive momentum) or going to the left (negative momentum); the transformation element $\langle q \mid E \rangle$ takes no cognizance of the momentum $p$ and is thus a superposition of the amplitude for each possibility.

### D. Example 2: Clebsch-Gordan Coefficients

A less trivial application of the classical-limit transformation relations in (2.50) is that which arises in coupling two angular momenta together to form eigenstates of total angular momentum. This application is also important since it is not possible to obtain the classical limit of the Clebsch-Gordan, or vector-coupling coefficients by any other procedure. Historically most semiclassical approximations (e.g., the one-dimensional wave function, spherical harmonics, and rotation matrices) have been obtained by applying the WKB approximation to the differential equation for the particular quantity. This is not possible in the present case, of course, for there is no such differential equation.

Let $(l, m_l)$ and $(j, m_j)$ be the two angular momenta and their projections onto a space-fixed axis; $(J, M)$ is the total angular momentum and its projection. The Clebsch-Gordan coefficients are the elements of the unitary transformation that connects the uncoupled, or direct product, states $|jm_jlm_l\rangle \equiv |jm_j\rangle\,|lm_l\rangle$ and the coupled states $|JMjl\rangle$ which are eigenstates of total angular momentum:

$$|JMjl\rangle = \sum_{m_l, m_j} |jm_jlm_l\rangle\langle jm_jlm_l \mid JMjl\rangle \tag{2.66}$$

Rose[18] uses the notation

$$C(jlJ; m_jm_lM) \equiv \langle jm_jlm_l \mid JMjl\rangle \tag{2.67}$$

for the Clebsch-Gordan coefficients. Since the Clebsch-Gordan coefficients are the elements of a unitary transformation, the general correspondence relations in (2.50) apply directly: the classical-limit approximation for them is given by (2.50d) where $F_4(p, P)$ is the generator of the canonical transformation from the "old momenta" $p$ to the "new momenta" $P$, where

$$\begin{aligned} p_1 &= j & P_1 &= j \\ p_2 &= l & P_2 &= l \\ p_3 &= m_j & P_3 &= M \\ p_4 &= m_l & P_4 &= J \end{aligned} \tag{2.68, 2.69}$$

As a first step one needs to express the new momenta in terms of the old variables. Clearly

$$\begin{aligned} P_1 &= p_1 \\ P_2 &= p_2 \\ P_3 &= p_3 + p_4 \end{aligned} \tag{2.70}$$

and to find $P_4 \equiv J$ in terms of the old variables note that

$$J^2 = l^2 + j^2 + 2\mathbf{l} \cdot \mathbf{j} \tag{2.71}$$

and

$$\mathbf{l} \cdot \mathbf{j} = l_x j_x + l_y j_y + l_z j_z \tag{2.72}$$

But

$$j_z = m_j$$
$$j_y = (j^2 - m_j^2)^{1/2} \sin q_{m_j}$$
$$i_x = (j^2 - m_j^2)^{1/2} \cos q_{m_j}$$

and

$$l_z = m_l$$
$$l_y = (l^2 - m_l^2)^{1/2} \sin q_{m_l}$$
$$l_x = (l^2 - m_l^2)^{1/2} \cos q_{m_l}$$

where $q_{m_j}$ and $q_{m_l}$ are the coordinates conjugate to $m_j$ and $m_l$; physically, they are the angles of precession of $j$ and $l$, respectively, about the space-fixed $z$-axis. Equation (2.72) thus becomes

$$\mathbf{l} \cdot \mathbf{j} = m_j m_l + [(j^2 - m_j^2)(l^2 - m_l^2)]^{1/2} \cos (q_{m_j} - q_{m_l}) \tag{2.73}$$

and $J \equiv P_4$ is thus expressed in terms of the old variables as

$$P_4^2 = p_1^2 + p_2^2 + 2p_3p_4 + 2[(p_1^2 - p_3^2)(p_2^2 - p_4^2)]^{1/2} \cos (q_3 - q_4) \tag{2.74}$$

Equations (2.70) and (2.74) thus give the new momenta in terms of the old variables.

Because of the conservation relation

$$M = m_j + m_l \tag{2.75}$$

the variables $m_j$ and $m_l$ are not independent. To eliminate this redundancy it is useful to transform to an intermediate set of variables $(p', q')$, where the intermediate momenta are defined in terms of the old by

$$\begin{aligned} p_1' &= p_1 \\ p_2' &= p_2 \\ p_3' &= p_3 + p_4 \\ p_4' &= p_4 \end{aligned} \tag{2.76}$$

This transformation is effected by the $F_3$-type generator

$$F_3(p, q') = -p_1q_1' - p_2q_2' - (p_3 + p_4)q_3' - p_4q_4' \tag{2.77}$$

From the derivative relations

$$q = -\frac{\partial F_3(p, q')}{\partial p}$$

$$p' = -\frac{\partial F_3(p, q')}{\partial q'}$$

one easily verifies that (2.76) are fulfilled, and in addition that

$$\begin{aligned} q_3 &= q_3' \\ q_4 &= q_3' + q_4' \end{aligned} \tag{2.78}$$

The new momenta $P$ are now given in terms of the intermediate variables by

$$\begin{aligned} P_1 &= p_1' \\ P_2 &= p_2' \\ P_3 &= p_3' \\ P_4^2 &= p_1'^2 + p_2'^2 + 2p_4'(p_3' - p_4') \\ &\quad + 2(p_2'^2 - p_4'^2)^{1/2}[p_1'^2 - (p_3' - p_4')^2]^{1/2} \cos q_4' \end{aligned} \tag{2.79}$$

Furthermore, since

$$F_4(p, P) = F_4(p, p') + F_4(p', P) \tag{2.80}$$

the desired generator $F_4(p, P)$ is identical to $F_4(p', P)$ because the $(p, q) \leftrightarrow (p', q')$ is a *point momentum* transformation and $F_4(p', p)$ is thus zero. The problem, then, is to construct the $F_4$ generator for the $(p', q') \leftrightarrow (P, Q)$ transformation. Since $p_i' = P_i$ for $i = 1, 2, 3$, [(2.79)], the $(p', q') \leftrightarrow (P, Q)$ transformation is essentially a one-dimensional transformation $(p_4', q_4') \leftrightarrow (P_4, Q_4)$, with $p_1' = P_1 \equiv j$, $p_2' = P_2 \equiv l$, and $p_3' = P_3 \equiv M$ appearing simply as constant parameters. The $F_4$ generator for this one-dimensional transformation can be obtained by integration of the derivative relation for $F_4$, (2.18a),

$$F_4(p_4', P_4) = -\int dp_4' q_4'(p_4', P_4) \tag{2.81}$$

where $q_4'$ is given in terms of $p_4'$ and $P_4$ from (2.79):

$$q_4'(p_4', P_4) = \pm\cos^{-1}\left\{\frac{P_4^2 - p_1'^2 - p_2'^2 - 2p_4'(p_3' - p_4')}{2(p_2'^2 - p_4'^2)^{1/2}[p_1'^2 - (p_3' - p_4')^2]^{1/2}}\right\} \tag{2.82}$$

the two signs indicating a double-valuedness of $q_4'$.

The most difficult task now is evaluating the integral indicated by (2.81) and (2.82); this can be done, however, and gives

$$F_4(p_4', P_4) = \pm\chi \tag{2.83}$$

where

$$\chi = j \cos^{-1}\left[\frac{-M(j^2 + l^2 - J^2) - m_l(j^2 + J^2 - l^2)}{\alpha[j^2 - (M - m_l)^2]^{1/2}}\right]$$

$$+ l \cos^{-1}\left[\frac{(M - m_l)(l^2 + J^2 - j^2) + M(l^2 + j^2 - J^2)}{\alpha(l^2 - m_l^2)^{1/2}}\right]$$

$$+ J \cos^{-1}\left[\frac{m_l(J^2 + j^2 - l^2) - (M - m_l)(J^2 + l^2 - j^2)}{\alpha(J^2 - M^2)^{1/2}}\right]$$

$$- (M - m_l) \cos^{-1}\left[\frac{j^2 + J^2 - l^2 - 2M(M - m_l)}{2(J^2 - M^2)^{1/2}[j^2 - (M - m_l)^2]^{1/2}}\right]$$

$$+ m_l \cos^{-1}\left[\frac{l^2 + J^2 - j^2 - 2m_l M}{2(J^2 - M^2)^{1/2}(l^2 - m_l^2)^{1/2}}\right] \tag{2.84}$$

with

$$\alpha = [(J + l + j)(-J + l + j)(J - l + j)(J + l - j)]^{1/2}$$
$$= (-J^4 - l^4 - j^4 + 2J^2j^2 + 2J^2l^2 + 2l^2j^2)^{1/2} \tag{2.85}$$

where $p_1'$, $p_2'$, $p_3'$, $p_4'$, and $P_4$ have been replaced by $j$, $l$, $M$, $m_l$, and $J$, respectively. [The simplest way to verify that (2.83) and (2.84) are correct is to differentiate with respect to $p_4' \equiv m_l$.] The preexponential factor of the transformation element is easily found to be

$$\frac{\partial^2 F_4(p_4', P_4)}{\partial p_4' \, \partial P_4} = \pm \frac{\partial \chi}{\partial m_l \, \partial J} = \pm \frac{2J}{\beta} \tag{2.86}$$

where

$$\beta = [-J^4 - l^4 - j^4 + 2J^2l^2 + 2J^2j^2 + 2l^2j^2$$
$$- 4m_l^2J^2 - 4M^2l^2 + 4m_lM(J^2 + l^2 - j^2)]^{1/2} \tag{2.87}$$

The Clebsch-Gordan coefficient is then obtained from (2.50d), there being two terms because of the double-valued nature of $q_4'$, (2.82):

$$C(jlJ; M - m_l, m_l) = \left(\frac{J}{\pi\beta}\right)^{1/2}\left[\exp\left(\frac{i\pi}{4} + i\chi\right) + \exp\left(-\frac{i\pi}{4} - i\chi\right)\right]$$
$$= 2\left(\frac{J}{\pi\beta}\right)^{1/2} \cos\left(\frac{\pi}{4} + \chi\right) \tag{2.88}$$

with $\chi$ and $\beta$ given by (2.84) and (2.87).

The final result takes a more symmetrical form if one relabels $(j, l, J, m_j, m_l, M)$ as $(j_1, j_2, j_3, m_1, m_2, m_3)$, respectively, and also replaces

$m_3$ by $-m_3$. The general expression for the classical limit of the Clebsch-Gordan coefficient is then

$$C(j_1j_2j_3; m_1m_2, -m_3) = 2\left(\frac{j_3}{\pi\beta}\right)^{1/2} \cos\left(\frac{\pi}{4} + \chi\right) \tag{2.89a}$$

where $-m_3 = m_1 + m_2$, and

$$\begin{aligned}
\chi = {} & j_1 \cos^{-1}\left[\frac{m_3(j_1^2 + j_2^2 - j_3^2) - m_2(j_1^2 + j_3^2 - j_2^2)}{\alpha\lambda_1}\right] \\
& + j_2 \cos^{-1}\left[\frac{m_1(j_2^2 + j_3^2 - j_1^2) - m_3(j_2^2 + j_1^2 - j_3^2)}{\alpha\lambda_2}\right] \\
& + j_3 \cos^{-1}\left[\frac{m_2(j_3^2 + j_1^2 - j_2^2) - m_1(j_3^2 + j_2^2 - j_1^2)}{\alpha\lambda_3}\right] \\
& - m_1 \cos^{-1}\left[\frac{\lambda_1^2 + \lambda_3^2 - \lambda_2^2}{2\lambda_1\lambda_3}\right] \\
& + m_2 \cos^{-1}\left[\frac{\lambda_2^2 + \lambda_3^2 - \lambda_1^2}{2\lambda_2\lambda_3}\right]
\end{aligned} \tag{2.89b}$$

with

$$\lambda_i = (j_i^2 - m_i^2)^{1/2}$$

$$\alpha = [(j_1 + j_2 + j_3)(-j_1 + j_2 + j_3)(j_1 - j_2 + j_3)(j_1 + j_2 - j_3)]^{1/2}$$

$$\beta = [(\lambda_1 + \lambda_2 + \lambda_3)(-\lambda_1 + \lambda_2 + \lambda_3)(\lambda_1 - \lambda_2 + \lambda_3)(\lambda_1 + \lambda_2 - \lambda_3)]^{1/2}$$

In applications, of course, one would want to make the usual semiclassical replacement $j_i \to j_i + \frac{1}{2}$.

Equation (2.89) is a new result, only the case $m_3 = 0$ having been given before.[8a] It is not difficult to show that (2.89) reduces to previously obtained results[19,20] in cases where some of the angular momenta are much larger than others.

### E. Example 3: Dynamical Transformations

The most important canonical transformation for our purposes is the *dynamical* transformation, that is, the time evolution of the coordinates and momenta that takes place according to the classical equations of motion. It is well-known in classical mechanics[15] that the values of the coordinates and momenta at time $t_2$, $p_2$ and $q_2$, are related to the values at time $t_1$, $p_1$ and $q_1$, by a canonical transformation.

The transition amplitude for motion from coordinate $q_1$ to coordinate $q_2$ in time increment $(t_1, t_2)$ is an element of the dynamical transformation; in the classical limit it is thus given by (2.50a),

$$\langle q \mid Q \rangle = \left[ -\frac{\partial^2 F_1(q, Q)/\partial q\, \partial Q}{2\pi i \hbar} \right]^{1/2} \exp\left[ \frac{iF_1(q, Q)}{\hbar} \right] \tag{2.90}$$

where $F_1(q, Q)$ is the generator of the dynamical transformation. In quantum mechanics the "initial" state is conventionally written on the right, the "final" state on the left, so to conform with this the "old" variables $(p, q)$ are chosen to be the values at $t_2$, and the "new" variables $(P, Q)$ are the values at $t_1$; that is, one has

$$(P, Q) \equiv (p_1, q_1) \tag{2.91a}$$

$$(p, q) \equiv (p_2, q_2) \tag{2.91b}$$

For illustrative purposes consider first the short-time limit of dynamics so that the solution of the classical equations of motion is given by

$$q_2 = q_1 + p_1 \frac{\Delta t}{m} \tag{2.92a}$$

$$p_2 = p_1 \tag{2.92b}$$

where $\Delta t = t_2 - t_1$, or inversely,

$$q_1 = q_2 - p_2 \frac{\Delta t}{m} \tag{2.93a}$$

$$p_1 = p_2 \tag{2.93b}$$

For this dynamical transformation the "new" coordinate $Q$ is thus given in terms of the old variables by

$$Q(p, q) = q - p\frac{\Delta t}{m} \tag{2.94}$$

which can be solved for $p$, giving

$$p(q, Q) = \frac{(q - Q)m}{\Delta t} \tag{2.95}$$

The generator $F_1(q, Q)$ can be obtained by integrating the derivative relation

$$p(q, Q) = \frac{\partial F_1(q, Q)}{\partial q}$$

Using (2.95), this gives

$$F_1(q, Q) = \frac{(\frac{1}{2}q^2 - qQ)m}{\Delta t} + C(Q) \tag{2.96}$$

where $C(Q)$ is an unknown function of $Q$. To determine it one appeals to the other derivative relative for $F_1$,

$$P = -\frac{\partial F_1(q, Q)}{\partial Q}$$

which leads to

$$\frac{(q - Q)m}{\Delta t} = \frac{qm}{\Delta t} - C'(Q)$$

and one thus finds

$$C(Q) = \tfrac{1}{2}Q^2 \frac{m}{\Delta t}$$

so that (2.96) finally gives

$$F_1(q, Q) = \tfrac{1}{2}(q - Q)^2 \frac{m}{\Delta t} \tag{2.97}$$

With $F_1(q, Q)$ given by (2.97) the transformation element of (2.90) becomes

$$\langle q \mid Q \rangle = \left(2\pi i\hbar \frac{\Delta t}{m}\right)^{-1/2} \exp\left[\frac{im(q_2 - q_1)^2}{2\hbar\,\Delta t}\right] \tag{2.98}$$

where $q$ and $Q$ have been replaced by $q_2$ and $q_1$.

The notation $\langle q \mid Q \rangle$ for the transition amplitude from initial point $q_1 = Q$ to final position $q_2 = q$ is unorthodox, the normal quantum mechanical notation for this quantity being

$$\langle q_2| \exp\left[-\frac{iH(t_2 - t_1)}{\hbar}\right] |q_1\rangle \tag{2.99}$$

That is, $|q\rangle$ is the ordinary Schrödinger state $|q_2\rangle$, but $|Q\rangle$ is the Heisenberg, or time-evolved state

$$|Q\rangle = \exp\left[-\frac{iH(t_2 - t_1)}{\hbar}\right] |q_1\rangle$$

that was the Schrödinger state $|q_1\rangle$ at $t_1$ but which has evolved dynamically to time $t_2$. Equations (2.98) and (2.99) thus give the classical limit for matrix elements of the quantum mechanical propagator, or time-evolution operator

$$\langle q|_2 \exp\left[-\frac{iH(t_2-t_1)}{\hbar}\right]|q_1\rangle = \left[\frac{2\pi i\hbar(t_2-t_1)}{m}\right]^{-1/2} \exp\left[\frac{im(q_2-q_1)^2}{2\hbar(t_2-t_1)}\right] \tag{2.100}$$

which is valid for short-time increments; (2.100) is also the correct quantum mechanical result for short times.

In general, of course, it is not possible to solve the classical equations of motion analytically and construct the generator of the dynamical transformation as was done above for the short-time case. One can appeal to the well-known result of classical mechanics,[15] however, that the generator of the dynamical transformation is the classical action, the time integral of the Lagrangian:

$$\phi(q_2, q_1) = \int_{t_1}^{t_2} dt\{p(t)\dot{q}(t) - H[p(t), q(t)]\} \tag{2.101}$$

the symbol $\phi$ now being used rather than $F_1$. The classical limit for matrix elements of the propagator are then given by the general correspondence relation,

$$\langle q_2| \exp\left[-\frac{iH(t_2-t_1)}{\hbar}\right]|q_1\rangle$$
$$= \left[-\frac{\partial^2\phi(q_2,q_1)/\partial q_2\,\partial q_1}{2\pi i\hbar}\right]^{1/2} \exp\left[\frac{i\phi(q_2,q_1)}{\hbar}\right] \tag{2.102}$$

The action integral $\phi(q_2, q_1)$ is understood to be a function of $q_2$ and $q_1$ in the following sense: the initial conditions $(p_1, q_1)$ determine the classical trajectory $p(t), q(t)$ from which the integrand of (2.101) is constructed: the value of the action integral is thus an explicit function of $(p_1, q_1)$. The final position $q_2$ is also a function of $(p_1, q_1)$, however, and one imagines inverting the equation

$$q_2 = q_2(p_1, q_1)$$

to eliminate $p_1$ in favor of $q_2$ as the other independent variable in addition to $q_1$; that is, $q_1$ and $q_2$ are taken as the two boundary conditions that determine the classical trajectory rather than the initial conditions $(p_1, q_1)$. The "double-ended" boundary conditions $(q_1, q_2)$ thus are an example of the "one old variable and one new variable" that are the independent variables for the generator of a canonical transformation. As in the general situation, too, $p_1(q_2, q_1)$ is not necessarily a single-valued function; that is, there may be more than one value of $p_1$—and therefore more than one classical trajectory—that goes from $q_1$ to $q_2$ in the time increment

$(t_1, t_2)$. The transition amplitude in (2.102) is thus understood to be a sum of similar terms over all classical trajectories that go from $q_1$ at time $t_1$ to $q_2$ at time $t_2$.

The action function $\phi(q_2, q_1)$ obeys the same derivative relations as any $F_1$-type generator, (2.14), but it is illustrative to demonstrate this explicitly for the case of a dynamical transformation. Considering the initial conditions $p_1$ and $q_1$ as the independent variables, one has

$$\left(\frac{\partial}{\partial q_1}\right)_{p_1} \phi[q_2(p_1, q_1), q_1] = \frac{\partial \phi(q_2, q_1)}{\partial q_2}\frac{\partial q_2(p_1, q_1)}{\partial q_1} + \frac{\partial \phi(q_2, q_1)}{\partial q_1} \tag{2.103}$$

In (2.101) the trajectory $p(t)$, $q(t)$ is a function of the initial conditions, that is,

$$p(t) \equiv p(t; p_1, q_1)$$
$$q(t) \equiv q(t; p_1, q_1)$$

Differentiating (2.101) with respect to $q_1$ (holding $p_1$ constant) thus gives

$$\left(\frac{\partial}{\partial q_1}\right)_{p_1} \phi[q_2(p_1, q_1), q_1]$$
$$= \int_{t_1}^{t_2} dt \left[\frac{\partial p(t)}{\partial q_1}\dot{q}(t) + p(t)\frac{d}{dt}\frac{\partial q(t)}{\partial q_1} - \frac{\partial H}{\partial p}\frac{\partial p(t)}{\partial q_1} - \frac{\partial H}{\partial q}\frac{\partial q(t)}{\partial q_1}\right] \tag{2.104}$$

Because of the equations of motion,

$$\dot{q}(t) = \frac{\partial H}{\partial p}$$
$$\dot{p}(t) = -\frac{\partial H}{\partial q}$$

the first and third terms in (2.104) cancel, and integration of the second term by parts gives

$$\left(\frac{\partial}{\partial q_1}\right)_{p_1} \phi[q_2(p_1, q_1), q_1] = p(t)\frac{\partial q(t)}{\partial q_1}\bigg|_{t_1}^{t_2} + \int_{t_1}^{t_2} dt \frac{\partial q(t)}{\partial q_1}\left[-\dot{p}(t) - \frac{\partial H}{\partial q}\right] \tag{2.105}$$

The second term vanishes because of the equations of motion, and (2.105) thus becomes

$$\left(\frac{\partial}{\partial q_1}\right)_{p_1} \phi[q_2(p_1, q_1), q_1] = p_2\frac{\partial q_2(p_1, q_1)}{\partial q_1} - p_1 \tag{2.106}$$

From (2.106) and (2.103) one thus identifies the standard result, (2.14):

$$\frac{\partial \phi(q_2, q_1)}{\partial q_2} = p_2 \tag{2.107a}$$

$$\frac{\partial \phi(q_2, q_1)}{\partial q_1} = -p_1 \tag{2.107b}$$

One consequence of these derivative relations is that the preexponential factor in (2.102) can be expressed in the following form, which usually is more convenient in applications:

$$-\frac{\partial^2 \phi(q_2, q_1)/\partial q_2\, \partial q_1}{2\pi i\hbar} = \left[2\pi i\hbar\left(\frac{\partial q_2}{\partial p_1}\right)_{q_1}\right]^{-1} \tag{2.108}$$

Equation (2.108) provides the simple physical interpretation of the preexponential factor that was discussed in Section II.B: with $q_1$ fixed, for example, $q_2$ and $p_1$ are functionally related by the classical equations of motion, and their probability distributions are thus related by

$$\text{Prob}\,(q_2)\, dq_2 = \text{Prob}\,(p_1)\, dp_1$$

With $q_1$ fixed, however, $p_1$ is a random variable (after the uncertainty principle) so that Prob $(p_1) = h^{-1}$, and one thus obtains

$$\text{Prob}\,(q_2) \equiv \left|\langle q_2|\exp\left[-\frac{iH(t_2 - t_1)}{\hbar}\right]|q_1\rangle\right|^2 = \frac{h^{-1}}{|(\partial q_2/\partial p_1)_{q_1}|}$$

which is (2.108).

In many applications it is more convenient to use the momentum representation of the propagator than the coordinate representation in (2.102). This is given by the general expression (2.50d) with the $F_4$-type generator $\phi(p_2, p_1)$, which is related to $\phi(q_2, q_1)$ of (2.101) in the usual manner, (2.15c):

$$\phi(p_2, p_1) = \phi(q_2, q_1) + p_1 q_1 - p_2 q_2 \tag{2.109}$$

Integrating the first term in (2.101) by parts and using (2.109) gives

$$\phi(p_2, p_1) = \int_{t_1}^{t_2} dt\{-q(t)\dot{p}(t) - H[p(t), q(t)]\} \tag{2.110}$$

The momentum representation of the propagator is then given by

$$\langle p_2|\exp\left[-\frac{iH(t_2 - t_1)}{\hbar}\right]|p_1\rangle$$
$$= \left[-\frac{\partial^2 \phi(p_2, p_1)/\partial p_2\, \partial p_1}{2\pi i\hbar}\right]^{1/2} \exp\left[\frac{i\phi(p_2, p_1)}{\hbar}\right] \tag{2.111}$$

In (2.110) and (2.111) $p_1$ and $p_2$ are the two doubled-ended boundary conditions that determine the appropriate classical trajectory. Just as for the coordinate representation, $p_1$ and $p_2$ do not necessarily determine just one classical trajectory, so that (2.111) may be a sum of several such terms. In addition, making use of the derivative relation

$$\frac{\partial \phi(p_2, p_1)}{\partial p_1} = q_1$$

allows one to write the preexponential factor in (2.111) in the equivalent form

$$-\frac{\partial^2 \phi(p_2, p_1)/\partial p_2\, \partial p_1}{2\pi i \hbar} = \left[-2\pi i \hbar \left(\frac{\partial p_2}{\partial q_1}\right)_{p_1}\right]^{-1} \tag{2.112}$$

Another important point worth emphasizing is that the form of (2.111), for example, is invariant to a canonical transformation; that is, (2.111) is valid for *any* momentum representation: ordinary Cartesian momenta, angular momenta, or any generalized momenta. To see this explicitly, suppose one transforms (2.111) to another momentum representation of momenta $P$:

$$\begin{aligned} &\langle P_2 | \exp\left[-\frac{iH(t_2 - t_1)}{\hbar}\right] | P_1 \rangle \\ &\quad = \int dp_2 \int dp_1 \langle P_2 \mid p_2 \rangle \langle p_2 | \exp\left[-\frac{iH(t_2 - t_1)}{\hbar}\right] | p_1 \rangle \langle p_1 \mid P_1 \rangle \end{aligned} \tag{2.113}$$

If $\phi(P_2, P_1)$ is the phase of the LHS of (2.113), then stationary phase evaluation of the integrals over $p_1$ and $p_2$ leads to

$$\phi(P_2, P_1) = \phi(p_2, p_1) - F_4(p_2, P_2) + F_4(p_1, P_1) \tag{2.114}$$

where $p_1$ and $p_2$ are determined by the appropriate stationary phase conditions and $F_4(p, P)$ is the generator of the $(p, q) \leftrightarrow (P, Q)$ transformation. But

$$\begin{aligned} -F_4(p_2, P_2) + F_4(p_1, P_1) &= \int_{t_1}^{t_2} dt\, -\frac{d}{dt} F_4(p, P) \\ &= \int_{t_1}^{t_2} dt\, -\frac{\partial F_4(p, P)}{\partial p}\, \dot{p} - \frac{\partial F_4(p, P)}{\partial P}\, \dot{P} \\ &= \int_{t_1}^{t_2} dt [q(t)\dot{p}(t) - Q(t)\dot{P}(t)] \end{aligned}$$

so that with (2.110), (2.114) becomes

$$\phi(P_2, P_1) = \int_{t_1}^{t_2} dt[-Q(t)\dot{P}(t) - H(P, Q)]$$

which is of the same form as (2.110).

The classical limit of the propagator in (2.102) has been obtained before by a number of other approaches. Perhaps the most straightforward procedure is to note that the coordinate matrix elements of the propagator satisfy the time dependent Schrödinger equation in the variables $q_2$ and $t_2$. Solution of the Schrödinger equation via the WKB approximation leads to the Hamilton-Jacobi equation for the phase and the continuity equation for the square of the preexponential factor.[21] The solution to the Hamilton-Jacobi equation is the classical action[15] $\phi$, and the solution of the continuity equation is the Jacobian factor of (2.104).

Equation (2.102) is also obtained by the asymptotic evaluation of Feynman's path integral representation of the propagator.[22] This approach is conceptually illuminating and particularly useful in emphasizing that there may be several classical trajectories determined by the double-ended boundary conditions and that one must add the amplitudes from each such classical path; that is, exact quantum mechanics results if one superposes the amplitudes for all possible paths that connect $q_1$ and $q_2$ (a path integral), and classical-limit quantum mechanics is obtained if one superposes the amplitudes for all the *classical* paths that connect $q_1$ and $q_2$.

There has been a great deal of work on the semiclassical approximation for the propagator. Some of the earliest is that of Van Vleck[23]; he seems to be the first to have obtained the preexponential factor in the form of (2.102), and the determinant of the matrix of second derivatives is often referred to as the "Van Vleck determinant." Later work, particularly that of Morette,[24] Fujiwara,[25] and Schiller,[26] has also been important in establishing the properties of the semiclassical propagator. The derivation that has been given here is intellectually pleasing in showing that one does not need to appeal to the Schrödinger equation, path integrals, or anything other than classical mechanics itself, the stationary phase approximation, and the quantum principle of superposition.

## III. THE CLASSICAL S-Matrix

### A. General Formula

With the classical-limit approximation to the propagator established in Section II.E it is a relatively straightforward matter to construct the classical limit of $S$-matrix elements (i.e., the "classical $S$-matrix") which describe transitions between specific quantum states of the collision partners. Several independent derivations of the appropriate expressions have been given,[8a,9a] and here we present another one which may in some respects be more satisfactory; all the different approaches are useful in illuminating various aspects of the semiclassical theory.

First it is necessary to carry out some quantum mechanical manipulations in order to express the $S$-matrix explicitly in terms of the propagator, the classical $S$-matrix then being obtained by invoking the classical-limit approximation for the propagator. The necessary quantum mechanical expressions are those for the Green's function in terms of the $T$-operator and in terms of the propagator[27]:

$$G^+(E) = G_0^+(E) + G_0^+(E)T(E)G_0^+(E) \tag{3.1}$$

$$G^+(E) = (i\hbar)^{-1}\int_0^\infty dt \exp\left(\frac{iEt}{\hbar}\right) \exp\left(-\frac{iHt}{\hbar}\right) \tag{3.2}$$

$G_0^+(E)$ being the unperturbed Green's function.

The general system consists of $N$ degrees of freedom, one being relative translation of the collision partners, and $(N - 1)$ "internal" degrees of freedom which are quantized in the asymptotic regions; that is, the Hamiltonian operator is

$$H = \frac{P^2}{2\mu} + h_{int} + V \tag{3.3}$$

where $h_{int}$ in the Hamiltonian for the internal degrees of freedom and $V$ is an interaction which vanishes as the translational coordinate $R$ becomes large. The Schrödinger equation

$$h_{int}\,|\mathbf{n}\rangle = \varepsilon_{\mathbf{n}}\,|\mathbf{n}\rangle \tag{3.4}$$

defines the internal states $|\mathbf{n}\rangle$ and the internal energies $\varepsilon_{\mathbf{n}}$; there are $(N - 1)$ quantum numbers $n = n_1, n_2, \ldots, n_{N-1}$ for the $(N - 1)$ quantized degrees of freedom. The quantities of interest in a scattering problem are the on-shell $S$-matrix elements

$$S_{\mathbf{n}_2,\mathbf{n}_1}(E) \tag{3.5}$$

which are the transition amplitudes describing transitions between quantum states of the internal Hamiltonian; as indicated, the $S$-matrix elements depend on the value of the total energy.

The unperturbed Green's function—that is, the Green's function for the unperturbed Hamiltonian $H_0 \equiv H - V$—is diagonal in the internal quantum numbers and has the usual[27] free-particle structure in the translational coordinate:

$$\langle R_2\mathbf{n}_2|\, G_0^+(E)\, |R_1\mathbf{n}_1\rangle = \delta_{\mathbf{n}_2,\mathbf{n}_1}\left(\frac{-2\mu}{\hbar^2 k_1}\right) \sin(k_1 R_<) \exp(ik_1 R_>) \tag{3.6}$$

where

$$k_1 = \left[\frac{2\mu(E - \varepsilon_{\mathbf{n}_1})}{\hbar^2}\right]^{1/2}$$

and $R_<(R_>)$ is the smaller (larger) of $R_1$ and $R_2$. Forming matrix elements of (3.1), for $\mathbf{n}_1 \neq \mathbf{n}_2$, thus gives

$$\langle R_2\mathbf{n}_2| G^+(E) |R_1\mathbf{n}_1\rangle = \int_0^\infty dR_2' \int_0^\infty dR_1' \langle R_2\mathbf{n}_2| G_0^+(E) |R_2'\mathbf{n}_2\rangle \times \langle R_2'\mathbf{n}_2| T(E) |R_1'\mathbf{n}_1\rangle \langle \mathbf{n}_1 R_1'| G_0^+(E) |R_1\mathbf{n}_1\rangle$$

For $R_1$, $R_2 \rightarrow \infty$, this becomes

$$\lim_{R_1, R_2 \rightarrow \infty} \langle R_2\mathbf{n}_2| G^+(E) |R_1\mathbf{n}_1\rangle = \left(\frac{2\mu}{\hbar^2}\right)^2 \exp(ik_1R_1 + ik_2R_2)\langle k_2\mathbf{n}_2| T(E) |k_1\mathbf{n}_1\rangle \tag{3.7}$$

where the $T$-matrix element is

$$\langle k_2\mathbf{n}_2| T(E) |k_1\mathbf{n}_1\rangle = \int_0^\infty dR_2' \int_0^\infty dR_1' \frac{\sin(k_2R_2')}{k_2} \times \langle R_2'\mathbf{n}_2| T(E) |R_1'\mathbf{n}_1\rangle \frac{\sin(k_1R_1')}{k_1} \tag{3.8}$$

With this normalization for the radial wave functions in (3.8) the $T$- and $S$-matrices are related by

$$S_{\mathbf{n}_2,\mathbf{n}_1}(E) - \delta_{\mathbf{n}_2,\mathbf{n}_1} = -2i\left(\frac{2\mu}{\hbar^2}\right)(k_1k_2)^{1/2}\langle k_2\mathbf{n}_2| T(E) |k_1\mathbf{n}_1\rangle \tag{3.9}$$

so that from (3.7) and (3.9) one identifies the $S$-matrix in terms of the Green's function:

$$S_{\mathbf{n}_2,\mathbf{n}_1}(E) = \lim_{R_1, R_2 \rightarrow \infty} \left(-\frac{\hbar^2}{2\mu}\right) 2i(k_1k_2)^{1/2} \times \exp(-ik_1R_1 - ik_2R_2)\langle R_2\mathbf{n}_2| G^+(E) |R_1\mathbf{n}_1\rangle \tag{3.10}$$

From (3.2) $G^+(E)$ is given as the Fourier transform of the propagator, so that the final expression for the $S$-matrix explicitly in terms of the propagator is

$$S_{\mathbf{n}_2,\mathbf{n}_1}(E) = -\lim_{R_1, R_2 \rightarrow \infty} \left(\frac{\hbar^2 k_1 k_2}{\mu^2}\right)^{1/2} \exp(-ik_1R_1 - ik_2R_2) \times \int_0^\infty dt \exp\left(\frac{iEt}{\hbar}\right)\langle R_2\mathbf{n}_2| \exp\left(-\frac{iHt}{\hbar}\right) |R_1\mathbf{n}_1\rangle \tag{3.11}$$

Equation (3.11) is a formally exact quantum mechanical expression; its classical limit is obtained by introducing the classical-limit approximation for the matrix elements of the propagator in the appropriate representation. The classical coordinates and momenta that correspond to the matrix elements in (3.11) are the ordinary translational coordinate and momentum, $R$ and $P$, and the *action-angle variables*[15] $\{n_i\}$ and $\{q_i\}$, $i = 1, \ldots, N-1$, for the internal degrees of freedom. The action variables $\mathbf{n}$ are the classical counterpart of the quantum numbers for the internal degrees of freedom; often they are referred to simply as the quantum numbers, but one should keep in mind that they are just a particular set of generalized momenta and, for example, vary with time according to Hamilton's equations. The classical Hamiltonian function is given in terms of these variables by

$$H(P, R, \mathbf{n}, \mathbf{q}) = \frac{P^2}{2\mu} + \varepsilon(\mathbf{n}) + V(R, \mathbf{n}, \mathbf{q}) \tag{3.12}$$

where $\varepsilon(\mathbf{n})$ is the WKB eigenvalue function for the internal degrees of freedom. [It is clear that quantizing the internal degrees of freedom via the WKB quantum condition limits one in practice either to collisions involving diatomic molecules or to a separable (i.e., normal mode) description of the energy levels of a polyatomic molecule.] Since the interaction $V$ vanishes in the asymptotic regions, that is, $V \to 0$ as $R_1, R_2 \to \infty$, the classical equations of motion show that

$$\dot{\mathbf{n}}(t) = -\frac{\partial H}{\partial \mathbf{q}} \to 0$$

That is, the quantum numbers become constant before and after collision, and furthermore, the WKB quantum condition requires that their constant values be integers. It is integer values of the action variable $\mathbf{n}$, of course, that provide the semiclassical definition of the initial and final quantum states of the collision partners.

In terms of the canonical variables $(P, R, \mathbf{n}, \mathbf{q})$ the matrix elements of the propagator in (3.11) are recognized to be a mixed representation, a coordinate representation with regard to the translational degree of freedom, and a momentum representation with regard to the internal degrees of freedom. This causes no difficulty, however, and in the classical limit (3.11) becomes

$$\begin{aligned} S_{\mathbf{n}_2,\mathbf{n}_1}(E) = &-\lim_{R_1 R_2 \to \infty} \left(-\frac{P_1 P_2}{\mu^2}\right)^{1/2} \exp\left[\frac{i}{\hbar}(P_1 R_1 - P_2 R_2)\right] \\ &\times \int_0^\infty dt \left[\frac{\partial^2 \phi(R_2\mathbf{n}_2, R_1\mathbf{n}_1; t)/\partial(R_2\mathbf{n}_2)\,\partial(R_1\mathbf{n}_1)}{(-2\pi i\hbar)^N}\right]^{1/2} \\ &\times \exp\left\{\frac{i}{\hbar}[Et + \phi(R_2\mathbf{n}_2, R_1\mathbf{n}_1; t)]\right\} \end{aligned} \tag{3.13}$$

where the action integral in this mixed representation is

$$\phi(R_2\mathbf{n}_2, R_1\mathbf{n}_1; t) = \int_{t_1}^{t_2} dt'[P(t')\dot{R}(t') - \mathbf{q}(t')\dot{\mathbf{n}}(t') - H] \tag{3.14}$$

and where $t \equiv t_2 - t_1$; in going from (3.11) to (3.13) the replacement

$$k_1 = -\frac{P_1}{\hbar}$$

$$k_2 = \frac{P_2}{\hbar}$$

had also been made. As with all integrations involving oscillatory integrands, the time integral in (3.13) is to be evaluated by stationary phase; since (as proved in Section IV.C)

$$\frac{\partial}{\partial t}\phi(R_2\mathbf{n}_2, R_1\mathbf{n}_1; t) = -E(R_2\mathbf{n}_2, R_1\mathbf{n}_1; t) \tag{3.15}$$

where $E(R_2\mathbf{n}_2, R_1\mathbf{n}_1; t)$ is the value of the total energy for the trajectory with the indicated boundary conditions, the stationary phase approximation gives

$$S_{\mathbf{n}_2,\mathbf{n}_1}(E) = -\left[-\frac{P_1P_2}{\mu^2}\frac{\partial t}{\partial E}\frac{\partial^2\phi(R_2\mathbf{n}_2, R_1\mathbf{n}_1; t)}{\partial(R_2\mathbf{n}_2)\,\partial(R_1\mathbf{n}_1)}\Big/(-2\pi i\hbar)^{N-1}\right]^{1/2} \times \exp\left[\frac{i\Phi(\mathbf{n}_2, \mathbf{n}_1; E)}{\hbar}\right] \tag{3.16}$$

where the phase $\Phi$ is

$$\Phi(\mathbf{n}_2, \mathbf{n}_1; E) = P_1R_1 - P_2R_2 + Et + \int_{t_1}^{t_2} dt'[P(t')\dot{R}(t') - \mathbf{q}(t')\dot{\mathbf{n}}(t') - H]$$

which takes the simple form

$$\Phi(\mathbf{n}_2, \mathbf{n}_1; E) = \int_{-\infty}^{\infty} dt[-R(t)\dot{P}(t) - \mathbf{q}(t)\cdot\dot{\mathbf{n}}(t)] \tag{3.17}$$

The stationary phase condition simply expresses energy conservation:

$$-\frac{\partial\phi(R_2\mathbf{n}_2, R_1\mathbf{n}_1; t)}{\partial t} \equiv E(R_2\mathbf{n}_2, R_1\mathbf{n}_1; t) = E \tag{3.18}$$

which for fixed $R_2$, $R_1$, $\mathbf{n}_1$, and $\mathbf{n}_2$ defines the functional relation between $E$ and $t$.

The preexponential factor in (3.16) can be greatly simplified by expressing the action integral $\phi(R_2\mathbf{n}_2, R_1\mathbf{n}_1; t)$ in terms of $\Phi(\mathbf{n}_2, \mathbf{n}_1; E)$

$$\phi(\mathbf{n}_2R_2, \mathbf{n}_1R_1; t) = \Phi(\mathbf{n}_2, \mathbf{n}_1; E) + P_2R_2 - P_1R_1 - Et \tag{3.19}$$

where $E$ on the RHS of (3.19) is the function of $t$ defined by (3.18). If use is also made of the relation

$$t = R_2 \frac{\partial P_2}{\partial E} - R_1 \frac{\partial P_1}{\partial E} + \frac{\partial \Phi(\mathbf{n}_2, \mathbf{n}_1; E)}{\partial E} = \frac{\mu R_2}{P_2} - \frac{\mu R_1}{P_1} + \frac{\partial \Phi(\mathbf{n}_2, \mathbf{n}_1; E)}{\partial E} \tag{3.20}$$

which is obtained by differentiating (3.19) with respect to $t$ and using (3.18), then one finds (after a reasonable amount of algebra)

$$\begin{aligned}\frac{\partial^2 \phi(\mathbf{n}_2 R_2, \mathbf{n}_1 R_1; t)}{\partial \mathbf{n}_2\, \partial \mathbf{n}_1} = \frac{\partial^2 \Phi(\mathbf{n}_2, \mathbf{n}_1; E)}{\partial \mathbf{n}_2\, \partial \mathbf{n}_1} &- \left(\frac{\partial t}{\partial E}\right)^{-1} \left[\frac{\partial^2 \Phi(\mathbf{n}_2, \mathbf{n}_1; E)}{\partial \mathbf{n}_2\, \partial E} + \frac{\mu^2 R_2}{P_2{}^3} \frac{\partial \varepsilon(\mathbf{n}_2)}{\partial \mathbf{n}_2}\right] \\ &\times \left[\frac{\partial^2 \Phi(\mathbf{n}_2, \mathbf{n}_1; E)}{\partial \mathbf{n}_1\, \partial E} - \frac{\mu^2}{P_1{}^3} R_1 \frac{\partial \varepsilon(\mathbf{n}_1)}{\partial \mathbf{n}_1}\right]\end{aligned} \tag{3.21a}$$

$$\frac{\partial^2 \phi(\mathbf{n}_2 R_2, \mathbf{n}_1 R_1; t)}{\partial \mathbf{n}_2\, \partial R_1} = \left(\frac{\partial t}{\partial E}\right)^{-1} \frac{\mu}{P_1} \left[\frac{\partial^2 \Phi(\mathbf{n}_2, \mathbf{n}_1; E)}{\partial \mathbf{n}_2\, \partial E} + \frac{\mu^2 R_2}{P_2{}^3} \frac{\partial \varepsilon(\mathbf{n}_2)}{\partial \mathbf{n}_2}\right] \tag{3.21b}$$

$$\frac{\partial^2 \phi(\mathbf{n}_2 R_2, \mathbf{n}_1 R_1; t)}{\partial R_2\, \partial \mathbf{n}_1} = -\left(\frac{\partial t}{\partial E}\right)^{-1} \frac{\mu}{P_2} \left[\frac{\partial^2 \Phi(\mathbf{n}_2, \mathbf{n}_1; E)}{\partial \mathbf{n}_2 \partial E} - \frac{\mu^2 R_1}{P_1{}^3} \frac{\partial \varepsilon(\mathbf{n}_1)}{\partial \mathbf{n}_1}\right] \tag{3.21c}$$

$$\frac{\partial^2 \phi(\mathbf{n}_2 R_2, \mathbf{n}_1 R_1; t)}{\partial R_2\, \partial R_1} = \frac{\mu^2}{P_1 P_2} \left(\frac{\partial t}{\partial E}\right)^{-1} \tag{3.21d}$$

Thus the determinant

$$\left| \frac{\partial^2 \phi(R_2 \mathbf{n}_2, R_1 \mathbf{n}_1; t)}{\partial(R_2 \mathbf{n}_2, R_1 \mathbf{n}_1)} \right| \equiv \begin{vmatrix} \dfrac{\partial^2 \phi}{\partial \mathbf{n}_2\, \partial \mathbf{n}_1} & \dfrac{\partial^2 \phi}{\partial \mathbf{n}_2\, \partial R_1} \\ \dfrac{\partial^2 \phi}{\partial R_2\, \partial \mathbf{n}_1} & \dfrac{\partial^2 \phi}{\partial R_2\, \partial R_1} \end{vmatrix}$$

is found with the aid of (3.21) to be

$$= \frac{\mu^2}{P_1 P_2} \left(\frac{\partial t}{\partial E}\right)^{-1} \frac{\partial^2 \Phi(\mathbf{n}_2, \mathbf{n}_1; E)}{\partial \mathbf{n}_2\, \partial \mathbf{n}_1} \tag{3.22}$$

so that the preexponential factor in (3.16) simplifies to give the final expression for the classical $S$-matrix,

$$S_{\mathbf{n}_2, \mathbf{n}_1}(E) = i \left[\frac{\partial^2 \Phi(\mathbf{n}_2, \mathbf{n}_1; E)/\partial \mathbf{n}_2\, \partial \mathbf{n}_1}{(-2\pi i \hbar)^{N-1}}\right]^{1/2} \exp\left[\frac{i \Phi(\mathbf{n}_2, \mathbf{n}_1; E)}{\hbar}\right] \tag{3.23}$$

with $\Phi(\mathbf{n}_2, \mathbf{n}_1; E)$ given by (3.17).

The preexponential factor in (3.23) could actually have been constructed directly from the phase $\Phi(\mathbf{n}_2, \mathbf{n}_1; E)$ by invoking the unitarity condition for the on-shell $S$-matrix:

$$\int d\mathbf{n}_2 S_{\mathbf{n}_2,\mathbf{n}_1'}(E)^* S_{\mathbf{n}_2,\mathbf{n}_1}(E) = \delta(\mathbf{n}_1' - \mathbf{n}_1) \tag{3.24}$$

[see the discussion following (2.48)]. Thus a much shorter, though seemingly less rigorous, route to (3.17) for the phase [and then to (3.23) via the unitarity condition] begins with the formal expression for the $S$-operator as an infinite time limit of the propagator in the interaction representation[27]:

$$S = \lim_{\substack{t_1 \to -\infty \\ t_2 \to +\infty}} \exp\left[\frac{iH_0 t_2}{\hbar}\right] \exp\left[-\frac{iH(t_2 - t_1)}{\hbar}\right] \exp\left[-\frac{iH_0 t_1}{\hbar}\right] \tag{3.25}$$

$S$-matrix elements are then simply matrix elements of this operator between eigenstates of $H_0$, that is, the momentum states $|P\mathbf{n}\rangle$. Ignoring all preexponential factors (which contain an energy-conserving delta function) thus gives

$$\begin{aligned}\langle P_2\mathbf{n}_2|\, S\, |P_1\mathbf{n}_1\rangle &\sim S_{\mathbf{n}_2,\mathbf{n}_1}(E) \\ &\sim \exp\left[\frac{iE(t_2 - t_1)}{\hbar}\right]\langle P_2\mathbf{n}_2|\exp\left[-\frac{iH(t_2 - t_1)}{\hbar}\right]|P_1\mathbf{n}_1\rangle\end{aligned} \tag{3.26}$$

where $P_1$ and $P_2$ are given by energy conservation

$$P_i = \pm\{2\mu[E - \varepsilon(\mathbf{n}_i)]\}^{1/2}$$

and use has been made of the fact that

$$\exp\left(\pm\frac{iH_0 t}{\hbar}\right)|P\mathbf{n}\rangle = \exp\left(\pm\frac{iEt}{\hbar}\right)|P\mathbf{n}\rangle$$

With the classical-limit approximation for the matrix elements of the propagator in the momentum representation, (2.110) and (2.111), (3.26) becomes

$$S_{\mathbf{n}_2,\mathbf{n}_1}(E) \sim \exp\left\{\frac{i}{\hbar}E(t_2 - t_1) + \frac{i}{\hbar}\int_{t_1}^{t_2} dt[-R(t)\dot{P}(t) - q(t)\cdot\dot{\mathbf{n}}(t) - H]\right\}$$

or

$$S_{\mathbf{n}_2,\mathbf{n}_1}(E) \sim \exp\left[\frac{i\Phi(\mathbf{n}_2, \mathbf{n}_1; E)}{\hbar}\right] \tag{3.27}$$

where $\Phi$ is the same phase function obtained before, (3.17). With the preexponential factor determined by unitarity, (3.23) is thus reproduced (except for overall constant phase factors).

In practice the evaluation of the preexponential factor in (3.23) is simplified if one makes use of the derivative relations of the phase function. From (3.19), for example, it is easy to show (by differentiation with respect to $\mathbf{n}_1$) that

$$\begin{aligned}\frac{\partial\Phi(\mathbf{n}_2, \mathbf{n}_1; E)}{\partial\mathbf{n}_1} &= \mathbf{q}_1 + R_1\frac{\partial P_1}{\partial\mathbf{n}_1}\\ &= \mathbf{q}_1 - \frac{\mu R_1}{P_1}\frac{\partial\varepsilon(\mathbf{n}_1)}{\partial\mathbf{n}_1}\\ &\equiv \bar{\mathbf{q}}_1\end{aligned} \tag{3.28}$$

Since in the initial asymptotic region Hamilton's equations give

$$\mathbf{q}(t_1) = \text{constant} + \frac{\partial\varepsilon(\mathbf{n}_1)}{\partial\mathbf{n}_1}t_1$$

$$R(t_1) = \text{constant} + \frac{P_1 t_1}{\mu}$$

it is easy to see that $\bar{\mathbf{q}}_1$ defined by (3.28) is time independent in the initial asymptotic region; that is, it has the "free" time dependence of $\mathbf{q}(t)$ subtracted out. Further differentiation of (3.28) gives

$$\frac{\partial^2\Phi(\mathbf{n}_2, \mathbf{n}_1; E)}{\partial\mathbf{n}_2\,\partial\mathbf{n}_1} = \left(\frac{\partial\bar{\mathbf{q}}_1}{\partial\mathbf{n}_2}\right)_{\mathbf{n}_1} = \left[\left(\frac{\partial\mathbf{n}_2}{\partial\bar{\mathbf{q}}_1}\right)_{\mathbf{n}_1}\right]^{-1} \tag{3.29}$$

so that the classical $S$-matrix of (3.23) is equivalently (and more conveniently) expressed as

$$S_{\mathbf{n}_2,\mathbf{n}_1}(E) = i\left[(-2\pi i\hbar)^{N-1}\left(\frac{\partial\mathbf{n}_2}{\partial\bar{\mathbf{q}}_1}\right)_{\mathbf{n}_1}\right]^{-1/2}\exp\left[\frac{i\Phi(\mathbf{n}_2, \mathbf{n}_1; E)}{\hbar}\right] \tag{3.30}$$

with $\Phi$ still given by (3.17).

To apply (3.30) one must thus find the classical trajectory (or trajectories) with initial conditions

$$\begin{aligned}\mathbf{n}(t_1) &\equiv \mathbf{n}_1 = \text{specified integers}\\ R(t_1) &= \text{large}\\ P(t_1) &= -\{2\mu[E - \varepsilon(\mathbf{n}_1)]\}^{1/2}\\ \mathbf{q}(t_1) &= \bar{\mathbf{q}}_1 + \frac{\partial\varepsilon(\mathbf{n}_1)}{\partial\mathbf{n}_1}\frac{\mu R(t_1)}{P(t_1)}\end{aligned} \tag{3.31}$$

and final conditions

$$\begin{aligned} \mathbf{n}(t_2) &\equiv \mathbf{n}_2 = \text{specified integers} \\ R(t_2) &= \text{large} \\ P(t_2) &= +\{2\mu[E - \varepsilon(\mathbf{n}_2)]\}^{1/2} \\ \mathbf{q}(t_2) &= \text{anything} \end{aligned} \tag{3.32}$$

If there is more than one trajectory that satisfies these boundary conditions, then (3.30) is a sum of terms, one for each such trajectory. To find the trajectories that obey these boundary conditions it is useful to introduce the "quantum number function" $\mathbf{n}_2(\bar{\mathbf{q}}_1, \mathbf{n}_1; E)$, the final value of the quantum numbers, not necessarily integral, which result from a classical trajectory with the initial conditions in (3.31). For a given total energy $E$ and a given set of initial integer quantum numbers $\mathbf{n}_1$, the task is to find the particular values of the angle variables $\bar{\mathbf{q}}_1$ for which $\mathbf{n}_2(\bar{\mathbf{q}}_1, \mathbf{n}_1; E)$ turns out to be the specific integers $\mathbf{n}_2$; that is, suppressing the arguments $\mathbf{n}_1$ and $E$, one must solve the equations

$$\mathbf{n}_2(\bar{\mathbf{q}}_1) = \mathbf{n}_2 \tag{3.33}$$

where $\mathbf{n}_2$ on the RHS is a specific set of integers; this is a set of $N - 1$ nonlinear equations in $N - 1$ unknowns. The preexponential factor in (3.30) is now seen to be the determinant of the matrix of partial derivatives of $\mathbf{n}_2(\bar{\mathbf{q}}_1)$, evaluated at the root of (3.33).

Finally, it should be pointed out that the discussion in this section has tacitly assumed that the collision process under consideration is a nonreactive one; Section III.D discusses the modifications that are necessary to treat reactions.

### B. Example: Vibrational Excitation in Collinear $A + BC$

To illustrate the general results obtained above, it is useful to discuss first the simplest nontrivial example of a collision system that possesses internal degrees of freedom in addition to translation. The model system is that of an atom-diatom collision with all three atoms constrained to lie in a straight line. The classical Hamiltonian for the system is

$$H(P, R, p, r) = \frac{P^2}{2\mu} + \frac{p^2}{2m} + v(r) + V(r, R) \tag{3.34}$$

where $R$ and $P$ are ordinary Cartesian coordinates and momenta for translation of $A$ relative to the center of mass $BC$, and $r$ and $p$ are the Cartesian variables for the vibration $BC$; $\mu$ and $m$ are the reduced masses for these two degrees of freedom, respectively, $v(r)$ is the vibrational potential for $BC$, and $V(r, R)$ is the interaction potential that couples translation and vibration. This system has only one internal degree of freedom in addition to translation.

To proceed semiclassically one first transforms from the Cartesian variables $(p, r)$ to the action-angle variables $(n, q)$ of the oscillator. The action variable $n$ is the classical counterpart of the vibrational quantum number and is determined in terms of the old variables $(p, r)$ by (units being used such that $\hbar = 1$)

$$(n + \tfrac{1}{2})\pi = (2m)^{1/2} \int_{r_<}^{r_>} dr'[\varepsilon - v(r')]^{1/2} \tag{3.35a}$$

with

$$\varepsilon = \frac{p^2}{2m} + v(r) \tag{3.35b}$$

and where $r_<$ and $r_>$ are the zeroes of the integrand, that is, the classical turning points. The angle variable $q$ is given by

$$q = \frac{(2m)^{1/2}}{n'(\varepsilon)} \int_{r_<}^{r} dr'[\varepsilon - v(r')]^{-1/2} \tag{3.36}$$

with $\varepsilon$ given by (3.35b) and where $n(\varepsilon)$ is the function defined by (3.35a). In terms of the variables $(P, R, n, q)$ the Hamiltonian of (3.34) becomes

$$H(P, R, n, q) = \frac{P^2}{2\mu} + \varepsilon(n) + V(r(n, q), R) \tag{3.37}$$

where $\varepsilon(n)$ is the inverse function of $n(\varepsilon)$ of (3.35) and $r(n, q)$ is the vibrational coordinate expressed in terms of the action-angle variables of vibration; $\varepsilon(n)$ is the standard WKB approximation for the vibrational eigenvalues for $n = 0, 1, 2, \ldots$. It is clear that $n$ and $P$ are the constants of the motion of the unperturbed Hamiltonian

$$H_0(P, R, n, q) = \frac{P^2}{2\mu} + \varepsilon(n)$$

that is, that

$$\dot{P}(t) = -\frac{\partial H_0}{\partial R} = 0$$

$$\dot{n}(t) = -\frac{\partial H_0}{\partial q} = 0$$

For simple vibrational potentials it is possible to obtain explicit expressions for $\varepsilon(n)$ and $r(n, q)$. With a harmonic oscillator,

$$v(r) = \tfrac{1}{2}m\omega^2(r - r_0)^2 \tag{3.38}$$

for example, one finds

$$\varepsilon(n) = (n + \tfrac{1}{2})\omega \tag{3.39a}$$

and

$$r(n, q) = r_0 + \left(\frac{2n + 1}{\omega m}\right)^{1/2} \sin q \tag{3.39b}$$

It is also possible to construct explicit expressions for the case of a Morse vibrational potential.[8e] For the general case one can expand the potential as does Dunham[28]

$$v(r) = \tfrac{1}{2}m\omega^2(r - r_0)^2[1 + a_1\xi + a_2\xi^2 + \cdots] \tag{3.40}$$

where

$$\xi = \frac{(r - r_0)}{r_0}$$

and expansions for $\varepsilon(n)$ and $r(n, q)$ can be constructed:

$$\varepsilon(n) = (n + \tfrac{1}{2})\omega\{1 + (\tfrac{3}{2}a_2 - \tfrac{15}{8}a_1^2)\lambda(n + \tfrac{1}{2}) + 0(\lambda^2)\} \tag{3.41a}$$

$$\begin{aligned} r(n, q) = r_0\{1 &+ 2[\lambda(n + \tfrac{1}{2})]^{1/2} \sin q \\ &- a_1[3 + \cos(2q)]\lambda(n + \tfrac{1}{2}) \\ &- [(3a_2 - \tfrac{11}{4}a_1^2)\sin q + (\tfrac{1}{2}a_2 + \tfrac{3}{8}a_1^2)\sin(3q)][\lambda(n + \tfrac{1}{2})]^{3/2} \\ &+ 0(\lambda^2)\} \end{aligned} \tag{3.41b}$$

where

$$\lambda = (2m\omega r_0^2)^{-1}$$

For most diatomic molecules $\lambda$ is quite small, being the largest for $H_2$ ($\lambda_{H_2} = 0.014$), so that (3.41) should be adequate for any applications.

According to the general results of Section III.A the on-shell $S$-matrix elements, the transition amplitudes from initial vibrational state $n_1$ to final vibrational state $n_2$, are given by

$$S_{n2,n1}(E) = \left[-2\pi i\left(\frac{\partial n_2}{\partial \bar{q}_1}\right)_{n_1}\right]^{-1/2} \exp[i\Phi(n_2, n_1)] \tag{3.42a}$$

where

$$\Phi(n_2, n_1) = -\int_{-\infty}^{\infty} dt[R(t)\dot{P}(t) + q(t)\dot{n}(t)] \tag{3.42b}$$

The transition probability for the $n_1 \rightarrow n_2$ (or $n_2 \rightarrow n_1$) transition is

$$P_{n_2,n_1}(E) = |S_{n_2,n_1}(E)|^2 \tag{3.43}$$

The action variable $n$ is required semiclassically to be an integer in the asymptotic regions where $BC$ is a free oscillator; this is the semiclassical definition of the vibrational "state."

The appropriate double-ended boundary conditions for the trajectories relevant to (3.42) are

$$n_1 = \text{integer} \tag{3.44a}$$
$$P_1 = -\{2\mu[E - \varepsilon(n_1)]\}^{1/2} \tag{3.44b}$$
$$n_2 = \text{integer} \tag{3.44c}$$
$$P_2 = +\{2\mu[E - \varepsilon(n_2)]\}^{1/2} \tag{3.44d}$$

where $E$ is the fixed total energy. Classical trajectories for a system with several degrees, however, must in general be computed by numerically integrating the equations of motion with a specified set of *initial* conditions. To satisfy the above double-ended boundary conditions one thus begins a trajectory with the initial conditions

$$n_1 = \text{integer} \tag{3.45a}$$
$$R_1 = \text{large} \tag{3.45b}$$
$$P_1 = -\{2\mu[E - \varepsilon(n_1)]\}^{1/2} \tag{3.45c}$$
$$q_1 = \bar{q}_1 + \varepsilon'(n_1)\frac{\mu R_1}{P_1} \tag{3.45d}$$

and integrates the equations of motion forward in time until $R(t)$ is again large and $\dot{P}(t) \to 0$; "large $R$" means a value sufficiently large that $V(r, R)$ is negligible, and $n(t)$ and $P(t)$ thus constant. The final value of the action variable (not necessarily an integer) is thought of as a function of $\bar{q}_1$ and $n_1$, the functional value $n_2(\bar{q}_1, n_1)$ being determined by integrating the equations of motion with the initial conditions in (3.45) (the initial conditions of the translational degree of freedom are always determined implicitly by the scattering boundary condition and energy conservation). The double-ended boundary conditions in (3.44) can therefore be satisfied only if one chooses $\bar{q}_1$ to be the particular value for which $n_2(\bar{q}_1, n_1)$ is an integer; that is, one must solve the equation

$$n_2(\bar{q}_1, n_1) = n_2 \tag{3.46}$$

[When $n_2$ is written with arguments as in the LHS of (3.46) it denotes the final value of the action variable (not necessarily integral) that results from the classical trajectory with the indicated initial conditions; written without arguments, as in the RHS of (3.46), it denotes a specific integer value. For the $0 \to 1$ vibrational transition, for example, (3.46) reads $n_2(\bar{q}_1, 0) = 1$.]

Fig. 1 shows an example of the function $n_2(\bar{q}_1, n_1)$ for the case $n_1 = 1$; since the initial condition $\bar{q}_1 + 2\pi$ leads to precisely the same trajectory as the value $\bar{q}_1$, $n_2(\bar{q}_1, n_1)$ is a periodic function of $\bar{q}_1$. It is clear, therefore,

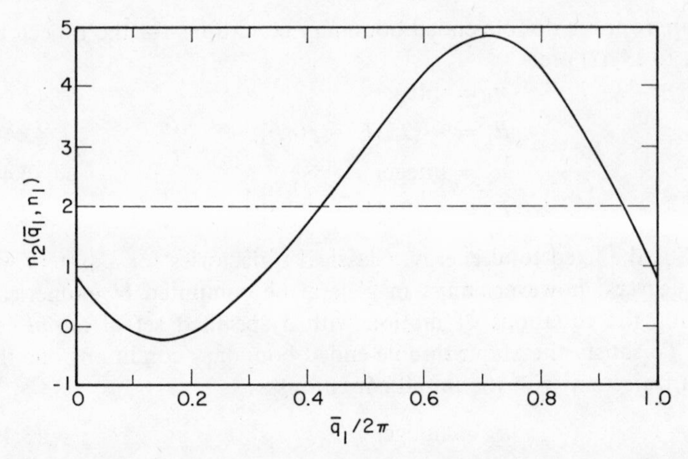


**Fig. 1.** An example of the quantum number function $n_2(\bar{q}_1, n_1)$, here with $n_1 = 1$. The ordinate is the final value of the vibrational quantum number as a function of the initial phase $\bar{q}_1$ of the oscillator, along a classical trajectory with the initial conditions in (3.45). The dotted line at $n_2 = 2$ indicates the graphical solution for the two roots of the equation $n_2(\bar{q}_1, 1) = 2$.

that there must be an *even number* of roots to (3.46); that is, the function $\bar{q}_1(n_2, n_1)$ is a multivalued function, meaning that there is more than one classical trajectory that obeys the double-ended boundary conditions of (3.44). For the simple case shown in Fig. 1, there are typically two roots to (3.44), so that the $S$-matrix element in (3.42a) is the sum of two such terms. If $\bar{q}_{\mathrm{I}}$ and $\bar{q}_{\mathrm{II}}$ are the two roots indicated in Fig. 1, then the $S$-matrix element is

$$S_{n_2, n_1} = P_{\mathrm{I}}^{1/2} \exp\left(\frac{i\pi}{4} + i\phi_{\mathrm{I}}\right) + p_{\mathrm{II}}^{1/2} \exp\left(-\frac{i\pi}{4} + i\phi_{\mathrm{II}}\right) \quad (3.47)$$

where the preexponential probability factors are

$$p_{\mathrm{I}} = [2\pi\, |n_2'(\bar{q}_{\mathrm{I}})|]^{-1}$$

$$p_{\mathrm{II}} = [2\pi\, |n_2'(\bar{q}_{\mathrm{II}})|]^{-1}$$

and the phases $\phi_{\mathrm{I}}$ and $\phi_{\mathrm{II}}$ are given by (3.42b) for trajectories I and II. The transition probability is

$$P_{n_2, n_1} = p_{\mathrm{I}} + p_{\mathrm{II}} + 2(p_{\mathrm{I}} p_{\mathrm{II}})^{1/2} \sin(\phi_{\mathrm{II}} - \phi_{\mathrm{I}}) \quad (3.48)$$

Fig. 2 shows the results for a model system chosen to represent He + $H_2$, a highly quantumlike system. The dotted lines connect the points that are the completely classical result,

$$P_{n_2, n_1}^{\mathrm{CL}} = p_{\mathrm{I}} + p_{\mathrm{II}} \quad (3.49)$$

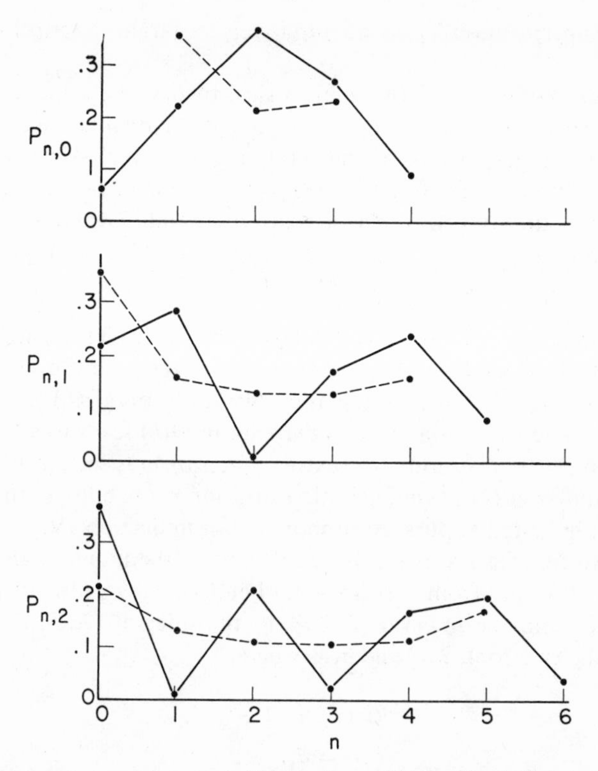


**Fig. 2.** Vibrational transition probabilities for collinear He + $H_2$ at total energy $E = 10\, h\omega$ for initial quantum states $n_1 = 0$ (top), 1, and 2 (bottom). The dotted lines connect results of the completely classical approximation (3.49), and the solid lines connect the exact quantum values of Ref. 29; on this scale there is essentially no difference between the exact quantum values and the uniform semiclassical results of Refs. 8b and 8c.

that results by discarding the interference term in (3.48). If interference is neglected, nature thus reverts to *classical* superposition, as in (3.49), simply the addition of the probabilities related to the two different trajectories that lead to the $n_1 \rightarrow n_2$ transition. The solid lines in Fig. 2 connect the semiclassical values, and it is seen that the interference structure is quite prominent. These semiclassical values are in excellent agreement (within a few percent) with essentially exact quantum mechanical calculations,[29] showing that the semiclassical theory is an accurate description of these quantum effects. In a certain sense, therefore, this system is highly quantumlike, in that the purely classical transition probabilities (the dotted lines in Fig. 2) are poor; the quantum effects are chiefly of an interference nature, however, so that "classical dynamics plus quantum superposition" is a quantitatively accurate description of the collision.

The reader familiar with the semiclassical theory of elastic scattering[1,2] will recognize an analogy with many of the above qualitative effects; this analogy, in fact, greatly aids in understanding the semiclassical description of inelastic scattering. The function $n_2(\bar{q}_1)$ plays the same role as the classical deflection function, $\Theta(b)$, of elastic scattering, and (3.46) corresponds to the equation

$$\Theta(b) = \pm\theta \tag{3.50}$$

which is central to the analysis of elastic scattering. There are multiple roots to (3.50) which lead to a superposition of terms in the semiclassical scattering amplitude, just as the multiple roots of (3.46) do so for a specific inelastic transition. Just as extrema in $\Theta(b)$ leads to a "rainbow" structure in elastic scattering, the extrema in $n_2(\bar{q}_1)$ (as seen in Fig. 1) give rise to a similar effect in the inelastic transition probabilities; this feature is seen in Fig. 2 and is pursued in more detail in Section IV.

An interesting feature is illustrated if one for the moment allows $\bar{q}_1$ to to take on all values from $-\infty$ to $+\infty$. Let $\bar{q}_{\text{I}}$ and $\bar{q}_{\text{II}}$ be the two roots to (3.46) that lie in the interval $(0, 2\pi)$. In the interval $(2\pi, 4\pi)$ there are therefore the two roots $\bar{q}_{\text{III}}$ and $\bar{q}_{\text{IV}}$, where

$$\bar{q}_{\text{III}} = \bar{q}_{\text{I}} + 2\pi$$

$$\bar{q}_{\text{IV}} = \bar{q}_{\text{II}} + 2\pi$$

Trajectories I and III are essentially the same, of course; in fact,

$$n_{\text{III}}(t) = n_{\text{I}}(t)$$

$$q_{\text{III}}(t) = q_{\text{I}}(t) + 2\pi$$

$$R_{\text{III}}(t) = R_{\text{I}}(t)$$

$$P_{\text{III}}(t) = P_{\text{I}}(t)$$

for all $t$. From (3.42b) the phases $\phi_{\text{III}}$ and $\phi_{\text{I}}$ are thus seen to be related by

$$\phi_{\text{III}} = \int_{-\infty}^{\infty} dt[-R_{\text{III}}\dot{P}_{\text{III}}(t) - q_{\text{III}}(t)\dot{n}_{\text{III}}(t)]$$

$$= \int_{-\infty}^{\infty} dt[-R_{\text{I}}(t)\dot{P}_{\text{I}}(t) - (q_{\text{I}}(t) + 2\pi)\dot{n}_{\text{I}}(t)]$$

$$= \phi_{\text{I}} - 2\pi\,\Delta n$$

where

$$\Delta n = n_2 - n_1$$

Similarly, $\phi_{IV} = \phi_{II} - 2\pi\,\Delta n$, so that superposition of the four roots in the internal $(0, 4\pi)$ gives

$$S = S_0[1 + e^{-2\pi i\,\Delta n}]$$

where $S_0$ is the $S$-matrix resulting from the two roots in the primary interval $(0, 2\pi)$. Superposing the two roots from the intervals $(4\pi, 6\pi)$, $(6\pi, 8\pi), \ldots$, and $(-2\pi, 0), (-4\pi, -2\pi), \ldots$, thus gives the following:

$$S = S_0 \sum_{k=-\infty}^{\infty} e^{-2\pi ik\,\Delta n} \tag{3.51}$$

According to the Poisson sum formula,[30] however,

$$\sum_{k=-\infty}^{\infty} e^{-2\pi ik\,\Delta n} = \sum_{k=-\infty}^{\infty} \delta(k - \Delta n) \tag{3.52}$$

so that $S$ in (3.51) contains a delta function factor which requires $\Delta n$ to be an integer. Superposition, therefore, leads to quantization of $n_2$ in the final asymptotic region if $n_1$ is an integer initially. The $S$-matrix element $S_0$, which comes from the two roots in the primary interval $(0, 2\pi)$, is thus the $S$-matrix element on the "quantum number shell."

Finally, it is interesting to see how the strictly classical result, (3.49), that neglects interferences can be cast in the popular Monte Carlo framework.[11] Equation (3.49) gives the strictly classical result as

$$P^{\mathrm{CL}}_{n_2,n_1} = (2\pi)^{-1}[|n_2'(\bar{q}_{\mathrm{I}})|^{-1} + |n_2'(\bar{q}_{\mathrm{II}})|^{-1}] \tag{3.53}$$

Averaging this over a quantum number width about $n_2$ gives

$$\bar{P}_{n_2,n_1} \equiv \int_{n2-1/2}^{n2+1/2} dn_2 P^{\mathrm{CL}}_{n2,n1} \tag{3.54a}$$

$$= (2\pi)^{-1} \int_{n2-1/2}^{n2+1/2} dn_2 \left[ \left| \frac{\partial n_2}{\partial \bar{q}_1} \right|_{\mathrm{I}^+}^{-1} + \left| \frac{\partial n_2}{\partial \bar{q}_1} \right|_{\mathrm{II}}^{-1} \right] \tag{3.54b}$$

and changing variables of integration from $n_2$ to $\bar{q}_1$ eliminates the Jacobian from the integrand of (3.54b):

$$\bar{P}_{n2,n1} = \frac{1}{2\pi} \int d\bar{q}_1 \tag{3.55a}$$

The limits of the $\bar{q}_1$ integral are those values for which

$$n_2 - \tfrac{1}{2} \leq n_2(\bar{q}_1) \leq n_2 + \tfrac{1}{2} \tag{3.55b}$$

The Monte Carlo way of evaluating (3.55) is, with $n_1$ fixed, to choose $\bar{q}_1$

$$\bar{q}_1 = 2\pi\xi_1$$

where $\xi_1$ is a random number in (0, 1); if (3.55b) is satisfied for the trajectory with this $\bar{q}_1$, a "success" is scored. This process is repeated many times, and the value of $P_{n_2,n_1}$ is the number of "successes" divided by the total number of trajectories run.

One may recognize that this averaging procedure in (3.54a) destroys the microscopic reversibility of $P_{n_2,n_1}$; that is, with $\bar{P}_{n_2,n_1}$ defined in (3.54a),

$$\bar{P}_{n_2,n_1} \neq \bar{P}_{n_1,n_2}$$

This is obvious because $n_2$ has been averaged over, but $n_1$ precisely quantized. The only consistent approach is thus to average over a quantum number width of $n_1$ also:

$$\bar{\bar{P}}_{n2,n1} \equiv \int_{n2-1/2}^{n2+1/2} dn_2 \int_{n1-1/2}^{n1+1/2} dn_1 P^{\text{CL}}_{n2,n1} \tag{3.56}$$

Proceeding as above, this can be reduced to

$$\bar{\bar{P}}_{n2,n1} = \frac{1}{2\pi} \int_{n1-1/2}^{n1+1/2} dn_1 \int d\bar{q}_1 \tag{3.57}$$

with the limits of $\bar{q}_1$ restricted as in (3.55b). The Monte Carlo procedure for evaluating this "doubly averaged" transition probability in (3.57) is to choose $n_1$ and $\bar{q}_1$ as

$$\bar{q}_1 = 2\pi\xi_1$$
$$n_1 = n_1 - \tfrac{1}{2} + \xi_2$$

where $\xi_1$ and $\xi_2$ are random numbers in (0, 1); success is scored if (3.55b) is fulfilled for this trajectory, etc. It should be clear that the "doubly averaged" transition probability does obey microscopic reversibility.

### C. Selection Rules

In rotational excitation of a homonuclear diatomic molecule such as $H_2$ it is known quantum mechanically that the change in rotational quantum number must be even, that is, $\Delta j = 0, 2, 4, \ldots$ ; classical mechanics, of course, allows any continuous change in $j$. It will be seen, however, that classical-limit quantum mechanics—that is, classical dynamics plus quantum superposition—properly accounts for these selection rules. Selection rules, which were first discussed in this semiclassical framework by Marcus,[9a] are thus a direct result of quantum superposition, so that any semiclassical theory that incorporates superposition can correctly describe them.

To illustrate the semiclassical origin of selection rules, consider the following model system: a nonvibrating diatomic molecule $BC$ (i.e., a rigid rotor) and an atom $A$ all confined to a plane, and with the center of mass motion of $A$ relative to $BC$ confined to a line. The Hamiltonian for this system is

$$H(P, R, j, q) = \frac{P^2}{2\mu} + Bj^2 + V(q, R) \tag{3.58}$$

where $(P, R)$ are the ordinary translational variables, $j$ is the angular momentum of the rotor, $q$, the coordinate conjugate to $j$, is the angle of rotation of the rotor, $B$ is the rotation constant of the rotor, and $V(q, R)$ is the interaction potential. Classical $S$-matrix elements for rotational transitions $j_1 \to j_2$ are given by essentially the same formulas as in Section III.B; thus

$$S_{j_2,j_1} = \left[-2\pi i\left(\frac{\partial j_2}{\partial \bar{q}_1}\right)_{j_1}\right]^{-1/2} \exp\left[i\phi(j_2, j_1)\right] \tag{3.59}$$

where

$$\phi(j_2, j_1) = -\int_{-\infty}^{\infty} dt[R(t)\dot{P}(t) + q(t)\dot{j}(t)] \tag{3.60}$$

To find the appropriate classical trajectories one solves the equation

$$j_2(\bar{q}_1, j_1) = j_2 \tag{3.61}$$

for given integers $j_1$ and $j_2$; $j_2(\bar{q}_1, j_1)$ on the LHS of (3.38) is the final value of $j(t)$, not necessarily integral, that results from the trajectory with the indicated initial conditions.

Just as for the example of vibrational excitation in Section III.B, $j_2(\bar{q}_1)$ is a periodic function of $\bar{q}_1$ with period $2\pi$, for an initial orientation angle $\bar{q}_1 + 2\pi$ is physically the same as initial angle $\bar{q}_1$. Equation (3.61) must therefore have an even number of roots, and (3.59) will consist of several terms, one for each such root.

If the molecule $BC$ is *homonuclear*, however, then the function $j_2(\bar{q}_1)$ will be periodic with period $\pi$; that is, the orientation $\bar{q}_1 + \pi$ is related to the orientation $\bar{q}_1$ simply by the interchange of atoms $B$ and $C$, so for a homonuclear molecule ($B = C$), the initial orientations $\bar{q}_1$ and $\bar{q}_1 + \pi$ are equivalent. The function $j_2(\bar{q}_1)$ in the interval $(\pi, 2\pi)$ is thus the same as that in the interval $(0, \pi)$. Equation (3.61) must, therefore, have an even number of roots in the interval $(0, \pi)$. If $\bar{q}_{\text{I}}$ and $\bar{q}_{\text{II}}$ are the two roots, say, of (3.61) in the interval $(0, \pi)$, then there are two other roots $\bar{q}_{\text{III}}$ and $\bar{q}_{\text{IV}}$ in the interval $(\pi, 2\pi)$, where

$$\bar{q}_{\text{III}} = \bar{q}_{\text{I}} + \pi$$

$$\bar{q}_{\text{IV}} = \bar{q}_{\text{II}} + \pi$$

The preexponential functions for roots III and IV are the same as those for roots I and II, respectively, and it is easy to see that the phases are related by

$$\phi_{III} = \phi_{I} - \pi\,\Delta j$$
$$\phi_{IV} = \phi_{II} - \pi\,\Delta j$$

where $\Delta j = j_2 - j_1$. The $S$-matrix element is the sum of terms from all the roots of (3.61) in the interval $(0, 2\pi)$:

$$S = S_0[1 + e^{-i\pi\,\Delta j}] \tag{3.62}$$

where $S_0$ is the sum of the amplitudes from the two roots in the interval $(0, \pi)$. Since $\Delta j$ is an integer, (3.62) thus becomes

$$S = S_0[1 + (-1)^{\Delta j}] \tag{3.63}$$

showing that the transition probability for odd $\Delta j$ is zero.

More generally, suppose the rotor has an $m$-fold symmetry; for example, $m = 2$ corresponds to the homonuclear diatomic molecule, and $m = 3$ would correspond to planar $CH_3$. The function $j_2(\bar{q}_1)$ would in this case be periodic in the interval $(0, 2\pi/m)$; if there were two roots, say, of (3.61) in this subinterval, then there would be $2m$ roots in the complete $(0, 2\pi)$ interval. By following the same arguments used above the $S$-matrix element for a particular $j_1 \to j_2$ transition would be given by

$$\begin{aligned} S &= S_0 \sum_{k=1}^{m} \exp\left(\frac{-2\pi i\,\Delta jk}{m}\right) \\ &= S_0 \frac{\sin(\pi\,\Delta j)}{\sin(\pi\,\Delta j/m)} \exp\left[i\pi\,\Delta j\left(\frac{1}{m} - 1\right)\right] \end{aligned} \tag{3.64}$$

Since $\Delta j$ is an integer, one sees that $S$ is zero if $\Delta j/m$ is not also an integer; that is, the selection rule is

$$\Delta j = 0, \pm m, \pm 2m, \ldots \tag{3.65}$$

Just as in ordinary quantum mechanics, therefore, selection rules are related to symmetries of the system. In classical-limit quantum mechanics these symmetries appear in the dynamical functions such as $j_2(\bar{q}_1)$ that relate initial and final values of the canonical variables. Superposition of amplitudes that come from symmetrically related trajectories then leads to the selection rules.

### D. Reactive Processes

To describe rearrangement, or reactive collisions, such as

$$A + BC \to AB + C \tag{3.66}$$

one must modify somewhat the formal development in Section III.A; the $S$-operator of (3.25), for example, must be supplied with arrangement labels. For the atom-diatom reaction in (3.66) the label is $a$, $b$, or $c$ to denote arrangements $A + BC$, $B + AC$, or $C + AB$. The scattering operator for (3.66), therefore, is

$$S_{c,a} = \lim_{\substack{t_2 \to +\infty \\ t_1 \to -\infty}} \exp\left[\frac{iH_c t_2}{\hbar}\right] \exp\left[\frac{-iH(t_2 - t_1)}{\hbar}\right] \exp\left[\frac{-iH_a t_1}{\hbar}\right] \tag{3.67}$$

where $H$ is the total Hamiltonian, and $H_a$ and $H_c$ are the unperturbed Hamiltonians for arrangements $a$ and $c$, respectively. If $(P^a, R^a)$ and $(\mathbf{n}^a, \mathbf{q}^a)$ are the translational and action-angle variables for arrangement $a$, and $(P^c, R^c)$ and $(\mathbf{n}^c, \mathbf{q}^c)$ the analogous quantities for arrangement $c$, then the momentum states $|P^a\mathbf{n}^a$ and $|P^c\mathbf{n}^c\rangle$ are the eigenstates of the asymptotic Hamiltonians $H_a$ and $H_c$, and the $S$-matrix elements describing (3.66) are given by

$$\begin{aligned} S_{\mathbf{n}_2c,\,\mathbf{n}_1{}^a}(E) &\simeq \langle P_2{}^c \mathbf{n}_2{}^c | \, S_{c,a} \, | P_1{}^a \mathbf{n}_1{}^a \rangle \\ &\simeq \lim_{(t_2 - t_1) \to \infty} \exp\left[\frac{iE(t_2 - t_1)}{\hbar}\right] \langle P_2{}^c \mathbf{n}_2{}^c | \\ &\quad \times \exp\left[\frac{-iH(t_2 - t_1)}{\hbar}\right] | P_1{}^a \mathbf{n}_1{}^a \rangle \end{aligned} \tag{3.68}$$

where the fact has been used that

$$\exp\left(\frac{\pm iH_\gamma t}{\hbar}\right) |P^\gamma \mathbf{n}^\gamma\rangle = \exp\left(\frac{\pm iEt}{\hbar}\right) |P^\gamma \mathbf{n}^\gamma\rangle$$

for $\gamma = a$ or $c$. In the spirit of the discussion preceding (3.26), all pre-exponential factors have been omitted in (3.68). [As a formal note, the $S$-operator in (3.67) should actually be modified according to $S_{c,a} \to \mathscr{P}_c S_{c,a} \mathscr{P}_a$, where $\mathscr{P}_a$ and $\mathscr{P}_c$ are the projection operators onto the channel states of Hamiltonians $H_a$ and $H_c$.[31] Since $\mathscr{P}_a |P^a\mathbf{n}^a\rangle = |P^a\mathbf{n}^a\rangle$ and $\mathscr{P}_c |P^c\mathbf{n}^c\rangle = |P^c\mathbf{n}^c\rangle$, however, it is clear that this has no consequence regarding the $S$-matrix elements in (3.68).]

With the classical-limit approximation for the propagator (3.68) becomes

$$S_{\mathbf{n}_{2c},\mathbf{n}_1{}^a}(E) \sim \exp\left[\frac{i\Phi(\mathbf{n}_2{}^c, \mathbf{n}_1{}^a; E)}{\hbar}\right] \tag{3.69}$$

but since $(P^a, \mathbf{n}^a)$ and $(P^c, \mathbf{n}^c)$ are the canonical momenta of *different* representations some care is needed in defining the phase $\Phi(\mathbf{n}_2{}^c, \mathbf{n}_1{}^a; E)$.

One way to express it is

$$\Phi(\mathbf{n}_2{}^c, \mathbf{n}_1{}^a; E) = F_4(P_2{}^c\mathbf{n}_2{}^c, P_2{}^a\mathbf{n}_2{}^a) + E(t_2 - t_1) + \phi(P_2{}^a\mathbf{n}_2{}^a, P_1{}^a\mathbf{n}_1{}^a; t_2 - t_1) \quad (3.70)$$

where $F_4(P^c\mathbf{n}^c, P^a\mathbf{n}^a)$ is the generator of the canonical transformation $(P^cR^c\mathbf{n}^c\mathbf{q}^c) \Leftrightarrow (P^aR^a\mathbf{n}^a\mathbf{q}^a)$; since the initial and final momenta in the third term of (3.70) refer to the same canonical variables, this term is given by (2.110), so that (3.70) becomes

$$\Phi(\mathbf{n}_2{}^c, \mathbf{n}_1{}^a; E) = F_4(P_2{}^c\mathbf{n}_2{}^c, P_2{}^a\mathbf{n}_2{}^a) - \int_{t_1}^{t_2} dt[R^a(t)\dot{P}^a(t) + \mathbf{q}^a(t) \cdot \mathbf{n}^a(t)] \quad (3.71)$$

Equation (3.71) is the phase that describes propagation beginning at $t_1$ in arrangement $a$ to time $t_2$ in arrangement $a$, rearrangement at $t_2$ from arrangement $a$ to $c$. There is nothing special about the time $t_2$, however, and one can imagine carrying out the "rearrangement transformation" at any intermediate time $\bar{t}$, so that one equivalently has

$$\begin{aligned}\Phi(\mathbf{n}_2{}^c, \mathbf{n}_1{}^a; E) = &-\int_{t_1}^{\bar{t}} dt[R^a(t)\dot{P}^a(t) + \mathbf{q}^a(t) \cdot \dot{\mathbf{n}}^a(t)] \\ &+ F_4[P^c(\bar{t})\mathbf{n}^c(\bar{t}), P^a(\bar{t})\mathbf{n}^a(\bar{t})] \\ &- \int_{\bar{t}}^{t_2} dt[R^c(t)\dot{P}^c(t) + \mathbf{q}^c(t) \cdot \dot{\mathbf{n}}^c(t)] \end{aligned} \quad (3.72)$$

Equation (3.72) has a pictorial interpretation: the system evolves in arrangement $a$ from time $t_1$ to $\bar{t}$, the arrangement is changed from $a$ to $c$ at time $\bar{t}$, and the system then evolves in arrangement $c$ from time $\bar{t}$ to $t_2$. By making use of the derivative relations for an $F_4$-type generator, (2.18), one can show explicitly that $\Phi(\mathbf{n}_2{}^c, \mathbf{n}_1{}^a; E)$ in (3.72) is independent of $\bar{t}$, that is, that

$$\frac{d}{d\bar{t}}\Phi(\mathbf{n}_2{}^c, \mathbf{n}_1{}^a; E) = 0$$

Equation (3.71) corresponds to the specific choice $\bar{t} = t_2$.

A simpler way in practice to evaluate the phase of the $S$-matrix for reactive processes is to make use of the fact that *the rearrangement transformation is a point transformation of the Cartesian coordinates*; that is, if $(\mathbf{p}^a, \mathbf{x}^a)$ and $(\mathbf{p}^c, \mathbf{x}^c)$ denote the ordinary Cartesian variables of arrangements $a$ and $c$, respectively, then the coordinates $\mathbf{x}^c$ are simple linear combinations of the coordinates $\mathbf{x}^a$ and do not involve the momenta $\mathbf{p}^a$.

If one considers the identity

$$\langle P_2{}^c\mathbf{n}_2{}^c|\, S\,|P_1{}^a\mathbf{n}_1{}^a\rangle = \int d\mathbf{x}_2{}^c \int d\mathbf{x}_2{}^a \int d\mathbf{x}_1{}^a \langle P_2{}^c\mathbf{n}_2{}^c \mid \mathbf{x}_2{}^c\rangle\langle \mathbf{x}_2{}^c \mid \mathbf{x}_2{}^a\rangle \times \langle \mathbf{x}_2{}^a|\, S\,|\mathbf{x}_1{}^a\rangle\langle \mathbf{x}_1{}^a \mid P_1{}^a\mathbf{n}_1{}^a\rangle \quad (3.73)$$

then it is clear that the phases of the various factors in (3.73) are related by

$$\Phi(\mathbf{n}_2{}^c, \mathbf{n}_1{}^a; E) = -F_2(\mathbf{x}_2{}^c, P_2{}^c\mathbf{n}_2{}^c) - F_1(\mathbf{x}_2{}^a, \mathbf{x}_2{}^c) + F_2(\mathbf{x}_1{}^a, P_1{}^a\mathbf{n}_1{}^a) + \int_{t_1}^{t_2} dt\,\mathbf{p}^a(t)\cdot\dot{\mathbf{x}}^a(t) \quad (3.74)$$

The second term in (3.74) is zero [since $(\mathbf{p}^a, \mathbf{x}^a) \leftrightarrow (\mathbf{p}^c, \mathbf{x}^c)$ is a point coordinate transformation], and for Cartesian coordinates one has

$$\mathbf{p}^a(t)\cdot\dot{\mathbf{x}}^a(t) = \mathbf{p}^c(t)\cdot\dot{\mathbf{x}}^c(t) = 2T$$

where $T$ is the total kinetic energy, so that (3.74) becomes

$$\Phi(\mathbf{n}_2{}^c, \mathbf{n}_1{}^a; E) = -F_2(\mathbf{x}_2{}^c, P_2{}^c\mathbf{n}_2{}^c) + F_2(\mathbf{x}_1{}^a, P_1{}^a\mathbf{n}_1{}^a) + \int_{t_1}^{t_2} dt(2T) \quad (3.75)$$

The pictorial interpretation of (3.75) is that the system begins at $t_1$ in a momentum state of the action-angle variables of arrangement $a$ and first undergoes a transformation to a Cartesian coordinate representation; the system then evolves from $t_1$ to $t_2$ in the Cartesian coordinate representation (in which the rearrangement is a point transformation), and at $t_2$ the system is transformed from Cartesian coordinates to the final action-angle variables of arrangement $c$. Since for a reactive collision system it is usually most convenient to carry out the numerical integration of the equations of motion in Cartesian coordinates, (3.75) is the most convenient for obtaining the phase of the reactive $S$-matrix elements.

The discussion in Section II.A regarding boundary conditions for the appropriate classical trajectories carries over to the present case of reactive processes with obvious modifications. To describe the process

$$A + BC(n_1{}^a) \rightarrow AB(n_2{}^c) + C$$

for example, one begins trajectories in arrangement $a$ with the initial conditions

$$\begin{aligned} \mathbf{n}^a(t_1) &= \mathbf{n}_1{}^a = \text{specified integers} \\ R^a(t_1) &= \text{large} \\ P^a(t_1) &= -\{2\mu[E - \varepsilon_a(\mathbf{n}_1{}^a)]\}^{1/2} \\ \mathbf{q}^a(t_1) &= \bar{\mathbf{q}}_1{}^a + \frac{\partial\varepsilon_a(\mathbf{n}_1{}^a)}{\partial\mathbf{n}_1{}^a}\frac{\mu R^a(t_1)}{P^a(t_1)} \end{aligned} \quad (3.76)$$

and must find the values of $\bar{\mathbf{q}}_1{}^a$ that satisfy the equations

$$\mathbf{n}_2{}^c(\bar{\mathbf{q}}_1{}^a, \mathbf{n}_1{}^a) = \mathbf{n}_2{}^c \tag{3.77}$$

where the meaning of these quantities is analogous to those in Section III.A. The only qualitative difference of the reactive situation from the nonreactive case of Section III.A is that the reactive quantum number function, $\mathbf{n}_2{}^c(\bar{\mathbf{q}}_1{}^a, \mathbf{n}_1{}^a)$, is not defined for all $\bar{\mathbf{q}}_1{}^a$ in the interval $(0, 2\pi)$; it is defined only over the part of this interval for which the classical trajectory with these initial conditions is reactive. Correspondingly, the nonreactive quantum number function $\mathbf{n}_2{}^a(\bar{\mathbf{q}}_1{}^a, \mathbf{n}_1{}^a)$ is defined only for those values of $\bar{\mathbf{q}}_1{}^a$ for which the classical trajectory is nonreactive. Classically, that is, the trajectory with the initial conditions of (3.76) is either reactive or nonreactive and thus defines a value of $\mathbf{n}_2{}^c(\bar{\mathbf{q}}_1, \mathbf{n}_1{}^a)$ or $\mathbf{n}_2{}^a(\bar{\mathbf{q}}_1{}^a, \mathbf{n}_1{}^a)$, but not both of them. One of the consequences of this segmented nature of the reactive function $\mathbf{n}_2{}^c(\bar{\mathbf{q}}_1{}^a, \mathbf{n}_1{}^a)$ is that there do not need to be an even number of roots to (3.77); there may be just one root, for example, and there can thus be a classical contribution to the $\mathbf{n}_1{}^a \to \mathbf{n}_2{}^c$ transition without any interference structure. This was not possible for the nonreactive system discussed in Section III.B, for example.

Just as for the nonreactive case, the preexponential factor of the classical $S$-matrix for reactive processes is most conveniently expressed in terms of the quantum number function of (3.77), that is,

$$S_{\mathbf{n}_2{}^c, \mathbf{n}_1{}^a}(E) = \left[(-2\pi i\hbar)^{N-1}\left(\frac{\partial \mathbf{n}_2{}^c}{\partial \bar{\mathbf{q}}_1{}^a}\right)_{\mathbf{n}_1{}^a}\right]^{-1/2} \exp\left[\frac{i\Phi(\mathbf{n}_2{}^c, \mathbf{n}_1{}^a; E)}{\hbar}\right]$$

with $\Phi$ given by (3.75) and with the derivatives of $\mathbf{n}_2{}^c$ with respect to $\bar{\mathbf{q}}_1{}^a$ evaluated at the root of (3.77). Since $\Phi(\mathbf{n}_2{}^c, \mathbf{n}_1{}^a; E)$ satisfies the relation

$$\frac{\partial \Phi(\mathbf{n}_2{}^c, \mathbf{n}_1{}^a; E)}{\partial \mathbf{n}_1{}^a} = \bar{\mathbf{q}}_1{}^a \tag{3.79}$$

the classical $S$-matrix can also be expressed in the more symmetrical (but less useful) form analogous to (3.23):

$$S_{\mathbf{n}_2{}^c, \mathbf{n}_1{}^a}(E) = \left[\frac{\partial^2 \Phi(\mathbf{n}_2{}^c, \mathbf{n}_1{}^a; E)/\partial \mathbf{n}_2{}^c\, \partial \mathbf{n}_1{}^a}{(-2\pi i\hbar)^{N-1}}\right]^{1/2} \exp\left[\frac{i\Phi(\mathbf{n}_2{}^c, \mathbf{n}_1{}^a; E)}{\hbar}\right] \tag{3.80}$$

Finally, as a practical note, the reactive quantum number function of (3.77) is normally[8e] computed in a manner which is essentially a literal interpretation of (3.75); beginning with the initial conditions of the action-angle variables, (3.76), one generates the corresponding initial values of the Cartesian coordinates, carries out the numerical integration of the equations of motion in Cartesian coordinates, and at the end of the trajectory transforms back to the action-angle variables of arrangement $c$ if the trajectory has been reactive, or to those of arrangement $a$ if it is nonreactive. The phase $\Phi$ of (3.75) is computed by adding to the equations of motion an additional differential equation for the function $\chi(t)$,

$$\dot{\chi}(t) = 2T$$

where $T$ is the total kinetic energy, with initial condition

$$\chi(t_1) = F_2(\mathbf{x}_1^a, P_1^a\mathbf{n}_1^a)$$

In terms of the final value of $\chi(t)$ the phase of the $S$-matrix is

$$\Phi(\mathbf{n}_2^\gamma, \mathbf{n}_1^a; E) = \chi(t_2) - F_2(\mathbf{x}_2^\gamma, P_2^\gamma\mathbf{n}_2^\gamma)$$

for a reactive ($\gamma = c$) or nonreactive ($\gamma = a$) trajectory.

Further discussion of reactive processes is deferred until Section IV.E where examples of reactive tunneling are discussed.

### E. Three-Dimensional Collision Systems; Quenching of Interference Effects

To conclude Section III we discuss the three-dimensional $A + BC$ collision system in more detail.[8a,8d,8f,8g] The differential cross-section is given in terms of the $S$-matrix elements by

$$\sigma_{n_2j_2m_2\leftarrow n_1j_1m_1}(\theta) = |f_{n_2j_2m_2\leftarrow n_1j_1m_1}(\theta)|^2 \tag{3.81}$$

where

$$f_{n_2j_2m_2\leftarrow n_1j_1m_1}(\theta) = \frac{1}{2ik_1}\sum_J (2J+1)d^J_{m_2,m_1}(\theta)S_{n_2j_2m_2,n_1j_1m_1}(J) \tag{3.82}$$

and the integral cross-sections by

$$\sigma_{n_2j_2m_2\rightarrow n_1j_1m_1} \equiv 2\pi\int_0^\pi d\theta \sin\theta\sigma_{n_2j_2m_2\leftarrow n_1j_1m_1}(\theta) \tag{3.83}$$

$$= \frac{\pi}{k_1^2}\sum_J (2J+1)\,|S_{n_2j_2m_2,n_1j_1m_1}(J)|^2 \tag{3.84}$$

where $n$, $j$, and $m$ are the vibrational, rotational, and helicity quantum numbers for $BC$; $d_{m_2,m_1}(\theta)$ are the rotation matrices [e.g., $d_{00}{}^J(\theta) = P_J(\cos\theta)$], and the $S$-matrix elements in the helicity, or $m$-representation are related to those in the $l$-representation by a vector-coupling transformation:

$$S_{n_2j_2m_2,n_1j_1m_1}(J) = \sum_{l_2,l_1} C(j_2l_2J; m_2, -m_2)i^{l_2-l_1} \times S_{n_2j_2l_2,n_1j_1l_1}(J)C(j_1l_1J; m_1, -m_1) \tag{3.85}$$

One is usually interested in integral cross-sections that are summed and averaged over $m_2$ and $m_1$, respectively,

$$\sigma_{n_2j_2 \leftarrow n_1j_1} = \frac{\pi}{k_1^2(2j_1+1)} \sum_J (2J+1) \sum_{m_2,m_1} |S_{n_2j_2m_2,n_1j_1m_1}(J)|^2 \quad (3.86)$$

and this can also be expressed directly in terms of $S$-matrix elements of the $l$-representation:

$$\sigma_{n_2j_2 \leftarrow n_1j_1} = \frac{\pi}{k_1^2(2j_1+1)} \sum_J (2J+1) \sum_{l_2,l_1} |S_{n_2j_2l_2,n_1j_1l_1}(J)|^2 \quad (3.87)$$

The classical $S$-matrix elements for the $l$-representation are given by the expressions of Section III.A:

$$S_{n_2j_2l_2,n_1j_1l_1}(J,E) = \left[(-2\pi i\hbar)^3 \frac{\partial(n_2j_2l_2)}{\partial(q_{n_1}q_{j_1}q_{l_1})}\right]^{-1/2} \times \exp\left[\frac{i\Phi(n_2j_2l_2,\, n_1j_1l_1)}{\hbar}\right] \quad (3.88)$$

where

$$\Phi(n_2j_2l_2,\, n_1j_1l_1) = \frac{\pi}{2}(l_2+l_1) - \int_{-\infty}^{\infty} dt \left[R(t)\frac{dP(t)}{dt} + q_n(t)\frac{dn(t)}{dt} + q_j(t)\frac{dj(t)}{dt} + q_l(t)\frac{dl(t)}{dt}\right] \quad (3.89)$$

The reduced Hamiltonian for the system, expressed in terms of the action-angle variables, is

$$H_J(P,R,n,q_n,l,q_l,j,q_j) = \frac{(P^2 + l^2/R^2)}{2\mu} + \varepsilon(n,j) + V(r,R,\cos\gamma) \quad (3.90)$$

where

$$\cos\gamma = \cos q_l \cos(q_j - \Delta q_j) + \left(\frac{J^2 - l^2 - j^2}{2lj}\right) \sin q_l \sin(q_j - \Delta q_j) \quad (3.91)$$

and $r = r(n, q_n, j)$ and $\Delta q_j = \Delta q_j(n, q_n, j)$ are functions to be discussed below. The Hamiltonian in (3.90) is labeled by the total angular momentum $J$ which is a conserved quantity and thus appears in the Hamiltonian only as a parameter.

The application of (3.88) and (3.89) is similar to the collinear $A + BC$ system discussed in Section III.B, but now there are three internal (i.e., quantized) degrees of freedom, $n$, $j$, and $l$. Holding $n_1$, $j_1$, and $l_1$ constant, therefore, one varies $\bar{q}_{n_1}$, $\bar{q}_{j_1}$, and $\bar{q}_{l_1}$ to find the root (or roots) of the equations

$$\begin{aligned} n_2(\bar{q}_{n_1}, \bar{q}_{j_1}, \bar{q}_{l_1}) &= n_2 \\ j_2(\bar{q}_{n_1}, \bar{q}_{j_1}, \bar{q}_{l_1}) &= j_2 \\ l_2(\bar{q}_{n_1}, \bar{q}_{j_1}, \bar{q}_{l_1}) &= l_2 \end{aligned} \tag{3.92}$$

where the interpretation of these quantities is the same as before: that is, the values on the RHS of (3.92) are specific integers, and the functions of the LHS are the final values of these quantities that result by integrating Hamilton's equations with the Hamiltonian in (3.90) and initial conditions

$$\begin{aligned} n(t_1) &= n_1 \\ j(t_1) &= j_1 \\ l(t_1) &= l_1 \\ R(t_1) &= \text{large} \\ P(t_1) &= -\{2\mu[E - \varepsilon(n_1, j_1)]\}^{1/2} \\ q_n(t_1) &= \bar{q}_{n_1} + \frac{\partial \varepsilon(n_1, j_1)}{\partial n_1} \frac{\mu R(t_1)}{P(t_1)} \\ q_j(t_1) &= \bar{q}_{j_1} + \frac{\partial \varepsilon(n_1, j_1)}{\partial j_1} \frac{\mu R(t_1)}{P(t_1)} \\ q_l(t_1) &= \bar{q}_{l_1} \end{aligned} \tag{3.93}$$

For each root of the simultaneous equations (3.92) there is a contribution to the $S$-matrix element given by (3.88) with the phase computed from (3.89). [The phase $(\pi/2)(l_2 + l_1)$ in (3.89) comes from the fact that the radial momentum $P$ is actually not a constant of the motion of $H_0$, the Hamiltonian in (3.90) with the interaction $V$ omitted; the phases $\pi l_1/2$ and $\pi l_2/2$ come from the trivial transformation at $t_1$ and $t_2$ from the radial momentum to a linear momentum that is a conserved quantity of $H_0$.]

To complete the specification of the Hamiltonian, (3.90), in terms of the action-angle variables one must express the vibrational coordinate $r$ and the centrifugal distortion correction $\Delta q_j$ in terms of the action-angle variables. The general expressions for these quantities have been given[8a] but are so cumbersome as to be impractical for applications. It is possible, however, to obtain explicit expressions for these quantities within the framework of a Dunham-like expansion[28]; since the Dunham expansion for the energy levels of most diatomic molecules converges quite rapidly, these expansions should be adequate for essentially all cases of non-reactive $A + BC$ collisions. If the vibrational potential of $BC$ is expanded as

$$v(r) = \tfrac{1}{2}m\omega^2(r - r_0)^2[1 + a_1\xi + a_2\xi^2 + \cdots] \tag{3.94}$$

with

$$\xi = \frac{(r - r_0)}{r_0} \tag{3.95}$$

and if $B$ is the rotation constant

$$B = (2mr_0^2)^{-1} \tag{3.96}$$

then the appropriate expansion parameter is

$$\lambda = \frac{B}{\omega} \tag{3.97}$$

With remainders of order $\lambda^2$, the quantities $\varepsilon(n, j)$, $r(n, q_n, j)$, and $\Delta q_j(n, q_n, j)$ are given by

$$\begin{aligned}\varepsilon(n, j) = (n + \tfrac{1}{2})\omega\{(1 + A) + (n + \tfrac{1}{2})\lambda[\tfrac{3}{2}a_2 - \tfrac{15}{8}a_1^2 \\ + 6A(1 + a_1) - 4A^2]\}\end{aligned} \tag{3.98}$$

$$\begin{aligned}r(n, q_n, j) = r_0\{1 &+ 2[\lambda(n + \tfrac{1}{2})]^{1/2} \sin q_n \\ &+ \lambda(n + \tfrac{1}{2})[4A - a_1(3 + \cos 2q_n)] \\ &- 2[\lambda(n + \tfrac{1}{2})]^{3/2}[3A(1 + a_1) \sin q_n \\ &+ \tfrac{1}{2}(3a_2 - \tfrac{11}{4}a_1^2) \sin q_n \\ &+ \tfrac{1}{4}(a_2 + \tfrac{3}{4}a_1^2) \sin 3q_n]\}\end{aligned} \tag{3.99}$$

$$\begin{aligned}\Delta q_j(n, q_n, j) = j\lambda\{-8[\lambda(n + \tfrac{1}{2})]^{1/2} \cos q_n + 2\lambda(n + \tfrac{1}{2}) \\ (3 - a_1) \sin 2q_n\}\end{aligned} \tag{3.100}$$

where

$$A = \frac{\lambda j^2}{n + \frac{1}{2}} = \frac{Bj^2}{(n + \frac{1}{2})\omega}$$

Table I gives the values of $\lambda^2$ for some common diatomic molecules, which indicate that these expansions should be more than adequate. With (3.98) through (3.100) the Hamiltonian in (3.90) is expressed explicitly in terms of the action-angle variables. Numerical integration of the equations of motion in terms of the action-angle variables is highly desirable, of course, because there are only four degrees of freedom in this case, as opposed to six if the numerical integration is carried out in Cartesian coordinates, and because the action-angle variables should be more slowly varying functions of time than the Cartesian variables.

If direct numerical integration of the equations of motion in action-angle variables is not convenient, then it is also possible to proceed as follows: specifying the initial conditions in terms of the action-angle variables, (3.93), one generates the corresponding initial values of the

TABLE I
Dunham Expansion Parameters for Some Common Diatomic Molecules

| Molecule | $\lambda^{2a}$ |
|---|---|
| $H_2$ | $1.9 \times 10^{-4}$ |
| $N_2$ | $7.3 \times 10^{-7}$ |
| $O_2$ | $8.4 \times 10^{-7}$ |
| HCl | $1.4 \times 10^{-5}$ |
| HF | $2.6 \times 10^{-5}$ |
| $Cl_2$ | $1.9 \times 10^{-7}$ |
| NO | $8.0 \times 10^{-7}$ |

[a] $\lambda^2 = (B/\omega)^2$, where $B$ is the rotational constant and $\omega$ the vibrational frequency; also, $\lambda^2 = (2mr_0{}^2\omega)^{-2}$, where $m$ is the reduced mass and $r_0$ the bond length. Molecular parameters are from G. Herzberg, *Spectra of Diatomic Molecules*, Van Nostrand, Princeton, N.J., 1950.

Cartesian coordinates and carries out the numerical integration in these variables; at the end of the trajectory one transforms back to action-angle variables to obtain the final quantum numbers. Either approach, of course, must give the same result for the classical $S$-matrix elements. Several applications[8d,8q,8h] to three-dimensional atom-diatom systems have been made, and comparison of individual classical $S$-matrix elements with quantum mechanical values indicates the semiclassical theory to be an accurate description of the collision dynamics even at this most detailed level.

Because there are three internal degrees for the present system there are often more than two roots to (3.92) for a given $n_1 j_1 l_1 \to n_2 j_2 l_2$ transition. If the three functions in (3.92) were simple harmonic functions of each of the three angle variables, then there would be $2^3 = 8$ roots to (3.92); that is, an $S$-matrix element would be the sum of eight terms of the form given by (3.88). With eight terms in an $S$-matrix element, however, there are $8 \times \frac{7}{2} = 28$ interference terms in $|S|^2$! With so many interference terms it is doubtful that any interference structure can survive—particularly so in light of the fact that the cross-sections of interest almost always involve sums and averages over quantum states for some of the degrees of freedom [see (3.87)]. In cases of strong coupling between translation and the various internal degrees of freedom, therefore, it seems that the interference structure predicted by the semiclassical theory will not be significant; this point has been borne out in numerical calculations.[8d] If, on the other hand, there is only weak coupling between various internal degrees of freedom, then it is quite reasonable to expect interference features in the internal state distribution to survive; this possibility is discussed more explicitly regarding "partial averaging" in Section IV.E.

## IV. CLASSICALLY FORBIDDEN PROCESSES

Often the process of interest lies outside the framework of classical mechanics; in thermal energy atom-diatom collisions, for example, there may be no classical trajectories for which the vibrational quantum number changes by an entire quantum; in such cases vibrationally inelastic transitions are said to be classically forbidden, in the sense that there are no classical trajectories that contribute to the appropriate classical $S$-matrix elements. Other practically important examples of classically forbidden processes are tunneling in reactive systems that have an activation barrier (see Section IV.D) and electronic transitions between different adiabatic electronic states (see Section V).

The ability to analytically continue classical mechanics within the framework of classical $S$-matrix theory and thus describe these classically forbidden, or weak, processes is perhaps the most practically important contribution of the semiclassical theory; this is because ordinary classical trajectory methods are unable to describe such processes at all. For classically allowed processes, on the other hand, the purely classical Monte Carlo methods are often completely adequate, missing only the interference structure which, as discussed in Section III.E, would often be quenched even if it were originally included in the $S$-matrix elements.

### A. Example: Vibrational Excitation in Collinear $A + BC$

To illustrate the concept of classically forbidden processes and their description within the framework of classical $S$-matrix theory, it is useful to consider first the model system discussed before in Section III.B. With reference to Fig. 1, which corresponds to $n_1 = 1$, it is clear that there are no roots to the equation

$$n_2(\bar{q}_1) = 5 \tag{4.1}$$

for the maximum of $n_2(\bar{q}_1)$ is less than 5; the $1 \to 5$ vibrational transition is thus classically forbidden at this collision energy. One might ask if vibrational state 5 is a closed channel, that is, if $E - \varepsilon(5) < 0$, where $E$ is the total energy; this is not so in this case, vibrational states up to $n = 9$ being energetically allowed at the collision energy in Fig. 1. The $1 \to 5$ transition is thus *energetically* allowed, but *dynamically* forbidden, that is, it does not take place via the classical equations of motion. In general, therefore, the term "classically forbidden" is quite distinct from "energetically forbidden"; it has meaning only with reference to the full classical dynamics and is related more to the strength of the coupling between translational and vibrational degrees of freedom than it is to the energy of the system.

Within the simple classical-limit theory the $1 \rightarrow 5$ transition probability is thus zero, which in practice means that it is small; the $1 \rightarrow 6$ transition is "more forbidden" classically, and is thus even smaller. To describe these processes semiclassically one notes that though there are no *real* values of $\bar{q}_1$ that satisfy (4.1), there are complex values that do so; to see this explicitly note that $n_2(\bar{q}_1)$ may be approximated by a Taylor's series expansion about its maximum

$$n_2(\bar{q}_1) \simeq n_2^{\max} + \tfrac{1}{2}n_2''(\bar{q}_{\max})(\bar{q}_1 - \bar{q}_{\max})^2 \tag{4.2}$$

so that the roots to (4.1) are found to be

$$\bar{q}_1 \simeq \bar{q}_{\max} \pm i\left[\frac{2(5 - n_2^{\max})}{|n_2''(\bar{q}_{\max})|}\right]^{1/2} \tag{4.3}$$

Equation (4.3) is, of course, only an approximation to the complex roots of (4.1), and to proceed more rigorously one needs a general way of finding these complex roots more accurately. The precise definition of $n_2(\bar{q}_1)$ is the final value of $n(t)$ for the trajectory with the initial conditions $n_1$, $\bar{q}_1$, $P_1$, $R_1$ of (3.45), and this definition also applies *even if* $\bar{q}_1$ *is complex*; that is, it is possible to integrate Hamilton's equations numerically with complex initial conditions and in this way generate $n_2(\bar{q}_1)$ for any complex value of $\bar{q}_1$. Complex-valued classical trajectories may sound "unphysical," but as discussed in the next section, all the physically meaningful quantities—$n_1$, $R_1$, $P_1$, $n_2$, $R_2$, $P_2$—are real in the asymptotic regions, and since $\bar{q}_1$ and $\bar{q}_2$ are not observable it is irrelevant that they are complex.

Along a complex-valued trajectory the action integral $\Phi(n_2, n_1)$, (3.42b), is also complex, so that the *S*-matrix element of (3.42a) has an exponential damping factor $\exp(-\text{Im}\,\Phi/\hbar)$. If $\bar{q}_{\text{I}}$ is a complex root of (4.1), then it is easy to see by complex conjugation of (4.1) that the other root $q_{\text{II}}$ is $q_{\text{I}}^*$; the phases $\phi_{\text{I}}$ and $\phi_{\text{II}}$ are also complex conjugates of each other, $\phi_{\text{II}} = \phi_{\text{I}}^*$. If the root $\bar{q}_{\text{I}}$ corresponds to exponential damping, that is, $\text{Im}\,\phi_{\text{I}} > 0$, then it is clear that $\bar{q}_{\text{II}}$ corresponds to an exponential *enhancement* $\exp(+\text{Im}\,\Phi/\hbar)$. It is intuitively clear on physical grounds, and can be shown more rigorously,[8,9] that the exponentially large term should not be included, and the *S*-matrix element thus has just the one term

$$S_{n_2,n_1} = [-2\pi i n_2'(\bar{q}_{\text{I}})]^{-1/2} \exp\left(\frac{i\Phi_{\text{I}}}{\hbar}\right) \tag{4.4}$$

and the transition probability is

$$P_{n2,n1} = [2\pi\,|n_2'(\bar{q}_{\rm I})|]^{-1} \exp\left(\frac{-2\,|{\rm Im}\,\Phi_{\rm I}|}{\hbar}\right) \quad (4.5)$$

Equation (4.5), having an exponential damping factor involving a classical action integral, is quite reminiscent of the WKB tunneling probability for barrier penetration in one-dimensional systems.[17] Classically forbidden processes are, in fact, a generalization of one-dimensional tunneling to dynamical systems with several degrees of freedom.

In applications the "primitive" classical-limit formulas in (3.48) and (4.5) are not accurate if the transition is "just forbidden," or "just allowed," that is, if $n_2$ is too close to $n_2^{\rm max}$. Uniform semiclassical formulas[8b,9g] may be invoked in such cases, however, and they are able to describe the intermediate case quite well. The reader familiar with the semiclassical theory of elastic scattering will recognize that the situation $n_2$ near $n_2^{\rm max}$ is analogous to the rainbow effect in elastic scattering[1,2]; the purely classical probability

$$P_{n2,n1}^{\rm CL} = [2\pi\,|n_2'(\bar{q}_{\rm I})|]^{-1} + [2\pi\,|n_2'(\bar{q}_{\rm II})|]^{-1} \quad (4.6)$$

becomes infinite as $n_2 \to n_2^{\rm max}$, $n_2 < n_2^{\rm max}$, (the "bright" side of the rainbow angle), and becomes exactly zero for $n_2 > n_2^{\rm max}$ (the "dark" side of the rainbow angle). The primitive semiclassical expression describes the interference on the bright side of the rainbow ($n_2 < n_2^{\rm max}$) but also goes to infinity as $n_2 \to n_2^{\rm max}$; the uniform semiclassical formula, however, describes the transition through the forbidden/allowed boundary.

Calculations of the type outlined above—that is, the numerical integration of complex-valued classical trajectories to obtain the complex root of (3.46)—have been carried out[8g,9f] for the nonreactive, collinear $A + BC$ collision system, and agreement with essentially exact quantum mechanical values is excellent (within a few percent) even for highly forbidden transitions with probabilities as small as $10^{-11}$. Since for low-energy collisions it is typically the case that *all* vibrationally inelastic transitions are classically forbidden, strictly classical Monte Carlo trajectory methods are of highly questionable reliability in such cases; the ability to analytically continue classical mechanics in such a way as to describe these processes accurately is thus an important advance.

### B. Boundary Conditions for Analytically Continued (i.e., Complex-Valued) Classical Trajectories; Origin of Complex Time

To see that complex-valued classical trajectories are indeed physically meaningful within the context of classical $S$-matrix theory, it is useful to discuss more explicitly the boundary conditions which are appropriate for them. For the general system of $N$ degrees of freedom there are $(N - 1)$ internal, quantized degrees of freedom characterized by the action-angle variables $\mathbf{n}$ and $\mathbf{q}$, $(\mathbf{n}, \mathbf{q}) \equiv (\mathbf{n}_i, \mathbf{q}_i)$, $i = 1, \ldots, N - 1$; the translational degree of freedom is characterized by radial coordinate and momentum $R$ and $P$.

The classical trajectory related to a specific $\mathbf{n}_1 \rightarrow \mathbf{n}_2$ transition corresponds to the initial conditions

$$\mathbf{n}_1 = \text{specified integers} \tag{4.7a}$$

$$R_1 = \text{real and large} \tag{4.7b}$$

$$P_1 = -\{2\mu[E - \varepsilon(\mathbf{n}_1)]\}^{1/2} \tag{4.7c}$$

$$\mathbf{q}_1 = \bar{\mathbf{q}}_1 + \frac{\partial \varepsilon(\mathbf{n}_1)}{\partial \mathbf{n}_1} \frac{\mu R_1}{P_1} \tag{4.7d}$$

Since the quantum numbers $\mathbf{n}_1$ are real, one sees that the translational momentum $P_1$, which is determined implicitly by energy conservation, is also real. It is clear that $R_1$ and $P_1$ must be real-valued for they are physical quantities, the distance and velocity between the collision partners in the initial asymptotic region. The quantum numbers $\mathbf{n}_1$ are required semiclassically to be integers (and thus real), but the conjugate angle variables $\mathbf{q}_1$ are completely unrestricted (cf. the uncertainty principle). The final values of the variables should be similar to (4.7):

$$\mathbf{n}_2 = \text{specified integers} \tag{4.8a}$$

$$R_2 = \text{real and large} \tag{4.8b}$$

$$P_2 = +\{2\mu[E - \varepsilon(\mathbf{n}_2)]\}^{1/2} \tag{4.8c}$$

$$\mathbf{q}_2 = \text{anything} \tag{4.8d}$$

As discussed in Section IV.A, however, there may be no real values of $\bar{\mathbf{q}}_1$ for which $\mathbf{n}_2(\bar{\mathbf{q}}_1, \mathbf{n}_1)$ achieves the desired integer values in (4.8a); that is, the transition may be classically forbidden, and one must then allow $\bar{\mathbf{q}}_1$ to take on complex values. If one integrates Hamilton's equations with the initial conditions in (4.7) with $\bar{\mathbf{q}}_1$ complex, however, the coupling of the variables during the trajectory means that *all* the coordinates and momenta become complex-valued; thus in general, the final values of $\mathbf{n}$, $\mathbf{q}$, $R$, and $P$ are all complex-valued for a trajectory with complex initial conditions. Equation (4.8a), which is actually the $2N - 2$ equations

$$\text{Re}\,\mathbf{n}_2 = \text{specified integers}$$

$$\text{Im}\,\mathbf{n}_2 = 0$$

can be satisfied, however, by the appropriate choice of the $2N - 2$ variables $\text{Re}\,\bar{\mathbf{q}}_1$, $\text{Im}\,\bar{\mathbf{q}}_1$. With (4.8a) satisfied by the choice of $\bar{\mathbf{q}}_1$, (4.8c) is automatically fulfilled. Finally, the condition that $R_2$ be real, (4.8b), can be achieved by the appropriate choice of the imaginary part of the *time*. To see this, note that the solution of the equations of motion in the final asymptotic region gives

$$\mathbf{n}(t) = \mathbf{n}_2 \tag{4.9a}$$

$$P(t) = P_2 \tag{4.9b}$$

$$\mathbf{q}(t) = \mathbf{q}(\tilde{t}_2) + \frac{\partial \varepsilon(\mathbf{n}_2)}{\partial \mathbf{n}_2}(t - \tilde{t}_2) \tag{4.9c}$$

$$R(t) = R(\tilde{t}_2) + \frac{P_2(t - \tilde{t}_2)}{\mu} \tag{4.9d}$$

where $\tilde{t}_2$ is real, say. $R(\tilde{t}_2)$ will not in general be real but can be made so if one chooses the final time $t_2$ to be

$$t_2 = \tilde{t}_2 - i\frac{\mu}{P_2}\,\text{Im}\,R(\tilde{t}_2) \tag{4.10}$$

Since $\mathbf{n}(t)$ and $P(t)$ are constant in the asymptotic region, this final imaginary time increment does not affect their values.

To have consistent, physically appropriate boundary conditions that treat initial and final states on an equal footing, it is thus necessary to allow the time increment $t_2 - t_1$ to be complex-valued. This is not unreasonable, however, for the $S$-matrix corresponds to a specific value of the total energy $E$, so that the uncertainty principle allows the variable conjugate to energy, the time, to be anything and thus possibly complex; this is analogous to the fact that $S$-matrix elements in a quantum number representation demand that the quantum numbers $\mathbf{n}$ be specific values initially and finally, so that the conjugate angle variables $\mathbf{q}$ can be anything (and thus complex). [Perhaps more familiar than the present use of complex time in an energy representation is the use of complex energy when one is dealing with time evolution; that is, the time evolution of a metastable state is often described by a formalism that makes use of the complex energy $(E_0 - i\Gamma/2)$.] A specific physical meaning of the imaginary part of the complex time, which is uniquely determined by the boundary conditions in (4.8) and (4.9), is discussed in Sections IV.D and VI.A: there $\text{Im}\,(t_2 - t_1)$ is found to be related to $\hbar\beta$, where $\beta^{-1} = kT$ is a temperature.

Another way to see complex time enter the theory is to return to the derivation of the general expression for the classical $S$-matrix in Section

III.A. The value of $t \equiv t_2 - t_1$ which is the point of stationary phase in (3.13) is the time required for the classical trajectory which leads to the $\mathbf{n}_1 \rightarrow \mathbf{n}_2$ transition at total energy $E$; (3.20) expresses it as

$$t \equiv t_2 - t_1 = \frac{\mu R_2}{P_2} - \frac{\mu R_1}{P_1} + \frac{\partial \Phi(\mathbf{n}_2, \mathbf{n}_1; E)}{\partial E}$$

[$\mu R_2/P_2 - \mu R_1/P_1$ is the "free" transit time, and $\partial \Phi(\mathbf{n}_2, \mathbf{n}_1; E)/E$ is the "time delay."[32]] If there is no classical trajectory at energy $E$ that gives rise to the $\mathbf{n}_1 \rightarrow \mathbf{n}_2$ transition—that is, if the transition is classically forbidden—then there is no real value of $t$ that satisfies the stationary phase condition, (3.18), and the value of the integral is exponentially small. To evaluate the integral in such cases one resorts to the method of steepest descent[33]: that is, one looks for complex roots of (3.18) and, finding one, distorts the path of the time integral from the real time axis to a contour that passes through this "complex point of stationary phase"; in order to do this one must be able to analytically continue the integrand into the complex $t$-plane. Although the mathematical arguments pertaining to the method of steepest descent are somewhat different from those that apply to the method of stationary phase, the resulting integral formula is precisely the same, so that the classical $S$-matrix is still given by (3.23), but where $\Phi(\mathbf{n}_2, \mathbf{n}_1; E)$ is evaluated along the appropriate analytically continued classical trajectory. Since $\Phi(\mathbf{n}_2, \mathbf{n}_1; E)$ is complex, it is easy to see that the value of $\bar{\mathbf{q}}_1$, which leads to the appropriate trajectory,

$$\bar{\mathbf{q}}_1 = \frac{\partial \Phi(\mathbf{n}_2, \mathbf{n}_1; E)}{\partial \mathbf{n}_1}$$

is also complex. Thus the value of $t$ which is the "complex point of stationary phase" is identical to the complex value of $t_2 - t_1$ determined by the boundary conditions in (4.8) and (4.9).

It should not be disturbing that the quantities $R(t)$, $P(t)$, and $\mathbf{n}(t)$, though real in the limit $t \rightarrow \pm\infty$, are actually complex-valued *during* the collision, for the numerical integration of the equations of motion with complex initial conditions is simply an operational procedure for analytically continuing the quantum number function $\mathbf{n}_2(\bar{\mathbf{q}}_1, \mathbf{n}_1)$ and action integral $\Phi(\mathbf{n}_2, \mathbf{n}_1; E)$ from which classical $S$-matrix elements are constructed. The $S$-matrix, the quantum mechanical description of a scattering process, however, refers only to the initial and final states; that is, it makes no sense quantum mechanically to ask what is happening *during* the collision. Since quantum mechanics does not allow one to observe the system during the collision, there is thus nothing inconsistent or unphysical about $R(t)$, $P(t)$, and $\mathbf{n}(t)$ being complex-valued at these intermediate times. Only in the asymptotic regions are they properly observable, and here the boundary conditions in (4.8) and (4.9) show that they are indeed real.

A final point that should be noted is that since the coordinates and momenta are analytic functions of the complex time $t$, their values at time $t_2$ depend only on the initial conditions at time $t_1$ and the final value $t_2$—they do not depend on the particular path in the complex time plane along which the time variable is incremented from $t_1$ to $t_2$ in integrating the equations of motion. [In carrying out the numerical integration of complex-valued trajectories this fact is actually exploited and the path in the complex time plane is chosen to facilitate the calculation.] The particular paths that $R(t)$, $P(t)$, etc., trace out in their respective complex planes do, of course, depend on the particular time path that connects $t_1$ and $t_2$, but their final values $R(t_2)$, $P(t_2)$, etc., do not. Thus the functions $\mathbf{n}_2(\bar{\mathbf{q}}_1, \mathbf{n}_1)$ and $\Phi(\mathbf{n}_2, \mathbf{n}_1; E)$ (and, therefore, the classical $S$-matrix elements) are invariant to the complex time path, but what the trajectory actually "does" during the collision depends on it. This is another manifestation of the fact that a quantum description does not allow one to discuss what happens during the collision itself.

### C. Branch Points in Complex Time

Regarding the invariance of the final properties of a classical trajectory to the complex time path along which it is integrated (as discussed at the end of the preceding section) one must add one qualification, that one consider only time paths that pass on the same side of any branch points that may exist in the complex time plane. To illustrate the fact that there can be branch points in the complex time plane and to demonstrate their significance, it is helpful to discuss a simple one-dimensional model that is exactly solvable.

Consider the symmetrical Eckart potential barrier

$$V(x) = V_0 \operatorname{sech}^2\left(\frac{x}{a}\right) \tag{4.11}$$

and an energy $E \equiv (\frac{1}{2})mv^2 < V_0$, as shown in Fig. 3. If the particle is to the left of the barrier for real time, then one finds the trajectory to be

$$x(t) = -a \sinh^{-1}\left[\lambda \cosh\left(\frac{vt}{a}\right)\right] \tag{4.12}$$

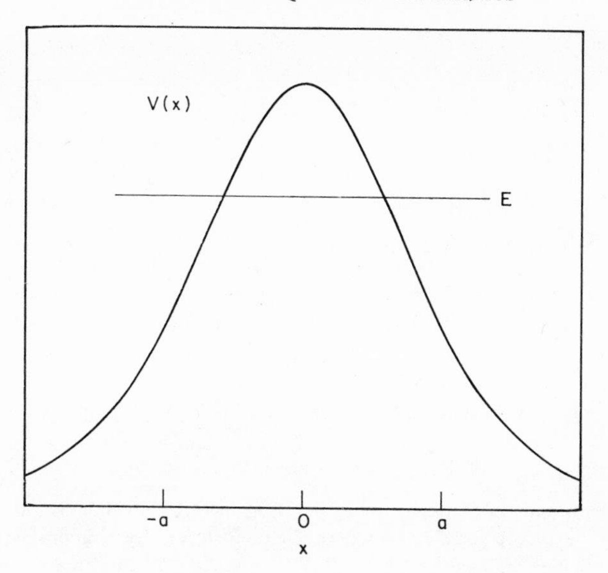


**Fig. 3.** Eckart potential barrier, $V(x) = V_0 \operatorname{sech}^2 (x/a)$.

where

$$\lambda = \left(\frac{V_0}{E} - 1\right)^{1/2} \tag{4.13}$$

and where the zero of time has been chosen (without restriction) so that $x(0)$ is the classical turning point. One can see easily from (4.12) that as $t$ moves from $-\infty$ to 0 to $+\infty$ the particle moves from $x = -\infty$ to the turning point $x(0)$ back to $-\infty$.

The analytic properties of $x(t)$ follow from those of the inverse hyperbolic sine function; since $\sinh^{-1}(z)$ has branch points at $z = \pm i$, $x(t)$ is seen to have branch points at values of $t$ for which

$$\lambda \cosh\left(\frac{vt}{a}\right) = \pm i$$

That is, the branch points in the complex time plane are

$$t_n = \pm t_* \pm \frac{i(n + \frac{1}{2})\pi a}{v} \tag{4.14}$$

where

$$n = 0, 1, 2, \ldots$$

$$t_* = \left(\frac{a}{v}\right) \tanh^{-1}\left[\left(\frac{E}{V_0}\right)^{1/2}\right] \tag{4.15}$$

The layout of these branch points is shown in Fig. 4.

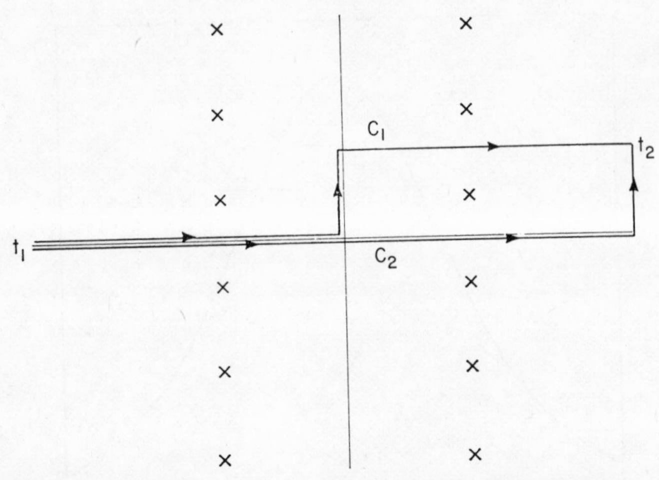


**Fig. 4.** Branch points of the function $x(t)$ defined in (4.12); their locations are given algebraically by (4.14) and (4.15). Curve $C_1$ consists of the straight-line segments $t_1 \equiv -\infty \rightarrow 0, 0 \rightarrow i\pi a/v \rightarrow +\infty + i\pi a/v \equiv t_2$, and curve $C_2$ the straight line segments $t_1 \equiv -\infty \rightarrow +\infty, +\infty \rightarrow +\infty + i\pi a/v \equiv t_2$.

$x(t)$ is thus a multivalued function of $t$, and the *principal tunneling branch* corresponds to the time path $C_1$ in Fig. 4: as $t$ is incremented from $-\infty$ to 0, the particle moves from $x = -\infty$ to its classical turning point; as $t$ is incremented from 0 to $i\pi a/v$ the particle tunnels through the barrier from its left turning point to the right turning point; and as $t$ is incremented from $i\pi a/v$ to $\infty + i\pi a/v$, $x$ moves from the right turning point to $+\infty$. The probability associated with this tunneling transition is

$$P_{\text{tunnel}} = \exp\left(\frac{-2 \operatorname{Im} \Phi}{\hbar}\right) \tag{4.16}$$

where the imaginary part of the action integral $\Phi$ is the usual "barrier penetration" integral, which for this potential s

$$\operatorname{Im} \Phi = \pi a[(2mV_0)^{1/2} - (2mE)^{1/2}] \tag{4.17}$$

[Equations (4.16) and (4.17) give the standard WKB tunneling probability.[17]] This same final value of $x(t)$, and the probability associated with the transition, is obtained if one chooses the time path to be any distortion of the curve $C_1$ that does not move it through any of the branch points; that is, the curve $C_1$, and any distortion thereof that doesn't cross branch points, defines one of the single-valued branches of $x(t)$, the principal tunneling branch.

The time path $C_2$ in Fig. 4 connects the same values $t_1$ and $t_2$ as does curve $C_1$, but one can easily verify that following $x(t)$ continuously along curve $C_2$ leads to a reflective, nontunneling transition. In fact, the branch of $x(t)$ defined by curve $C_2$, and any distortion of it that doesn't cross branch points, is the principal nontunneling branch that corresponds to a purely real time path.

One can easily identify other branches of $x(t)$ by considering time paths that pass around the branch points in various ways: the time paths that pass above the branch points $t_* + i\frac{3}{2}\pi a/v$, $t_* + i\frac{5}{2}\pi a/v$, etc., correspond to a nontunneling trajectory that passes back and forth once through the barrier, a tunneling trajectory that makes one extra passage back and forth inside the barrier, etc., respectively. The probabilities associated with these branches is exp $(-4 \text{ Im } \Phi/\hbar)$, exp $(-6 \text{ Im } \Phi/\hbar)$, etc., respectively.

One sees in this example, therefore, that *it matters at what real value of time the imaginary time increments are made*. Imaginary time increments in the asymptotic regions (curve $C_2$) cause no tunneling, but tunneling does result if the imaginary time increment is made while the particle is in the "interaction region." For this particular example tunneling occurs only if the imaginary part of the time increment is made when the real part of the time is in the interval $(-t_*, t_*)$, where $t_*$ is defined by (4.15); for low energies, $E \to 0$, one thus has an increment of time

$$2t_* \simeq \left(\frac{a\mu}{2V_0}\right)^{1/2}$$

in which to make the imaginary time increment, and for energies near to top of the barrier, $E \to V_0$, $2t_*$ becomes infinite.

Another observation regarding this example is that the branch point times $t_n$ of (4.14) correspond to those times at which $x(t)$ is at a singularity of the potential $V(x)$; that is, one can easily show that

$$V[x(t_n)] = \infty$$

This should be a general feature of the analytic structure of the trajectory, since from the equation of motion

$$\mu\ddot{x}(t) = -V'(x)$$

it is clear that singular points of $x(t)$ correspond to positions $x(t)$ for which $V'(x)$ is singular.

In summary, the following features are expected to be characteristic of the general situation. (*1*) The final values of the coordinates and momenta of an analytically continued classical trajectory, and the classical $S$-matrix element associated with it, are independent of the particular complex time path along which time is incremented so long as one considers time paths that pass in the same manner through any branch points that may exist. (*2*) The existence of more than one energetically open asymptotic arrangement implies the existence of branch points in the finite time plane. (*3*) Branch points in the *t*-plane correspond to values of the coordinates for which derivatives of the potential energy are divergent. (*4*) In order to effect a classically forbidden rearrangement (i.e., tunneling) transition the imaginary part of the time increment should be made predominantly when the collision partners are closest together.

### D. Reactive Tunneling

One of the most important examples of a classically forbidden process is reactive tunneling in reactive systems that have activation barriers. The practical importance lies in the fact that the energy region near threshold, where the effects of tunneling are significant, often dominates the thermal energy kinetics, particularly if the reaction involves transfer of a hydrogen atom.

Calculations for reactive tunneling in the collinear $H + H_2$ reaction have been carried out,[8h,8i] and they demonstrate clearly the general features. The pertinent formulas are simple modifications of those in Section IV.A, and the discussion of the boundary conditions in Section IV.B also applies. The most novel aspect of the calculation is the necessity of "making" the trajectory react by the choice of the complex time path; that is, at collision energies below the classical threshold all ordinary real-valued trajectories are nonreactive, but by judicious choice of the complex time path one can find complex-valued trajectories that react. In particular, the time path is chosen predominantly along the real time axis until the turning point in the relative translation of $H_A$ and $H_BH_C$ is reached. If the time path were continued in the real direction $H_A$ and $H_BH_C$ would separate nonreactively, but by choosing predominantly imaginary time increments at this point one can force atoms $H_A$ and $H_B$ together. Choosing further time increments predominantly real then leads to separation of $H_AH_B$ and $H_C$, and a complex-valued reactive trajectory is thus achieved.

For the $H + H_2$ system (on the Porter-Karplus[34] potential surface) there appears to be just one complex-valued trajectory that contributes significantly to the ground state to ground state reaction at collision energies below the classical threshold; that is, no other complex-valued reactive trajectories could be found. Fig. 5 shows this reactive trajectory at two different collision energies; the most significant feature here is the degree to which the trajectory "cuts the corner" as it tunnels through the potential barrier. (Caution must be used in attaching physical significance

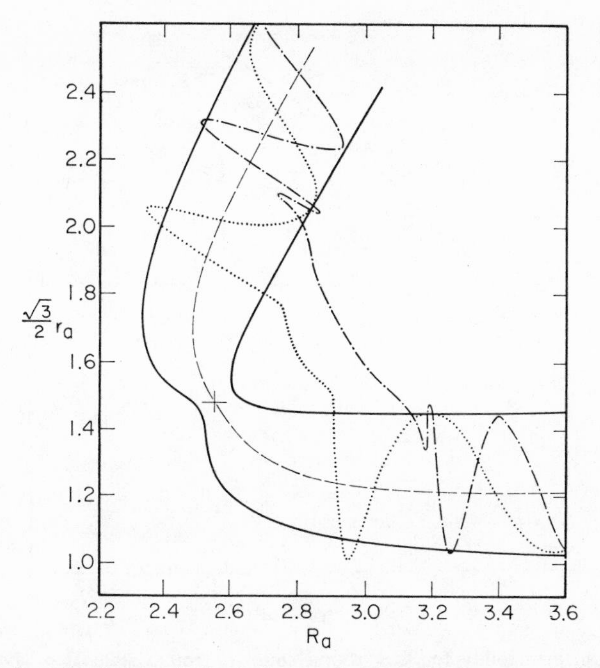


**Fig. 5.** Trajectories for reactive tunneling in the ground state to ground state $H + H_2$ reaction at a collision energy $E_0 = 0.20$ eV (dotted line) and $E_0 = 0.02$ eV (dash-dot line). For reference, the dashed line is the reaction coordinate, that is, the path of minimum potential energy, and the cross is the saddle point. $R_a$ and $r_a$ are the real parts of the complex translational and vibrational coordinates, respectively, of arrangement $a(A + BC)$.

to the trajectory, however, for as discussed in Section IV.B, it is not invariant to the choice of the complex time contour.)

Figs. 6 and 7 show the ground state to ground state reaction probability as a function of collision energy; also shown are the results of quantum mechanical[35,36] and purely classical[37] Monte Carlo calculations. It is seen that there is a substantial amount of reactive tunneling in this model system and that it is described quite accurately by classical *S*-matrix theory. Analogous to the discussion in Section IV.A one should be careful not to confuse the notion of "classically forbidden"—that is, forbidden by classical *dynamics*—and "energetically forbidden." For the present $H + H_2$ system, for example, reaction is energetically possible for collision energies above 0.12 eV, but classical dynamics does not lead to reaction for any collision energy below 0.21 eV; the dynamic threshold of 0.21 eV is obviously the relevant classical threshold, reaction taking place below it via tunneling.

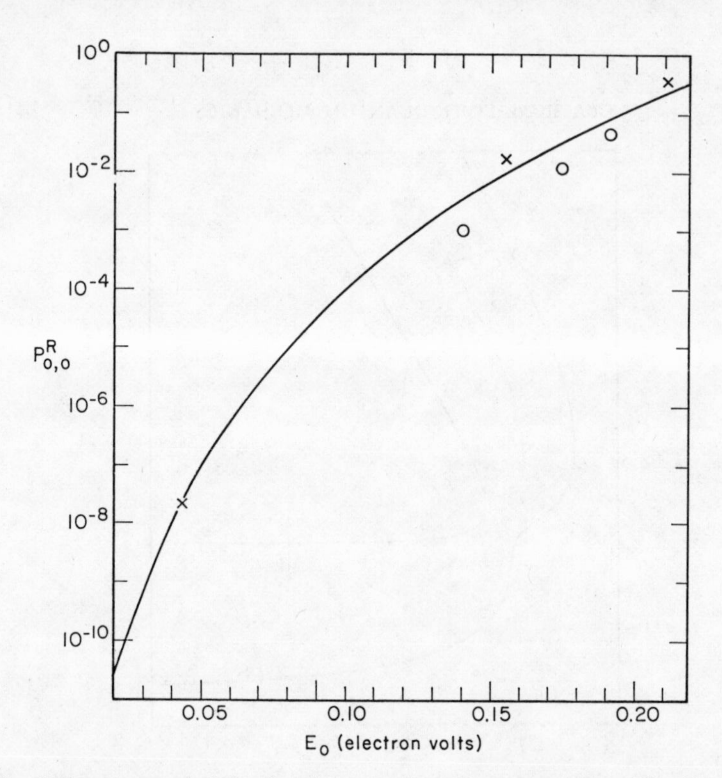


**Fig. 6.** Reaction probability for the ground state to ground state $H + H_2$ reaction (collinear) as a function of the relative collision energy $E_0$. The crosses and circles show the quantum mechanical values calculated by Diestler[35] and Wu and Levine,[36] respectively.

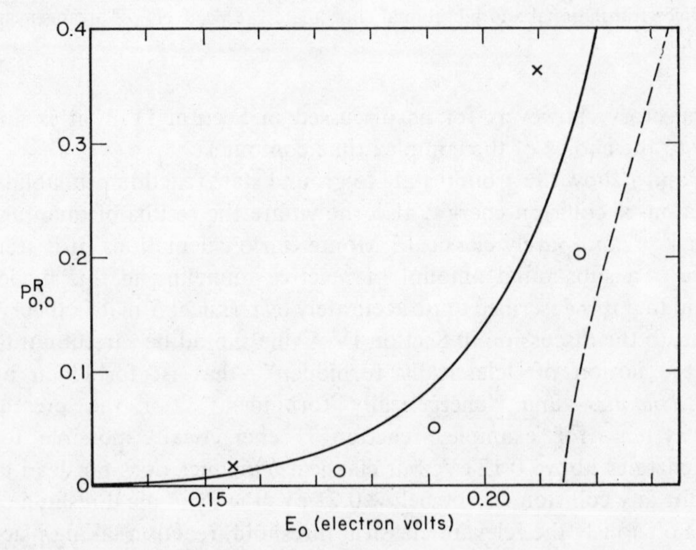


**Fig. 7.** Same as Fig. 6, but a more detailed picture of the energy region just below the classical threshold. The broken line is the purely classical Monte Carlo result[37] for the same (Porter-Karplus) potential surface.

To see even more clearly the significance of reactive tunneling for this system it is useful to compute the temperature dependent reaction probability

$$\bar{P}(T) = (kT)^{-1} \int_0^\infty dE \exp\left(\frac{-E}{kT}\right) P(E) \tag{4.18}$$

where $E$ is the initial collision energy and $P(E)$ the $0 \to 0$ reaction probability. Since there is just one complex-valued trajectory that contributes to the $0 \to 0$ reaction, $P(E)$ is given by

$$P(E) = \left(2\pi\hbar \left|\frac{\partial n_2}{\partial \bar{q}_1}\right|\right)^{-1} \exp\left(\frac{-2 \operatorname{Im} \Phi(E)}{\hbar}\right) \tag{4.19}$$

so that

$$\bar{P}(T) = (kT)^{-1} \int_0^\infty dE \left[2\pi\hbar \left|\frac{\partial n_2}{\partial \bar{q}_1}\right|\right]^{-1} \exp\left[-\frac{E}{kT} - \frac{2 \operatorname{Im} \Phi(E)}{\hbar}\right] \tag{4.20}$$

Equation (4.20) is of the form for which a steepest descent[33] approximation to the integral is useful; this approximation gives

$$\int dx\, g(x) \exp\left[\frac{-f(x)}{\hbar}\right] \simeq g(x_0)\left[\frac{2\pi\hbar}{f''(x_0)}\right]^{1/2} \exp\left[\frac{-f(x_0)}{\hbar}\right] \tag{4.21a}$$

where $x_0$ is determined by the steepest descent

$$f'(x_0) = 0 \tag{4.21b}$$

[Equation (4.21) will be recognized as the usual stationary phase approximation for the case that the "phase" is pure imaginary.] Evaluation of (4.20) within this approximation thus gives

$$\bar{P}(T) = P(E)\left[2\pi \frac{dE}{d(kT)}\right]^{1/2} \exp\left(\frac{-E}{kT}\right) \tag{4.22}$$

with $E$ given as a function of $T$ by the equation

$$\hbar\beta = -2 \operatorname{Im} \Phi'(E) \tag{4.23}$$

where $\beta = (kT)^{-1}$. Since the energy derivative of the action is the time increment, however, one has

$$\operatorname{Im} \Phi'(E) = \operatorname{Im} (t_2 - t_1)_E$$

so that the steepest descent relation in (4.23) reads

$$-(\tfrac{1}{2})\hbar\beta = \operatorname{Im} (t_2 - t_1)_E \tag{4.24}$$

From the imaginary part of the time increment of the particular complex trajectory for energy $E$, therefore, one associates a temperature according to (4.24); the physical meaning of this temperature is that this value of $E$ is the one that contributes most significantly (in the steepest descent sense) to a Boltzmann average for this temperature.

Fig. 8 shows the energy versus temperature relation of (4.24) for the $H + H_2$ reaction; for a given temperature this curve shows the energy region that is most important for the Boltzmann average. Since the classical threshold for the reaction is just above $E = 0.21$ eV (Fig. 7), Fig. 8 shows that *for temperatures below* 1000°K *the energy region most important to the Boltzmann average is below the classical threshold*, a fact that dramatically illustrates the practical importance of reactive tunneling.

Fig. 9 shows the temperature averaged reaction probability given by (4.22). The marked curvature of the Arrhenius plot ($\log \bar{P}(T)$ versus $T^{-1}$) below 1000°K is consistent with the fact that in this temperature region the most important collision energies are in the tunneling region below the classical threshold; it has long been a tenet of kinetics that curvature of an Arrhenius plot is indicative of large tunneling corrections.[38]

There is one additional interpretation of the $E$ versus $T$ relation of (4.23) and (4.24) and shown in Fig. 8. The temperature dependent reaction

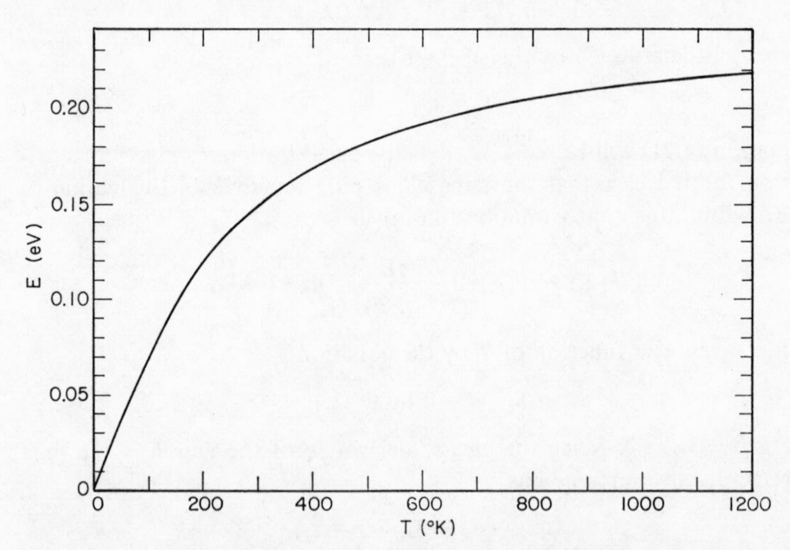


**Fig. 8.** The energy-temperature relation determined by the steepest descent condition, (4.24). For a given temperature $T$ the corresponding energy $E$ is the value that makes the dominant contribution to a Boltzmann average of the reaction probability for that temperature. Equation (4.27) shows that $E(T)$ also has an interpretation as the temperature dependent activation energy.

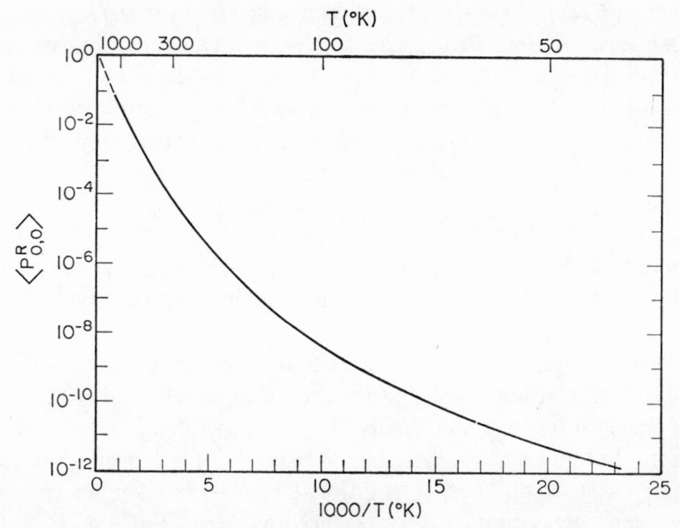


**Fig. 9.** The temperature dependent reaction probability for the ground state to ground state $H + H_2$ reaction, as given by (4.22) and (4.23). The semiclassical treatment is valid only in the present case for $T \lessapprox 1000°K$, so that the dashed part of the curve is simply an extrapolation to unit probability at infinite temperature.

probability of (4.22) has the form

$$\bar{P}(T) = A(T) \exp\left\{-\frac{E(T)}{kT} - \frac{2}{\hbar}\,\mathrm{Im}\,\Phi[E(T)]\right\} \tag{4.25}$$

where $A(T)$ is a slowly varying function of $T$ and where $E(T)$ is the function defined by (4.23). The *activation energy* is defined by

$$\begin{aligned} E_{\mathrm{act}} &= -\frac{d}{d(1/kT)} \ln \bar{P}(T) \\ &= kT^2 \frac{d \ln \bar{P}(T)}{dT} \end{aligned} \tag{4.26}$$

and is in general a function of temperature. If the temperature dependence of the preexponential factor $A$ is neglected, then from (4.25) one obtains

$$E_{\mathrm{act}} = E(T) - kT^2E'(T)\left[(kT)^{-1} + \frac{2\,\mathrm{Im}\,\Phi'(E)}{\hbar}\right]$$

Equation (4.23) shows that the second term vanishes, however, so that the final result is

$$E_{\mathrm{act}} = E(T) \tag{4.27}$$

*The function $E(T)$ defined by* (4.23) *and shown in Fig.* 8, *therefore, is seen to be the temperature dependent activation energy*. The activation energy, which is ordinarily considered to be a purely experimental quantity defined by (4.26), is thus seen to be expressible in terms of strictly *dynamical* quantities, the imaginary time increment of the relevant complex-valued trajectory playing a central role.

### E. Partial Averaging

To conclude the discussion of classically forbidden processes we consider more specifically the case of collision systems in three-dimensional space. The situation here, as discussed in Section III.E, is that there are several internal degrees of freedom and that in practice one is seldom concerned with transitions that specify quantum numbers for all these degrees of freedom initially and finally.

In an atom-diatom collision system, for example, it is typically the case at low energy that a number of rotational states are strongly coupled whereas only a few vibrational states are involved. For the $H + H_2$ reaction from ground state of $H_2$, $n_1 = j_1 = 0$, for example, it appears[39] that the final rotational states $j_2 = 0$ through at least $j_2 = 5$ have comparable cross-sections whereas $n_2 = 0$ is the only accessible vibrational state. In such cases it would be highly desirable to treat the rotational degrees of freedom in a completely classical-like Monte Carlo framework while simultaneously quantizing the vibrational degree of freedom within the semiclassical framework. Not only is this possible, but it greatly simplifies the practical aspects of applying classical $S$-matrix theory to three-dimensional collision systems.[8j–8l]

Suppose, for example, one is interested in the transition from initial vibrational-rotational state $(n_1, j_1)$ of $BC$ to final vibrational state $n_2$, summed over all final rotational states; from (3.87) this is (with $\hbar = 1$)

$$\sigma_{n_2 \leftarrow n_1 j_1} = \frac{\pi}{k_1^2(2j_1 + 1)} \int_0^\infty dJ(2J + 1) \int dj_2 \int dl_2 \int dl_1 (2\pi)^{-3} \times \left| \frac{\partial(n_2 j_2 l_2)}{\partial(\bar{q}_{n_1} \bar{q}_{j_1} \bar{q}_{l_1})} \right|^{-1} \exp\left[-2 \operatorname{Im} \Phi(n_2 j_2 l_2, n_1 j_1 l_1)\right] \quad (4.28)$$

where it is assumed that there are enough values of $J$, $l_2$, $l_1$, $j_1$ contributing to the sums to justify replacing the sums over them by integrals, interference terms have been discarded (they would presumably be quenched by the sums), and the transition is taken to be classically forbidden. The integration over final values $j_2$ and $l_2$ in (4.28) can be changed to integration over the conjugate initial conditions $\bar{q}_{j_1}$ and $\bar{q}_{l_1}$, and the two-dimensional Jacobian that comes from this change of integration variables cancels into the three-dimensional Jacobian in (4.28) to give

$$\sigma_{n_2\leftarrow n_1 j_1} = \frac{\pi}{k_1^2(2j_1+1)} \int_0^\infty dJ(2J+1) \int_0^1 d\left(\frac{\bar{q}_{j_1}}{2\pi}\right) \times \int_0^1 d\left(\frac{\bar{q}_{l_1}}{2\pi}\right) \int_{|J-j_1|}^{J+j_1} dl_1 P_{n_2,n_1}(j_1 l_1 \bar{q}_{j_1} \bar{q}_{l_1}; JE) \quad (4.29)$$

where $P_{n_2,n_1}$ has the form of a one-dimensional vibrational transition probability:

$$P_{n_2,n_1}(j_1 l_1 \bar{q}_{j_1} \bar{q}_{l_1}; JE) = \left[2\pi \left|\frac{\partial n_2}{\partial \bar{q}_{n_1}}\right|\right]^{-1} \exp\left[\frac{-2\,\mathrm{Im}\,\Phi}{\hbar}\right] \quad (4.30)$$

The one-dimensional-like vibrational transition probability of (4.30) is constructed in the following manner: beginning trajectories with the initial conditions of (3.93) one holds $j_1$, $l_1$, $\bar{q}_{j_1}$, and $\bar{q}_{l_1}$ (as well as $n_1$) fixed while only $\bar{q}_{n_1}$ is varied to satisfy the one final boundary condition

$$n_2(\bar{q}_{n_1}) = n_2 \quad (4.31)$$

That is, by not quantizing $l_2$ and $j_2$ in the final asymptotic region (which is valid because many integer values are accessible) one avoids the final boundary conditions associated with these degrees of freedom. The three-dimensional root search problem in (3.92), which is the most difficult aspect of applying classical *S*-matrix theory to three-dimensional systems, is thus reduced to a one-dimensional root search, a relatively simple problem. From the standpoint of the boundary conditions the calculation is thus like a collinear $A + BC$ collision, with the vibrational transition probabilities depending parametrically on the initial conditions of all the other internal degrees of freedom; this similarity to a collinear collision is only with regard to the "bookkeeping," of course, for the trajectories are still full three-dimensional trajectories, incorporating the dynamics of all degrees of freedom exactly.

The cross-section given by (4.29) involves an average over the phase space of initial conditions of the degrees of freedom which are not "state-selected" (i.e., quantized semiclassically). These integrals in (4.29) are probably most efficiently evaluated by Monte Carlo methods. Furthermore, one can obtain all the partial cross sections—that is, the distributions in final rotational quantum number $j_2$ and/or scattering angle (i.e., the differential cross-section)—in the usual Monte Carlo fashion by assigning the numerical value of the integrand of (4.29) to the appropriate "box" labeled by $j_2$ and scattering angle. The "partial averaging" approach thus provides the advantages of a Monte Carlo treatment for the classical-like degrees of freedom (i.e., those with many quantum states involved), but the quantumlike degrees of freedom (i.e., vibration) are quantized semiclassically in the usual manner, via the boundary conditions in the initial and final asymptotic regions.

Because of the potential utility of partial averaging it is worthwhile putting (4.29) in the form of a working formula. It is useful first to interchange the order of the $l_1$ and $J$ integrals,

$$\int_0^\infty dJ(2J+1)\int_{|J-j_1|}^{J+j_1} dl_1 = \int_0^\infty dl_1 \int_{|l_1-j_1|}^{l_1+j_1} dJ(2J+1) \qquad (4.32)$$

Defining $z_J$ by

$$J + \tfrac{1}{2} = [(l_1 - j_1)^2 + (2l_1+1)(2j_1+1)z_J]^{1/2} \qquad (4.33)$$

one has

$$(2J+1)\, dJ = (2l_1+1)(2j_1+1)\, dz_J$$

and introducing the impact parameter $b$,

$$l_1 + \tfrac{1}{2} = k_1 b$$

Equation (4.29) becomes

$$\sigma_{n_2 \leftarrow n_1 j_1} = \pi \int_0^\infty db\, 2b \int_0^1 dz_J \int_0^1 d\left(\frac{\bar{q}_{j_1}}{2\pi}\right) \int_0^1 d\left(\frac{\bar{q}_{l_1}}{2\pi}\right) \times P_{n_2, n_1}(j_1 l_1 \bar{q}_{j_1} \bar{q}_{l_1}; JE) \qquad (4.34)$$

Consider now just the impact parameter integral,

$$\sigma = \pi \int_0^\infty db\, 2bP(b) \qquad (4.35)$$

To cast it in a form more convenient for Monte Carlo integration it is useful to factor out as much of the $b$-dependence of the integrand as possible. Thus if $P_0(b)$ is an approximation to $P(b)$, then

$$\sigma = \pi \int_0^\infty db\, 2bP_0(b) \left[\frac{P(b)}{P_0(b)}\right] \qquad (4.36)$$

where the factor in brackets should be slowly varying. The new integration variable $z_b$ is now defined by

$$dz_b = db 2bP_0(b) \qquad (4.37)$$

A convenient choice for $P_0(b)$ that is typical of the impact parameter dependence of transition probabilities is

$$P_0(b) = B^{-2} \exp\left(\frac{-b^2}{B^2}\right) \qquad (4.38)$$

where $B$ is a scaling constant. From (4.37) and (4.38) one thus finds

$$z_b = 1 - \exp\left(\frac{-b^2}{B^2}\right) \tag{4.39}$$

so that (4.35) becomes

$$\sigma = \pi B^2 \int_0^1 dz_b (1 - z_b)^{-1} P(b) \tag{4.40a}$$

with

$$b = B[-\ln(1 - z_b)]^{1/2} \tag{4.40b}$$

If $P(b)$ is approximated reasonably well by $P_0(b)$, then the integrand of (4.40a) will be a slowly varying function of $z_b$.

Carrying out the impact parameter integration in the manner described above, (4.34) takes its final form

$$\sigma_{n2\leftarrow n1j1} = \pi B^2 \int_0^1 dz_b \int_0^1 dz_J \int_0^1 d\left(\frac{\bar{q}_{j_1}}{2\pi}\right) \int_0^1 d\left(\frac{\bar{q}_{l_1}}{2\pi}\right) \times (1 - z_b)^{-1} P_{n2,n1}(j_1 l_1 \bar{q}_{j_1} \bar{q}_{l_1}; JE) \tag{4.41a}$$

with

$$l_1 + \tfrac{1}{2} = k_1 B[-\ln(1 - z_b)]^{1/2} \tag{4.41b}$$

$$J + \tfrac{1}{2} = [(l_1 - j_1)^2 + (2l_1 + 1)(2j_1 + 1)z_J]^{1/2} \tag{4.41c}$$

Since all the integration variables in (4.41a) go from 0 to 1, the expression is directly amenable to Monte Carlo evaluation. It should be noted, too, that processes with extremely small cross-sections should not be any more difficult to evaluate via (4.41) than those with large cross-sections, for the magnitude of the cross-section is determined by the magnitude of $P_{n_2 n_1}$, not by the number of "successes" as in the usual Monte Carlo methods.

## V. ELECTRONIC TRANSITIONS IN LOW-ENERGY MOLECULAR COLLISIONS

If one employs a fully quantum mechanical description of the internal, quantized degrees of freedom—that is, via a coupled-channel expansion—then the formulation of a scattering problem is straightforward and self-consistent. Similarly, if a completely classical description of the dynamics is used for all the degrees of freedom, then the formulation of the problem is again internally consistent. The "classical $S$-matrix" theory developed in preceding sections is of the latter type—all dynamics is classical (with quantum superposition included), quantization of the internal degrees of freedom being achieved through the boundary conditions of the classical equations of motion; classical $S$-matrix theory is thus a dynamically exact, internally consistent description of collision processes.

In treating electronic transitions in molecular collisions it is clear that one *must* rely on a quantum mechanical description of the electronic degrees of freedom; that is, the electronic degrees of freedom must be described by their electronic states rather than by the classical orbits of the electrons. With a quantum state description required for the electronic degrees of freedom one is thus faced with a choice between two approaches: one possibility is to use a quantum state description for all quantized degrees of freedom, both electronic and nuclear (i.e., rotation and vibration), that is, a coupled-channel expansion in rotational, vibrational, and electronic states. The formulation of the problem is then straightforward, but rather unrealistic for a molecular system because of the large number of rotational and vibrational states that one would need to include in the expansion. The other alternative is to retain a classical description of the rotational and vibrational (and also translational) degrees of freedom within the framework of classical $S$-matrix theory, and this simplifies that aspect of the problem, but one is now faced with the task of combining classical and quantum dynamics per se, not just classical dynamics and quantum superposition.

There are a large number of semiclassical methods for dealing with a combination of a quantum state description of electronic degrees of freedom and a classical description of the heavy particle degrees of freedom. For high collision energies it is common to disregard the dynamics of the nuclei (e.g., by assuming a straight line, constant velocity trajectory) and concentrate on the electronic aspect of the problem. Most descriptions, too, have had the atom-atom case in mind and thus have not dealt with the special features that arise when there are several heavy particle degrees of freedom. The following sections outline a semiclassical model that is dynamically exact so far as the heavy particle dynamics is concerned, and it will be shown that the electronic transition probability inherent in the model actually contains several approximate results derived for special situations.

### A. Formulation of the Semiclassical Model

An extremely convenient description of electronically inelastic processes is the formulation by Pechukas,[10a] which follows Feynman's[40] discussion of the "path integral as functional." If $x$ and $\mathbf{q}$ denote all the electronic and nuclear coordinates of the system, respectively, then Feynman's path integral expression[22] for the coordinate representation of the propagator is

$$\langle \mathbf{q}_2 x_2 | \exp\left[\frac{-iH(t_2 - t_1)}{\hbar}\right] |\mathbf{q}_1 x_1\rangle = \int_{\mathbf{q}_1}^{\mathbf{q}_2} D\mathbf{q} \int_{x_1}^{x_2} Dx \exp\left\{\frac{i}{\hbar}\int_{t_1}^{t_2} dt[\tfrac{1}{2}\mu\dot{\mathbf{q}}^2 + T_x - V(x, \mathbf{q})]\right\} \quad (5.1)$$

where the path integral is over all electronic and nuclear paths that connect $(\mathbf{q}_1 x_1)$ and $(\mathbf{q}_2 x_2)$, and where $T_x$ is the electronic kinetic energy. Following Feynam,[40] one imagines performing the electronic path integral first, whereby (5.1) becomes

$$\langle \mathbf{q}_2 x_2 | \exp\left[-\frac{iH(t_2 - t_1)}{\hbar}\right] |\mathbf{q}_1 x_1\rangle = \int_{\mathbf{q}_1}^{\mathbf{q}_2} D\mathbf{q} K[x_2, x_1; \mathbf{q}(t)] \exp\left[\frac{i}{\hbar}\int_{t_1}^{t_2} dt\, \tfrac{1}{2}\mu\dot{\mathbf{q}}(t)^2\right] \quad (5.2)$$

where $K$ is the electronic propagator

$$K[x_2, x_1; \mathbf{q}(t)] = \int_{x_1}^{x_2} Dx \exp\left\{\frac{i}{\hbar}\int_{t_1}^{t_2} dt[T_x - V(x, \mathbf{q}(t))]\right\} \quad (5.3)$$

$K$ is the electronic propagator for the time dependent electronic Hamiltonian $T_x + V(x, \mathbf{q}(t))$, the time dependence entering through the fixed nuclear path $\mathbf{q}(t)$; this is the sense in which the electronic propagator is a functional of the nuclear path.

Following Pechukas, one takes matrix elements of (5.2) between the initial and final electronic states $\phi_1(x)$ and $\phi_2(x)$, giving

$$\langle 2\mathbf{q}_2 | \exp\left[-\frac{iH(t_2 - t_1)}{\hbar}\right] |1\mathbf{q}_1\rangle = \int_{\mathbf{q}_1}^{\mathbf{q}_2} D\mathbf{q} K_{2,1}[\mathbf{q}(t)] \exp\left[\frac{i}{\hbar}\int_{t_1}^{t_2} dt\, \tfrac{1}{2}\mu\dot{\mathbf{q}}(t)^2\right] \quad (5.4)$$

where $K_{2,1}[\mathbf{q}(t)]$ is the matrix element of the electronic propagator, (5.3), with respect to electronic states $\phi_1(x)$ and $\phi_2(x)$; that is, $K_{2,1}[\mathbf{q}(t)]$ is the probability amplitude for the $1 \to 2$ electronic transition with the nuclei constrained to follow the path $\mathbf{q}(t)$. $S$-matrix elements for the $1\mathbf{n}_1 \to 2\mathbf{n}_2$ transition are constructed from the propagator in the usual fashion [i.e., via (3.11)], where $\mathbf{n}_1$ and $\mathbf{n}_2$ are the initial and final values of the quantum numbers for the heavy particle degrees of freedom.

Equation (5.4) is the fundamental result of this formulation of the problem and is an exact quantum mechanical expression. It has a rather obvious interpretation in terms of a two-step procedure for solving the problem: one first chooses some trajectory $\mathbf{q}(t)$ for the nuclei; with this nuclear trajectory fixed one solves the time dependent electronic problem to obtain the electronic transition amplitude $K_{2,1}[\mathbf{q}(t)]$. One then multiplies this by the phase factor associated with nuclear kinetic energy and path integrates over all possible nuclear trajectories, or paths. The interest in the picture of (5.4) is that many approximate theories correspond to the first step of this procedure—that is, a nuclear trajectory is assumed and the time dependent electronic problem solved; the second step, that is, path integration over nuclear trajectories, is of course not carried out.

The general semiclassical model[8f] to be discussed below results from two approximations to (5.4): a semiclassical approximation to the electronic transition amplitude $K_{2,1}[\mathbf{q}(t)]$, and then an asymptotic (i.e., steepest descent) evaluation of the path integral. Under fairly general conditions the electronic transition amplitude is given by[41–43]

$$K_{2,1}[\mathbf{q}(t)] = \exp\left\{-\frac{i}{\hbar}\int_{t_1}^{t_*} dt W_1[\mathbf{q}(t)] - \frac{i}{\hbar}\int_{t_*}^{t_2} dt W_2[\mathbf{q}(t)]\right\} \tag{5.5}$$

where $W_1(\mathbf{q})$ and $W_2(\mathbf{q})$ are the adiabatic electronic potential energy surfaces for states 1 and 2, and $t_*$ is the complex time at which the surfaces intersect,

$$W_2[\mathbf{q}(t_*)] = W_1[\mathbf{q}(t_*)] \tag{5.6}$$

if there is more than one such value of $t_*$, then (5.5) is a sum of such terms. Section V.B shows that (5.5) reproduces a number of the commonly used semiclassical expressions: the Landau-Zener formula,[44,45] the Demkov model,[46,47] and Nikitin's[48] expression for the probability of fine structure transitions.

With (5.5) for the electronic transition amplitude (5.4) becomes

$$\langle 2\mathbf{q}_2| \exp\left[-\frac{iH(t_2-t_1)}{\hbar}\right]|1\mathbf{q}_1\rangle$$
$$= \int_{\mathbf{q}_1}^{\mathbf{q}_2} D\mathbf{q} \exp\left\{\frac{i}{\hbar}\int_{t_1}^{t_2} dt\, \tfrac{1}{2}\mu\dot{\mathbf{q}}(t)^2 - \frac{i}{\hbar}\int_{t_1}^{t_*} dt W_1[\mathbf{q}(t)] - \frac{i}{\hbar}\int_{t_*}^{t_2} dt W_2[\mathbf{q}(t)]\right\} \tag{5.7}$$

which leads to an extremely simple physical interpretation. Compare, for example, the ordinary path integral representation for the propagator of a molecular system with only one potential energy function $V(\mathbf{q})$,

$$\langle \mathbf{q}_2| \exp\left[-\frac{iH(t_2-t_1)}{\hbar}\right]|\mathbf{q}_1\rangle$$
$$= \int_{\mathbf{q}_1}^{\mathbf{q}_2} D\mathbf{q} \exp\left\{\frac{i}{\hbar}\int_{t_1}^{t_2} dt\, \tfrac{1}{2}\mu\dot{\mathbf{q}}(t)^2 - \frac{i}{\hbar}\int_{t_1}^{t_2} dt V[\mathbf{q}(t)]\right\}$$

By analogy with this expression it is clear in (5.7) that $W_1(\mathbf{q})$ is the potential energy function for the system from $t_1$ to the complex time $t_*$, and $W_2(\mathbf{q})$ is the potential function from $t_*$ to $t_2$. Steepest descent evaluation of the path integral in (5.7) thus leads in the usual way to classical motion on potential $W_1(\mathbf{q})$ from $t_1$ to $t_*$, and on $W_2(\mathbf{q})$ from $t_*$ to $t_2$:

$$\langle 2\mathbf{q}_2| \exp\left[-\frac{iH(t_2 - t_1)}{h}\right] |1\mathbf{q}_1\rangle$$

$$= \left[(-2\pi i\hbar)^N \left(\frac{\partial \mathbf{q}_2}{\partial \mathbf{p}_1}\right)_{q1}\right]^{-1/2} \exp\left[\frac{i\phi_{2,1}(\mathbf{q}_2, \mathbf{q}_1)}{\hbar}\right] \quad (5.8)$$

where $\phi_{2,1}$ is the action integral

$$\phi_{2,1}(\mathbf{q}_2, \mathbf{q}_1) = \int_{t1}^{t2} dt\, \tfrac{1}{2}\mu\dot{\mathbf{q}}(t)^2 - \int_{t1}^{t_*} dt W_1 - \int_{t_*}^{t2} dt W_2 \quad (5.9)$$

evaluated along the classical trajectory determined by double-ended boundary conditions $\mathbf{q}_1$ and $\mathbf{q}_2$ and which changes from potential $W_1$ to $W_2$ at time $t_*$.

The classical $S$-matrix elements are obtained from the classical limit of the propagator, (5.8), via the same procedure as in Section III.A, giving

$$S_{2\mathbf{n}_2, 1\mathbf{n}_1}(E) = \left[(-2\pi i\hbar)^{N-1} \left(\frac{\partial \mathbf{n}_2}{\partial \bar{\mathbf{q}}_1}\right)_{\mathbf{n}_1}\right]^{-1/2}$$

$$\times \exp\left\{-\frac{i}{\hbar}\int_{-\infty}^{\infty} dt[R(t)\dot{P}(t) + \mathbf{q}(t)\cdot\dot{\mathbf{n}}(t)]\right. \quad (5.10)$$

where $(P, R, \mathbf{n}, \mathbf{q})$ are the usual action-angle variables for the nuclear (i.e., rotational, vibrational, and translational) degrees of freedom. The classical trajectory relevant to (5.10) begins on potential surface $W_1(\mathbf{q})$ with the initial conditions of (3.31); one then integrates to the time $t_*$ at which (5.6) is satisfied and here changes to potential surface $W_2(\mathbf{q})$, all coordinates and momenta being continuous at $t_*$; from $t_*$ one integrates to $+\infty$ on potential surface $W_2(\mathbf{q})$. As for the case of only one potential surface, one must choose the initial angle variables $\bar{\mathbf{q}}_1$ so that

$$\mathbf{n}_2(\mathbf{q}_1, \mathbf{n}_1) = \mathbf{n}_2 \quad (5.11)$$

Since the switching time $t_*$ is complex, it is clear that the quantum number function in (5.11) is in general complex even if all the initial conditions are real. It is necessary, therefore, to choose $\bar{\mathbf{q}}_1$ complex in order to satisfy (5.11).

Transitions between adiabatic potential energy surfaces thus appear in classical $S$-matrix theory as a classically forbidden process in the sense that the exact nuclear dynamics must follow a complex-valued classical trajectory. As a result of this the action integral in (5.10) is complex, so that the $S$-matrix has the exponential damping factor characteristic of classically forbidden processes. Much of the discussion in Section IV regarding these processes thus applies directly to electronic transitions.

Application of the semiclassical theory would obviously be much simpler if complex-valued classical trajectories could be avoided. It is clear from the above discussion that this cannot be done in any rigorous fashion, but an approximate version of the theory has been extracted which does involve only real-valued trajectories.[8f] If interference features in the transition probabilities are discarded and the sums and averages over rotational/vibrational quantum states done by Monte Carlo, then this approximate version is essentially equivalent to the "trajectory surface-hopping" model used successfully by Tully and Preston[49] to treat transitions between potential surfaces in $H^+ + H_2$ collisions.

It is interesting to note that it is the adiabatic representation of the electronic states that emerges as the one most fundamental to the semiclassical description of electronic transitions; that is, to evaluate the expressions in (5.8) to (5.10) only the adiabatic potential surfaces (and their analytic continuations to complex coordinates) are required, there is no reference at all to nonadiabatic coupling terms which are the off-diagonal elements for this adiabatic representation. What this means is that all the information concerning the nonadiabatic coupling which is necessary for the semiclassical model is contained implicitly in the analytic structure of the adiabatic potential surfaces; Section V.C pursues this point of view further. It is clear, of course, that a fully quantum mechanical treatment of the problem would require explicit knowledge of the nonadiabatic coupling.

### B. Examples of the Electronic Transition Probability

The key element of the semiclassical model described in the previous section is the electronic transition amplitude given by (5.5); to the extent that this expression is accurate the general model should be accurate, for the steepest descent approximation used in the path integral evaluation of (5.7) should be quite adequate. This section thus discusses several simple examples to show that (5.5) can reproduce some previously obtained results for quite different situations.

To simplify the dynamical aspect of the problem, so as to concentrate attention on the electronic transition probability, the system is chosen to be an atom-atom collision (so that there is only one nuclear coordinate) and a constant velocity is assumed for times in the vicinity of $t_*$. One assumes further that the adiabatic electronic states, $W_1$ and $W_2$, result from diagonalization of a two by two electronic potential matrix:

$$W_i(r) = \tfrac{1}{2}[V_{11}(r) + V_{22}(r)] \pm \tfrac{1}{2}\{[V_{11}(r) - V_{22}(r)]^2 + 4V_{12}(r)^2\}^{1/2} \quad (5.12)$$

Equation (5.5) gives the electronic transition probability as

$$P_{2,1} = |K_{2,1}[r(t)]|^2 = e^{-2\delta} \quad (5.13)$$

where

$$\delta = -\frac{i}{\hbar}\int_{\mathrm{Re}\,t_*}^{t_*} dt\,\Delta W$$

or with (5.12) and the constant velocity approximation, this becomes

$$\delta = -\frac{i}{2\hbar v}\int_{r_*^*}^{r_*} dr\,\Delta W(r) \quad (5.14)$$

where

$$\Delta W(r) = \{[V_{22}(r) - V_{11}(r)]^2 + 4V_{12}(r)^2\}^{1/2} \quad (5.15)$$

and $r_*$ is the complex $r$ for which

$$\Delta W(r) = 0 \quad (5.16)$$

Equations (5.13) and (5.14) [or the more accurate version, (5.39) below] are essentially Stuckelberg's[41] general formula.

Consider first the Landau-Zener[44,45] model,

$$V_{22}(r) - V_{11}(r) = \Delta V'(r_0)(r - r_0) \quad (5.17a)$$

$$V_{12}(r) = V_{12}(r_0) \quad (5.17b)$$

which is often used to describe "avoided intersections" of adiabatic potential curves $W_1(r)$ and $W_2(r)$. This is the model treated in detail by Stuckelberg,[41] and in fact, it is Stuckelberg's solution of the avoided crossing problem that is the pattern for the general semiclassical model described in Section V.A. The complex crossing point $r_*$ is given by

$$r_* = r_0 + 2i\left|\frac{V_{12}(r_0)}{\Delta V'(r_0)}\right| \quad (5.18)$$

Letting

$$r = r_0 + 2ix\left|\frac{V_{12}(r_0)}{\Delta V'(r_0)}\right|$$

Equation (5.14) becomes

$$\delta = \frac{2V_{12}(r_0)^2}{\hbar v\,|\Delta V'(r_0)|}\int_{-1}^{1} dx(1 - x^2)^{1/2}$$

and elementary integration gives

$$\delta = \frac{\pi}{\hbar v} \frac{V_{12}(r_0)^2}{|\Delta V'(r_0)|} \tag{5.19}$$

Equation (5.13) thus gives the usual Landau-Zener formula,[44,45]

$$P_{2,1} = \exp\left[-\frac{2\pi}{\hbar v}\frac{V_{12}(r_0)^2}{|\Delta V'(r_0)|}\right] \tag{5.20}$$

Next consider Demkov's model,[46,47]

$$V_{22}(r) - V_{11}(r) = \Delta V \tag{5.21a}$$

$$V_{12}(r) = A \exp(-\lambda r) \tag{5.21b}$$

where $\Delta V$ is constant; this model had found considerable utility in describing nonresonant charge transfer.[47] The complex crossing point $r_*$ is thus the root of

$$\Delta V^2 + 4A^2 \exp(-2\lambda r) = 0$$

which in this case gives a sequence of roots

$$r_* = -\lambda^{-1} \ln\left(\frac{\Delta V}{2A}\right) + \frac{i\pi}{2\lambda}(1 + 2n), \qquad n = 0, 1, 2, \ldots$$

The principal root, that is, the one that leads to the largest transition probability, is

$$r_* = -\lambda^{-1} \ln\left(\frac{\Delta V}{2A}\right) + \frac{i\pi}{2\lambda} \tag{5.22}$$

Changing variables of integration

$$r = -\lambda^{-} \ln\left(\frac{\Delta V}{2A}\right) + \frac{i\pi x}{2\lambda}$$

(5.14) becomes

$$\delta = \frac{\pi\,\Delta V}{4\lambda\hbar v}\int_{-1}^{1} dx(1 + e^{-i\pi x})^{1/2}$$

and this gives

$$\delta = \frac{\pi\,\Delta V}{2\lambda\hbar v} \tag{5.23}$$

The electronic transition probability is thus given by

$$P_{2,1} = \exp\left(-\frac{\pi\,\Delta V}{\lambda\hbar v}\right) \tag{5.24}$$

which is essentially the Demkov formula[46] for a single crossing.

It is interesting to see that the Demkov result can also be obtained directly by considering the large $r$ expression for $\Delta W(r)$, the adiabatic energy difference itself. With the Demkov model of (5.19) one has

$$\Delta W(r) = [\Delta V^2 + 4A^2 \exp(-2\lambda r)]^{1/2} \simeq \Delta V + \frac{2A^2}{\Delta V} \exp(-2\lambda r) \tag{5.25}$$

From (5.25) one finds $r_*$ to be

$$r_* = -\lambda^{-1} \ln\left(\frac{\Delta V}{A\sqrt{2}}\right) + \frac{i\pi}{2\lambda} \tag{5.26}$$

the real part of $r_*$ in (5.26) is somewhat different from that in (5.22). With (5.25) and (5.26) it is then easy to show that (5.14) gives

$$\delta = \frac{\pi\Delta V}{2\lambda\hbar v} \tag{5.29}$$

which is the same as (5.23).

The above two paragraphs treat the cases where $\Delta W(r)$ is an exponential or a quadratic function of $r$. Another common example is an inverse power law,

$$\Delta W(r) = \Delta W(\infty) + \frac{C}{r^N} \tag{5.30}$$

It is not difficult to find the parameter $\delta$ for this case. The root $r_*$ with the smallest imaginary part is

$$r_* = \left[\frac{C}{\Delta W(\infty)}\right]^{1/N} e^{i\pi/N} \tag{5.31}$$

and with (5.30) and (5.31) it is easy to show that (5.14) gives

$$\delta = \frac{\Delta W(\infty)}{\hbar v} \frac{N}{N-1} \left[\frac{C}{\Delta W(\infty)}\right]^{1/N} \sin\left(\frac{\pi}{N}\right) \tag{5.32}$$

For large $N$ (5.32) takes the form

$$\delta = \frac{\pi}{2} \frac{\Delta W(\infty)}{\hbar v} \frac{\Delta W(r_0)}{\Delta W'(r_0)} \tag{5.33}$$

where $r_0 = \mathrm{Re}\, r_*$, which will be recognized as identical to Demkov's formula for the case where $\Delta W(r)$ is asymptotically an exponential.

As a final example consider the two by two potential matrix which characterizes fine structure transitions between the ${}^2\Pi_{1/2}$ and ${}^2\Sigma_{1/2}$ potential curves in collisions of a ${}^2$P atom with a rare gas atom.[48] If state "1" is the $j = \frac{1}{2}$ component and state "2" the $j = \frac{3}{2}$ component, then one has

$$V_{11}(r) = \tfrac{1}{3}\Sigma(r) + \tfrac{2}{3}\Pi(r) + \tfrac{2}{3}\Delta \tag{5.34a}$$

$$V_{22}(r) = \tfrac{2}{3}\Sigma(r) + \tfrac{1}{3}\Pi(r) - \tfrac{1}{3}\Delta \tag{5.34b}$$

$$V_{12}(r) = -\frac{\sqrt{2}}{3}[\Pi(r) - \Sigma(r)] \tag{5.34c}$$

where $\Sigma(r)$ and $\Pi(r)$ are the Born-Oppenheimer potential curves for the ${}^2$P and rare gas atoms, and $\Delta$ is the spin-orbit splitting of the $j = \frac{1}{2}$ and $j = 3/2$ levels of the isolated ${}^2$P atom. (For the formulas as written, $\Delta$ is positive for the F atom, say, and negative for the first excited term of an alkali atom.) The adiabatic energy splitting is thus given by

$$\Delta W(r) = \{[\Pi(r) - \Sigma(r)]^2 + \tfrac{2}{3}\Delta[\Pi(r) - \Sigma(r)] + \Delta^2\}^{1/2} \tag{5.35}$$

If the $\Sigma$–$\Pi$ separation is taken to be an exponential,

$$\Pi(r) - \Sigma(r) = C\exp(-\lambda r) \tag{5.36}$$

then one finds the complex root $r_*$ to be

$$r_* = -\lambda^{-1}\ln\left(\frac{\Delta}{C}\right) + i\lambda^{-1}\cos^{-1}(-\tfrac{1}{3}) \tag{5.37}$$

With (5.35) through (5.37) (5.14) becomes

$$\delta = \frac{\Delta}{2\lambda\hbar v}\int_{-x_0}^{x_0} dx[e^{-2ix} + \tfrac{2}{3}e^{-ix} + 1]^{1/2}$$

where $x_0 = \cos^{-1}(-\frac{1}{3})$; the value of the integral is $4\pi/3$, so that one finally has

$$\delta = \frac{2\pi}{3}\frac{\Delta}{\lambda\hbar v} \tag{5.38}$$

a result which [together with (5.13)] has been obtained previously by Nikitin[48] from a different approach.

In summary, therefore, one sees that a number of the commonly used simple formulas for electronic transition probabilities are special cases of the electronic transition amplitude, (5.15), which is a part of the general semiclassical model described in Section V.A. For quantitative applications it should be emphasized that the simple formulas discussed in this section may often be poor, however, not because of a failure of (5.5), but because of the additional *dynamical* approximations that are invoked. For an atom-atom collision, for example, correct treatment of the nuclear dynamics gives the following expression for $\delta$:

$$\delta = -\frac{i}{\hbar}\left(\frac{\mu}{2}\right)^{1/2} \int_{r_*^*}^{r_*} dr\{[E - W_1(r)]^{1/2} - [E - W_2(r)]^{1/2}\} \qquad (5.39)$$

where, as before, $r_*$ is the root of

$$W_2(r) - W_1(r) = 0$$

Equation (5.14) is a high-energy approximation to (5.39) which is valid only if

$$E \gg \Delta W(r_0)$$

The approximate expression (5.14) involves only the adiabatic potential difference $\Delta W(r)$, but it is seen that the dynamically correct expression (5.39) involves both $W_1(r)$ and $W_2(r)$, not just their difference.

## C. Analytic Structure of Adiabatic Electronic Potential Energy Surfaces

Adiabatic electronic eigenvalues, that is, Born-Oppenheimer potential curves or surfaces, are generally obtained by finding roots of the secular equation

$$\det |\mathbf{H} - W\mathbf{1}| = 0 \qquad (5.40)$$

where $\mathbf{H}$ is the matrix of the molecular electronic Hamiltonian in an electronic basis set; since $\mathbf{H}$ depends parametrically on the relative positions of the nuclei, so do the roots $W$. In terms of the notation of the previous section $\mathbf{H}$ is the potential matrix $\mathbf{V}(\mathbf{q})$, and the various roots $W$ are the adiabatic potential surfaces $W_i(\mathbf{q})$.

The $N \times N$ secular determinant in (5.40) can be written as an $N$th-order polynomial in $W$, and the $N$ roots of the secular equation are thus the $N$ roots of this polynomial. The $N$ different roots of an $N$th-order polynomial, however, are essentially the $N$ different branches of the $N$th root function: for example, the two roots for the case $N = 2$ are a result of the double-valuedness of the square root function, the three roots for the case $N = 3$ result from the triple-valuedness of the cube root function, etc. That is, *the N different roots of the secular equation are the N different branches, or Riemann sheets, of the same analytic function.* This is manifestly true for the two-dimensional case, for example,

$$W = \tfrac{1}{2}(H_{11} + H_{22}) \pm \tfrac{1}{2}[(H_{22} - H_{11})^2 + 4H_{12}^2]^{1/2} \qquad (5.41)$$

The multivalued nature of a function is characterized by its branch points; for the case $N = 2$ the branch points in coordinate space correspond to those coordinates for which the radicand in (5.41) is zero. If one traces the functional values along a path that encircles a branch point, then the final functional value differs from the initial value even though the independent variables (i.e., the coordinates) return to their original values; that is, one will have changed from one branch, or Riemann sheet, of the function to another.

In the physical context of electronic transitions, therefore, consider a classical trajectory that begins on potential surface $W_1(\mathbf{q})$ at $t_1$ with initial conditions $\mathbf{p}_1$ and $\mathbf{q}_1$; $\mathbf{q}(t)$ is then a function only of the (complex) value of $t$ and the contour in the $t$-plane along which one has incremented time from $t_1$ to $t$ in, say, numerically integrating Hamilton's equations. With the initial conditions fixed, $W_1[\mathbf{q}(t)]$ is thus a function of only the one variable $t$. Because of the multivalued nature of the adiabatic potential energy function, $W_1[\mathbf{q}(t)]$ will have branch points in the time plane corresponding to those complex times at which $\mathbf{q}(t)$ is at one of the branch points of $W(\mathbf{q})$. In the two state case, for example, branch points in the $t$-plane are those values of $\mathbf{q}(t)$ for which the radicand in (5.41) is zero. By choosing the complex time contour in various ways the trajectory can thus be terminated on various of its different branches; that is, a particular electronic transition is described by choosing the time contour to go around the branch points in such a way for the trajectory to terminate on the desired branch of the multivalued function $W(\mathbf{q})$.

In the present language, therefore, there is actually just one adiabatic potential function, $W(\mathbf{q})$; it happens, though, to be a multivalued, many-sheeted function, the multivaluedness arising from the root-nature of the secular equation, (5.40). The classical-limit approximation for matrix elements of the propagator given by (5.8) can thus be written as

$$\langle \beta \mathbf{q}_2 | \exp\left[ -\frac{iH(t_2 - t_1)}{\hbar} \right] | \alpha \mathbf{q}_1 \rangle = \left[ (2\pi i \hbar)^N \left( \frac{\partial \mathbf{q}_2}{\partial \mathbf{p}_1} \right)_{\mathbf{q}_1} \right]^{-1/2} \times \exp\left\{ \frac{i}{\hbar} \int_{C_{\alpha\beta}} dt [\tfrac{1}{2}\mu \dot{\mathbf{q}}(t)^2 - W(\mathbf{q}(t))] \right\} \quad (5.42)$$

where $C_{\alpha\beta}$ is a contour in the time plane which connects $t_1$ and $t_2$ and weaves through the branch points in the time plane in such a way that the trajectory beginning on Riemann sheet $\alpha$ (i.e., electronic state $\alpha$) concludes at $t_2$ on Riemann sheet $\beta$ (i.e., electronic state $\beta$); that is, it is clearly not necessary to choose the contour $C_{\alpha\beta}$ to go exactly through the branch point $t_*$ as is indicated in (5.8)—any contour that passes about the branch points in the same manner will suffice. Equation (5.42), which describes electronic transitions via the general semiclassical model of Section V.A, now has the form of the ordinary semiclassical propagator discussed in Section II.E, the only difference being that the potential function in (5.42) is a multivalued function.

The intellectual appeal of the present picture is that electronic transitions emerge naturally from the same expressions for the semiclassical propagator that apply to the ordinary case of just one potential energy surface, the transitions arising from the fact that the potential function defined by the electronic secular equation is multivalued. One of the difficulties in developing the theory along these lines, however, is that there appears to have been little mathematical development regarding the analytic properties of dynamical systems in the complex time plane. Regarding the possibility of carrying out practical calculations, the situation is rather optimistic, for much has been learned recently about how to numerically integrate Hamilton's equations with complex initial conditions and along complex time contours; all this technology can be carried over to the case of electronic transitions within this general semiclassical model.

### D. An Effective Potential for Weak Inelastic Transitions

To conclude the discussion of electronic transitions we consider an approach presented recently by Pechukas and Davis[50] which leads to an interesting insight into the process. The starting point is again the exact quantum mechanical relation in (5.4), and the path integral is again evaluated via steepest descent. The difference from the semiclassical model of Sections V.A to V.C is the choice for the electronic transition amplitude; rather than the expression in (5.5) Pechukas and Davis choose the result given by first-order time dependent perturbation theory:

$$K_{2,1}[\mathbf{q}(t)] = -\frac{i}{\hbar}\int_{t_1}^{t_2} dt H_{21}[\mathbf{q}(t)] \times \exp\left\{-\frac{i}{h}\int_{t_1}^{t} dt' H_{11}[\mathbf{q}(t')] - \frac{i}{\hbar}\int_{t}^{t_1} dt' H_{22}[\mathbf{q}(t')]\right\} \tag{5.43}$$

where $H_{ij}(\mathbf{q})$ is the matrix of the electronic Hamiltonian between the electronic states; thus

$$\begin{aligned} H_{11}(\mathbf{q}) &= \varepsilon_1 + V_{11}(\mathbf{q}) \\ H_{22}(\mathbf{q}) &= \varepsilon_2 + V_{22}(\mathbf{q}) \\ H_{12}(\mathbf{q}) &= V_{12}(\mathbf{q}) \end{aligned} \tag{5.44}$$

where $V_{ij}(\mathbf{q})$ vanishes for large separation of the collision partners. Equation (5.43) has an interesting physical interpretation in itself: the integrand

$$-\frac{i}{\hbar}H_{21}[\mathbf{q}(t)]\exp\left\{-\frac{i}{\hbar}\int_{t_1}^{t}dt'H_{11}-\frac{i}{\hbar}\int_{t}^{t_2}dt'H_{22}\right\} \tag{5.45}$$

clearly corresponds to the nuclei moving on the potential surface $H_{11}(\mathbf{q})$ from $t_1$ to $t$ and on potential surface $H_{22}(\mathbf{q})$ from $t$ to $t_2$; that is, the transition occurs at time $t$ with a probability given by

$$\left|\frac{H_{21}[\mathbf{q}(t)]}{\hbar}\right|^2$$

Since the transition can take place at any time between $t_1$ and $t_2$, (5.43) is the superposition of the amplitudes in (5.45) for all possible transition times $t$.

The path integral in (5.4) is now to be evaluated by steepest descent with the electronic transition amplitude given by (5.43). Writing (5.4) as

$$\langle 2\mathbf{q}_2|\exp\left[-\frac{iH(t_2-t_1)}{\hbar}\right]|1\mathbf{q}_1\rangle$$

$$=\int_{\mathbf{q}_1}^{\mathbf{q}_2}D\mathbf{q}\exp\left\{\frac{i}{\hbar}\int_{t_1}^{t_2}dt\,\tfrac{1}{2}\mu\dot{\mathbf{q}}^2+\ln K_{2,1}[\mathbf{q}(t)]\right\} \tag{5.46}$$

the steepest descent relation for the "classical path" is

$$\delta\left\{\int_{t_1}^{t_2}dt\,\tfrac{1}{2}\mu\dot{\mathbf{q}}(t)^2-i\hbar\ln K_{2,1}[\mathbf{q}(t)]\right\}=0 \tag{5.47}$$

Using the fact that

$$\delta\ln K_{2,1}[\mathbf{q}(t)]=\frac{\delta K_{2,1}[\mathbf{q}(t)]}{K_{2,1}[\mathbf{q}(t)]}$$

a reasonable amount of algebra shows that the Euler equation for the path of steepest descent is

$$\mu\ddot{\mathbf{q}}(t)=-\frac{\partial V_{\mathrm{eff}}(\mathbf{q},t)}{\partial\mathbf{q}} \tag{5.48}$$

where $V_{\mathrm{eff}}(\mathbf{q},t)$ the time dependent, complex-valued potential defined by

$$V_{\mathrm{eff}}(\mathbf{q},t)=[1-\lambda(t)]V_{11}(\mathbf{q})+\lambda(t)V_{22}(\mathbf{q})+\frac{i\hbar\lambda'(t)V_{12}(\mathbf{q})}{V_{12}[\mathbf{q}(t)]} \tag{5.49}$$

where

$$\lambda(t) = \frac{K_{2,1}(t)}{K_{2,1}(t_2)} \tag{5.50a}$$

$$K_{2,1}(t) \equiv -\frac{i}{\hbar}\int_{t_1}^{t} dt' V_{12}[\mathbf{q}(t')] \times \exp\left\{-\frac{i}{\hbar}\int_{t_1}^{t'} dt'' H_{11}[\mathbf{q}(t'')] - \frac{i}{\hbar}\int_{t'}^{t_2} dt'' H_{22}[\mathbf{q}(t'')]\right\} \tag{5.50b}$$

$K_{2,1}(t)$ is the amplitude for the $1 \rightarrow 2$ transition in the time interval $t_1 \rightarrow t$, and $\mathbf{q}(t)$ is the as yet unknown solution to (5.48).

An important property of the effective potential in (5.49) is that

$$\int_{t_1}^{t_2} dt \frac{\partial V_{\text{eff}}[\mathbf{q}(t), t]}{\partial t} = -(\varepsilon_2 - \varepsilon_1)$$

That is, the nuclear trajectory determined by (5.48) and (5.49) loses the amount of energy that is transferred to electronic degrees of freedom, so that the *total* energy is conserved. In addition, since

$$\lim_{t \to t_1} \lambda(t) = 0$$

$$\lim_{t \to t_2} \lambda(t) = 1$$

it is clear from (5.50) that

$$\lim_{t \to t_1} V_{\text{eff}}(\mathbf{q}, t) = V_{11}(\mathbf{q})$$

$$\lim_{t \to t_2} V_{\text{eff}}(\mathbf{q}, t) = V_{22}(\mathbf{q})$$

That is, the effective potential changes from the initial to final potential during the course of the transition *regardless of the magnitude of the transition probability itself.* This differs considerably from the often used ad hoc effective potential.

$$V_{11}(\mathbf{q})[1 - P_{2,1}(t)] + V_{22}(\mathbf{q})P_{2,1}(t) \tag{5.51}$$

where $P_{2,1}(t)$ is the probability that at $t$ the system is in state 2 (having begun in state 1):

$$P_{2,1}(t) = |K_{2,1}(t)|^2 \tag{5.52}$$

with $K_{2,1}(t)$ given by (5.50b). $P_{2,1}(t)$ may be small for all $t$—as it should be if perturbation theory is to be valid—and in this case the effective potential predicted by (5.51) is just the initial potential for all $t$. The correct potential in (5.49) is the effective potential for nuclear motion *given the fact* that the electronic state changes from 1 to 2.

As noted above, the value of $\lambda(t)$ in (5.50) requires knowledge of the trajectory $\mathbf{q}(t)$ at all times prior to and past $t$, so that the equation of motion for $\mathbf{q}(t)$ in (5.48) must in general be solved iteratively: some trajectory $\mathbf{q}_1(t)$ is assumed initially and used in (5.50) to compute $\lambda(t)$; (5.48) is then integrated with potential of (5.49) to obtain a better trajectory $\mathbf{q}_2(t)$; $\mathbf{q}_2(t)$ is then used to compute $\lambda(t)$, etc. Pechukas and Davis[50] have shown, however, that if the $V_{11}(\mathbf{q})$ and $V_{22}(\mathbf{q})$ of (5.44) are the *same* (which is often a good approximation), (5.49) simplifies to

$$\frac{\partial V_{\text{eff}}(\mathbf{q}, t)}{\partial \mathbf{q}} = \frac{\partial V_0(\mathbf{q})}{\partial \mathbf{q}} + i\hbar e^{i\omega t} \frac{\partial V_{12}(\mathbf{q})}{\partial \mathbf{q}} \Big/ \int_{t_1}^{t_2} dt e^{i\omega t} V_{12}[\mathbf{q}(t)] \tag{5.53}$$

where $V_0(\mathbf{q})$ is the common value of $V_{11}(\mathbf{q})$ and $V_{22}(\mathbf{q})$, and

$$\omega = \frac{(\varepsilon_2 - \varepsilon_1)}{\hbar} \tag{5.54}$$

In (5.53) the unknown trajectory $\mathbf{q}(t)$ enters in the effective potential only through a *constant*, the denominator of the second term, so that the iterative solution of the equation of motion would be considerably simpler than the general case.

The exact quantum mechanical expression in (5.4) is thus a useful starting point for any theory of electronic transitions in molecular collisions. The semiclassical model discussed in Sections V.A, V.B, and V.C follows from choosing (5.5) as the approximation to the electronic transition amplitude, whereas Pechukas and Davis's[50] work discussed in this section takes the electronic transition amplitude given by first-order perturbation theory, (5.43); in both cases the path integral pertaining to nuclear degrees of freedom is then evaluated via a steepest descent approximation. One can imagine any number of other semiclassical models based on various approximations for the electronic transition amplitude, followed by a steepest descent evaluation of the nuclear path integral. It is our feeling that the steepest descent approximation for the path integral is reliable under essentially all conditions of physical interest and that the weak link of any semiclassical theory of this type is the approximation to the electronic transition amplitude. Properties of the electronic transition amplitude functional $K_{2,1}[\mathbf{q}(t)]$ would thus seem a fruitful point of emphasis for further investigation.

## VI. SOME NONSCATTERING APPLICATIONS

### A. Classical-Path Approximation for the Boltzmann Density Matrix

Equilibrium statistical mechanics is completely described by the Boltzmann operator $\exp(-\beta H)$, $H$ being the Hamiltonian for the system;

matrix elements of it are the Boltzmann density matrix

$$\rho(\mathbf{q}_2, \mathbf{q}_1) \equiv \langle \mathbf{q}_2 | \exp(-\beta H) | \mathbf{q}_1 \rangle \tag{6.1}$$

Use is often made in statistical mechanics of the formal analogy of the Boltzmann operator and the quantum mechanical propagator,

$$\exp(-\beta H) = \exp\left(-\frac{iHt}{\hbar}\right) \tag{6.2}$$

by choosing time to be purely imaginary, $t = -i\hbar\beta$. The "classical-path approximation"[51] for the Boltzmann density matrix corresponds to using the classical-limit approximation for matrix elements of the quantum propagator and then replacing $t$ by $-i\hbar\beta$.

From (2.102) and (2.108) with $t = -i\hbar\beta$ one thus obtains

$$\langle \mathbf{q}_2 | \exp(-\beta H) | \mathbf{q}_1 \rangle \simeq \left[(2\pi\hbar)^N \left(\frac{\partial \mathbf{q}_2}{\partial \mathbf{p}_1}\right)_{\mathbf{q}_1}\right]^{-1/2} \exp\left\{-\frac{1}{\hbar}\int_0^{\hbar\beta} d\tau [\tfrac{1}{2}\mu \mathbf{q}'(\tau)^2 + V[\mathbf{q}(\tau)]]\right\} \tag{6.3}$$

where $\tau$ is the imaginary time variable, $\tau \equiv it$, and Cartesian coordinates have been chosen. Since

$$\frac{d^2\mathbf{q}}{dt^2} = -\frac{d^2\mathbf{q}}{d\tau^2}$$

the equation of motion for the classical trajectory $\mathbf{q}(\tau)$ is

$$\mu \mathbf{q}''(\tau) = +\frac{\partial V(\mathbf{q})}{\partial \mathbf{q}} \tag{6.4}$$

That is, the equation of motion in the $\tau$ variable is like classical motion in the potential with its sign reversed (the upside-down potential). If classical motion in $V(\mathbf{q})$ is periodic for real time, for example, motion with increasing $\tau$ will be aperiodic. The momenta $\mathbf{p}_1$ in (6.3) are defined with regard to the $\tau$-derivative,

$$\mathbf{p}_1 \equiv \mu \mathbf{q}'(\tau), \qquad \tau = 0$$

The boundary conditions that specify the trajectory $\mathbf{q}(\tau)$ in (6.3) are the usual double-ended ones, $\mathbf{q}_1$ and $\mathbf{q}_2$. A simpler expression that does not involve double-ended boundary conditions can be obtained, however, for diagonal elements of the density matrix, that is, the particle density. From the identity

$$\exp(-\beta H) = \exp(-\tfrac{1}{2}\beta H)\exp(-\tfrac{1}{2}\beta H)$$

one has

$$\rho(\mathbf{q}_1) \equiv \langle \mathbf{q}_1 | \exp(-\beta H) | \mathbf{q}_1 \rangle$$
$$= \int d\mathbf{q}_2 \langle \mathbf{q}_1 | \exp(-\tfrac{1}{2}\beta H) | \mathbf{q}_2 \rangle \langle \mathbf{q}_2 | \exp(-\tfrac{1}{2}\beta H) | \mathbf{q}_1 \rangle \tag{6.5}$$

But since

$$\langle \mathbf{q}_1 | \exp(-\tfrac{1}{2}\beta H) | \mathbf{q}_2 \rangle = \langle \mathbf{q}_2 | \exp(-\tfrac{1}{2}\beta H) | \mathbf{q}_1 \rangle$$

Equation (6.5) becomes

$$\rho(\mathbf{q}_1) = \int d\mathbf{q}_2 \, |\langle \mathbf{q}_2 | \exp(-\tfrac{1}{2}\beta H) | \mathbf{q}_1 \rangle|^2 \tag{6.6}$$

and with the classical-path approximation (6.3) for the integrand one obtains

$$\rho(\mathbf{q}_1) = h^{-N} \int d\mathbf{q}_2 \left[ \left( \frac{\partial \mathbf{q}_2}{\partial \mathbf{p}_1} \right)_{\mathbf{q}_1} \right]^{-1} \exp\left[ -\frac{2}{\hbar} \int_0^{\hbar\beta/2} d\tau H(\tau) \right] \tag{6.7}$$

where

$$H(\tau) = \tfrac{1}{2}\mu \mathbf{q}'(\tau)^2 + V[\mathbf{q}(\tau)]$$

With $\mathbf{q}_1$ held constant, however, a change of variables of integration from $\mathbf{q}_2$ to $\mathbf{p}_1$ in (6.7) gives

$$\rho(\mathbf{q}_1) = h^{-N} \int d\mathbf{p}_1 \exp\left[ -\frac{2}{\hbar} \int_0^{\hbar\beta/2} d\tau H(\tau) \right] \tag{6.8}$$

The partition function is the integral of $\rho(\mathbf{q}_1)$,

$$Z(\beta) \equiv \text{tr}\,[\exp(-\beta H)] = \int d\mathbf{q}_1 \rho(\mathbf{q}_1)$$

and with (6.8) it is given by

$$Z(\beta) = h^{-N} \int d\mathbf{p}_1 \int d\mathbf{q}_1 \exp\left[ -\frac{2}{\hbar} \int_0^{\hbar\beta/2} d\tau H(\tau) \right] \tag{6.9}$$

a phase space integral over initial conditions.

Equations (6.8) and (6.9) should be easy to use in applications, for not only has all reference to double-ended boundary conditions been removed, but the Jacobian factor is also eliminated by the change of variables. To evaluate the integrand of (6.9), for example, one integrates the equations of motion (6.4) with initial conditions $\mathbf{q}_1$ and $\mathbf{q}_1'(0) = \mathbf{p}_1/\mu$ from $\tau = 0$ to $\tau = \hbar\beta/2$, computing the integral over $H(\tau)$ simultaneously. Furthermore, since the integrand is positive, the integrals over $\mathbf{p}_1$ and $\mathbf{q}_1$ can be done by Monte Carlo methods.

The usual expressions of classical statistical mechanics are obtained if one makes a short time ($\hbar\beta \to 0$) approximation to the propagator; thus as $\hbar\beta \to 0$,

$$\frac{2}{\hbar}\int_0^{\hbar\beta/2} d\tau H(\tau) \simeq \left(\frac{2}{\hbar}\right)\left(\frac{\hbar\beta}{2}\right) H(0) = \beta H(\mathbf{p}_1, \mathbf{q}_1)$$

so that (6.9), for example, becomes

$$Z(\beta) = h^{-N}\int d\mathbf{p}_1 \int d\mathbf{q}_1 \exp\left[-\beta H(\mathbf{p}_1, \mathbf{q}_1)\right] \tag{6.10}$$

the standard classical expression. The classical-path approximation [(6.3), (6.8), and (6.9)] thus contains quantum effects in the density matrix: if the potential $V(\mathbf{q})$ were quadratic in all the coordinates, in fact, (6.3), (6.8), and (6.9) would give the *exact* quantum mechanical results. More generally, it has been shown[51] that an expansion of (6.8), for example, in powers of $\hbar$ gives the first quantum correction exactly correctly, with a small error appearing in the second quantum correction.

### B. Classical-Limit Quantization of Nonseparable Systems

It is interesting to see that quantization of bound dynamical systems is also a direct result of the quantum principle of superposition and can thus be described in the framework of classical-limit quantum mechanics. The most direct way of seeing this is to follow the approach of Gutzwiller.[52,53]

The density of quantum states per unit energy,

$$\rho(E) \equiv \text{tr}\left[\delta(E - H)\right] \tag{6.11}$$

can be expressed in terms of the quantum propagator by use of the Fourier transform identity:

$$\delta(E - H) = (2\pi\hbar)^{-1}\int_{-\infty}^{\infty} dt \exp\left(\frac{iEt}{\hbar}\right) \exp\left(-\frac{iHt}{\hbar}\right) \tag{6.12}$$

Thus

$$\rho(E) = h^{-1}\int_{-\infty}^{\infty} dt \exp\left(\frac{iEt}{\hbar}\right) \tilde{\rho}(t) \tag{6.13}$$

where

$$\tilde{\rho}(t) = \text{tr}\left[\exp\left(-\frac{iHt}{\hbar}\right)\right] \tag{6.13b}$$

That is, $\rho(E)$ and $\tilde{\rho}(t)$ are Fourier transforms of one another. If the trace in (6.11) is carried out using the eigenstates of $H$ as basis, then one sees that $\rho(E)$ is also expressed as

$$\rho(E) = \sum_n \delta(E - E_n) \tag{6.14}$$

where $\{E_n\}$ are the eigenvalues of $H$. Classical-limit eigenvalues are thus determined by using the classical-limit approximation for the propagator in (6.13b), and in view of (6.14) the values of $E$ for which $\rho(E)$ in (6.13a) is singular are identified as the eigenvalues.

If the trace in (6.13b) is carried out in the coordinate representation, one has

$$\tilde{\rho}(t) = \int d\mathbf{q} \langle \mathbf{q} | \exp\left(-\frac{iHt}{\hbar}\right) | \mathbf{q} \rangle \tag{6.15}$$

and with the classical-limit propagator this becomes

$$\tilde{\rho}(t) = \int d\mathbf{q} \left[ (2\pi i\hbar)^N \left(\frac{\partial \mathbf{q}_2}{\partial \mathbf{p}_1}\right)_{\mathbf{q}_1} \right]^{-1/2} \exp\left[\frac{i\phi(\mathbf{q}_2, \mathbf{q}_1; t)}{\hbar}\right] \tag{6.16}$$

with $\mathbf{q}_1 = \mathbf{q}_2 = \mathbf{q}$ in the integrand; that is, the classical trajectory (or trajectories) from which the integrand of (6.16) is constructed begins and ends at position $\mathbf{q}$. Within the framework of classical-limit quantum mechanics the integration over $\mathbf{q}$ in (6.16) is carried out via the stationary phase approximation, the points of stationary phase being determined by

$$\frac{\partial}{\partial \mathbf{q}} \phi(\mathbf{q}, \mathbf{q}; t) = 0$$

That is,

$$\left[\frac{\partial \phi(\mathbf{q}_2, \mathbf{q}_1; t)}{\partial \mathbf{q}_2} + \frac{\partial \phi(\mathbf{q}_2, \mathbf{q}_1; t)}{\partial \mathbf{q}_1}\right]_{\mathbf{q}_1 = \mathbf{q}_2 = \mathbf{q}} = 0 \tag{6.17}$$

Because of the derivative relations of the action integral, however, (6.17) becomes

$$\mathbf{p}_2 - \mathbf{p}_1 = 0$$

That is, the values of $\mathbf{q}$ that contribute (in a stationary phase sense) to the integral in (6.16) are those that lie on a trajectory beginning at $t = 0$ with initial condition $\mathbf{q}(0) = \mathbf{q}$, and $\mathbf{p}(0) = \mathbf{p}_1$, and ending at time $t$ with the *same* values $\mathbf{q}(t) = \mathbf{q}$, $\mathbf{p}(t) = \mathbf{p}_2 = \mathbf{p}_1$—that is, $\mathbf{q}$ must lie on a *periodic* trajectory whose period $\tau$, or an integer multiple of it, equals the time $t$. Apart from preexponential factors that will be ignored for the present simplified discussion, the trace of the propagator is thus given in the classical limit by

$$\tilde{\rho}(t) \sim \sum_n \exp\left[\frac{i\phi_n(t)}{\hbar}\right] \tag{6.19}$$

where $\phi_n(t)$ is the action integral for $n$ passes about the periodic trajectory with period $\tau_n$, where $n\tau_n = t$.

Consistent with the overall classical-limit theory, the time integral in (6.13a) is also evaluated by stationary phase with $\tilde{\rho}(t)$ given by (6.19):

$$\rho(E) \sim \sum_n \int_{-\infty}^{\infty} dt \exp\left\{\frac{i}{\hbar}[Et + \phi_n(t)]\right\} \tag{6.20}$$

The stationary phase condition is

$$E + \frac{\partial \phi_n(t)}{\partial t} = 0$$

but from (3.18) this is seen to be

$$E - E_n(t) = 0$$

or

$$t = t_n(E) \tag{6.21}$$

$t_n(E)$ being the time required for $n$ passes about the periodic trajectory with energy $E$. Again ignoring Jacobian factors, stationary phase evaluation of (6.20) thus gives

$$\rho(E) \sim \sum_n \exp\left[\frac{i}{\hbar}\int_0^{t_n(E)} dt\,\mathbf{p}(t)\cdot\dot{\mathbf{q}}(t)\right] \tag{6.22}$$

where $[\mathbf{q}(t), \mathbf{p}(t)]$ are the periodic trajectory with energy $E$. The time for $n$ passes about the periodic trajectory, however, is simply $n$ times the time for one pass, and the action integral is thus $n$ times the value for one pass, so that (6.22) becomes

$$\rho(E) \sim \sum_{n=-\infty}^{\infty} \exp\left[\frac{in\Phi(E)}{\hbar}\right] \tag{6.23}$$

where

$$\Phi(E) = \int_0^{\tau(E)} dt\,\mathbf{p}(t)\cdot\dot{\mathbf{q}}(t) \tag{6.24}$$

$\tau(E) \equiv t_1(E)$ being the period of the periodic trajectory with energy $E$.

Singularities in $\rho(E)$ can now be identified from (6.23). If

$$\frac{\Phi(E)}{\hbar} = 2\pi X(\text{integer}) \tag{6.25}$$

then

$$\exp\left[\frac{in\Phi(E)}{\hbar}\right] = 1$$

for all $n$, and the sum in (6.23) is divergent. [This also relies on the fact that the preexponential factors in (6.23) do not diminish too rapidly with increasing $n$.] An energy for which (6.25) is satisfied must therefore be an eigenvalue—(6.25) is thus the quantum condition, which is seen to be a direct result of the *constructive interference* of the amplitudes associated with successive orbits about the periodic trajectory with energy $E$. If (6.25) is not satisfied, then the interference will be destructive and the sum not divergent.

The quantum condition may thus be written as

$$n(E) = h^{-1}\Phi(E) \tag{6.26}$$

with $\Phi(E)$ being the action integral of (6.24) along the periodic trajectory with energy $E$. Equation (6.26) defines the quantum number as a function of energy; the eigenvalues are given by $E(n)$ for $n = 0, 1, 2, \ldots$, where $E(n)$ is the inverse function of $n(E)$. [A more careful derivation[52,53] modifies (6.26) by adding a certain fraction to the LHS; this comes from the phase that the preexponential Jacobian factor contributes to the summand in (6.23).]

If there are no constants of the motion (other than the energy itself) or discrete symmetries of the system, then it is clear that for this quantum condition to make sense *there must be one and only one periodic trajectory for every energy* $E$; otherwise, the definition of $n(E)$ in (6.26) is not unique, and one can not speak of *the* quantum number for energy $E$, or *the* energy level for quantum number $n$. From Whittaker's[54] discussion of ordinary and singular periodic trajectories one can infer that the periodic trajectories are *isolated* if there are no constants of the motion other than the energy; for the classical-limit quantum condition to be unique one must argue further that different isolated periodic trajectories for energy $E$ (if there are more than one) are associated with different discrete symmetries of the system. In general, therefore, (6.26) is replaced by

$$n(E, J, \ldots, \sigma) = h^{-1}\Phi(E, J, \ldots, \sigma) \tag{6.27}$$

where $\Phi(E, J, \ldots, \sigma)$ is the action integral of (6.24) along *the* periodic trajectory determined (uniquely) by energy $E$, the other constants of the motion $J, \ldots$, and the discrete symmetry index $\sigma$; $n(E, J, \ldots, \sigma)$ is the "counting quantum number" that orders the eigenvalues in the subspace characterized by $J, \ldots, \sigma$, the eigenvalues being $E(n, J, \ldots, \sigma)$, $n = 0, 1, \ldots$, the inverse function of $n(E, J, \ldots, \sigma)$ for fixed $J, \ldots, \sigma$. As they should be, therefore, the eigenvalues of the system are characterized by all the good quantum numbers of the system (which arise from discrete and continuous symmetries) and a counting index that simply orders the eigenvalues with the same values of the good quantum numbers. For one-dimensional systems it is easy to show that (6.26) reduces to the usual WKB quantum condition (provided the fraction $\frac{1}{2}$ is added to the LHS).

The periodic orbit theory thus provides a physically consistent description of quantization of multidimensional, nonseparable systems. There are some questions[55] regarding its validity, however, and regardless of its accuracy it is clear that it will be extremely difficult to apply in practice, due primarily to the difficulty of finding the periodic trajectories of a many-dimensional, nonseparable system. One way of circumventing the need for periodic trajectories is to use the same trick as in Section VI.A for dealing with diagonal elements of the density operator.

Diagonal matrix elements of the propagator can thus be expressed as

$$\langle \mathbf{q}_1 | \exp\left(-\frac{iHt}{\hbar}\right) | \mathbf{q}_1 \rangle = \int d\mathbf{q}_2 \langle \mathbf{q}_1 | \exp\left(-\frac{iHt}{2\hbar}\right) | \mathbf{q}_2 \rangle \langle \mathbf{q}_2 | \exp\left(-\frac{iHt}{2\hbar}\right) | \mathbf{q}_1 \rangle \quad (6.28)$$

and if the Hamiltonian satisfies time-reversal invariance, then

$$\langle \mathbf{q}_2 | \exp\left(-\frac{iHt}{2\hbar}\right) | \mathbf{q}_1 \rangle = \langle \mathbf{q}_1 | \exp\left(-\frac{iHt}{2\hbar}\right) | \mathbf{q}_2 \rangle$$

so that (6.28) becomes

$$\langle \mathbf{q}_1 | \exp\left(-\frac{iHt}{\hbar}\right) | \mathbf{q}_1 \rangle = \int d\mathbf{q}_2 \langle \mathbf{q}_2 | \exp\left(-\frac{iHt}{2\hbar}\right) | \mathbf{q}_1 \rangle^2 \quad (6.29)$$

and with the classical-limit propagator used in the integrand of (6.29) one has

$$\langle \mathbf{q}_1 | \exp\left(-\frac{iHt}{\hbar}\right) | \mathbf{q}_1 \rangle = (2\pi i\hbar)^{-N} \int d\mathbf{q}_2 \left[\left(\frac{\partial \mathbf{q}_2}{\partial \mathbf{p}_1}\right)_{\mathbf{q}_1}\right]^{-1} \exp\left[\frac{2i}{\hbar}\phi\left(\mathbf{q}_2, \mathbf{q}_1; \frac{t}{2}\right)\right] \quad (6.30)$$

Changing variables of integration from $\mathbf{q}_2$ to $\mathbf{p}_1$ and also integrating over $\mathbf{q}_1$ gives $\rho(t)$ of (6.15) as

$$\tilde{\rho}(t) = (ih)^{-N} \int d\mathbf{p}_1 \int d\mathbf{q}_1 \sigma(t) \exp\left[\frac{2i}{\hbar}\int_0^{t/2} dt' L(t')\right] \quad (6.31)$$

where $\sigma(t)$ is the sign of the Jacobian,

$$\sigma(t) = \text{Sign}\left[\frac{\partial \mathbf{q}(t/2)}{\partial \mathbf{p}_1}\right]_{\mathbf{q}_1}$$

and $L$ is the Lagrangian. The Fourier transform of $\tilde{\rho}(t)$ then gives $\rho(E)$.

The advantage of (6.31) is, of course, that it is not necessary to deal with double-ended boundary conditions to select certain trajectories, rather there is a phase space average over all initial conditions. If we proceed further, the Fourier transform in (6.13a) can be carried inside the phase space average so that (6.32) takes the form

$$\rho(E) = h^{-N} \int d\mathbf{p}_1 \int d\mathbf{q}_1 A(\mathbf{p}_1, \mathbf{q}_1; E) \tag{6.32}$$

where

$$A(\mathbf{p}_1, \mathbf{q}_1; E) = i^{-N} h^{-1} \int_{-\infty}^{\infty} dt \sigma(t) \exp\left\{\frac{i}{\hbar}\left[Et + 2\int_0^{t/2} dt' L(t')\right]\right\} \tag{6.33}$$

In practice the phase space average in (6.32) would probably be most efficiently evaluated for multidimensional systems by Monte Carlo methods.

An interesting question regarding (6.32) and (6.33)—assuming the phase space average to be carried out by Monte Carlo—is how many phase points $(\mathbf{p}_1, \mathbf{q}_1)$ must be sampled before quantum structure appears in $\rho(E)$. For the one-dimensional harmonic oscillator, for example, it is not difficult to show that the complete quantum structure appears in $A(\mathbf{p}_1, \mathbf{q}_1; E)$ for *every single* phase point; that is, the function $A(\mathbf{p}_1, \mathbf{q}_1; E)$ of (6.33) has delta function singularities for $E = \hbar\omega(n + \frac{1}{2})$, $n = 0, 1, 2, \ldots$, for all values of $(\mathbf{p}_1, \mathbf{q}_1)$. If this were true, in general, of course, it would not be necessary to carry out the phase space average in order to identify the eigenvalues. From a certain point of view this hope does not seem unreasonable, for if the system is ergodic—as it should be if the system has no discrete or continuous symmetries—then for sufficiently long times any initial phase point is equivalent to any other; it has been seen from Gutzwiller's periodic orbit theory that it is the long-time behavior that determines quantization.

From a more quantitative point of view one can show from (6.29) that the exact quantum expression for $\tilde{\rho}(t)$ is

$$\tilde{\rho}(t) = \int d\mathbf{q}_2 \int d\mathbf{q}_1 \Bigg\{ \sum_n \psi_n(\mathbf{q}_2)^2 \psi_n(\mathbf{q}_1)^2 \exp\left(-\frac{iE_n t}{\hbar}\right) + \sum_{n \neq n'} \psi_n(\mathbf{q}_2)\psi_{n'}(\mathbf{q}_2)\psi_n(\mathbf{q}_1)\psi_{n'}(\mathbf{q}_1) \exp\left[\frac{-i(E_n + E_{n'})}{2}\frac{t}{\hbar}\right]\Bigg\} \tag{6.34}$$

where $\psi_n(\mathbf{q})$ and $E_n$ are the eigenfunctions and eigenvalues; the functions $\psi_n$ are real since the energy levels are nondegenerate (because of the lack of degeneracy). Changing variables of integration and taking the Fourier transform gives $\rho(E)$ as (6.32), where the exact quantum expression for

$A(\mathbf{p}_1, \mathbf{q}_1; E)$ is seen to be

$$A(\mathbf{p}_1, \mathbf{q}_1; E) = h^{N-1} \int_{-\infty}^{\infty} dt \left| \frac{\partial \mathbf{q}(t/2)}{\partial \mathbf{p}_1} \right| \exp\left(\frac{iEt}{\hbar}\right)$$
$$\times \left\{ \sum_n \psi_n(t)^2 \psi_n(\mathbf{q}_1)^2 \exp\left(-\frac{iE_n t}{\hbar}\right) \right.$$
$$\left. + \sum_{\substack{n, n' \\ n \neq n'}} \psi_n(t) \psi_{n'}(t) \psi_n(\mathbf{q}_1) \psi_{n'}(\mathbf{q}_1) \exp\left[\frac{-i(E_n + E_{n'})t}{2\hbar}\right] \right\}$$
$$(6.35)$$

where $\psi_n(t) = \psi_n[\mathbf{q}(t/2)]$ and $\mathbf{q}(t/2)$ is the position at time $t/2$ for the classical trajectory with initial conditions $(\mathbf{p}_1, \mathbf{q}_1)$. Since the time average of $\psi_n(t)^2$ is nonvanishing and the Jacobian factor nonoscillatory, the first term in braces in (6.35) should lead to peaks at $E = E_n$; in the second term the factor $\psi_n(t)\psi_{n'}(t)$ will itself be oscillatory and thus diminish any contribution from these cross terms. (Integration over $\mathbf{p}_1$ or $\mathbf{q}_1$ will of course cause the cross terms to vanish rigorously via orthogonality.) It appears, therefore, that the full quantum spectrum may appear in the function $A(\mathbf{p}_1, \mathbf{q}_1; E)$ itself; if this is indeed true, then the classical-limit eigenvalues can be extracted from (6.33) without the need of a phase space average over $(\mathbf{p}_1, \mathbf{q}_1)$. This possibility clearly warrants further study.

Finally, it is easy to show that the statistical mechanical approximation to dynamics is obtained, as usual, if one makes a short-time approximation to the dynamical quantities; thus for short times

$$2\int_0^{t/2} dt' L(t') \simeq t \left[\frac{\mathbf{p}_1^{\,2}}{2\mu} - V(\mathbf{q}_1)\right]$$

so that (6.31) becomes

$$\tilde{\rho}(t) = (ih)^{-N} \int d\mathbf{p}_1 \int d\mathbf{q}_1 \exp\left\{\frac{it}{\hbar}\left[\frac{\mathbf{p}_1^{\,2}}{2\mu} - V(\mathbf{q}_1)\right]\right\} \tag{6.36}$$

and a Fourier transform of this can be carried out explicitly, giving

$$\rho(E) = (ih)^{-N} \int d\mathbf{p}_1 \int d\mathbf{q}_1\, \delta\left[E + \frac{\mathbf{p}_1^{\,2}}{2\mu} - V(\mathbf{q}_1)\right] \tag{6.37}$$

The change of variables $\mathbf{p}_1 \to i\mathbf{p}_1$ then gives the standard expression of classical statistical mechanics:

$$\rho(E) = h^{-N} \int d\mathbf{p}_1 \int d\mathbf{q}_1\, \delta[E - H(\mathbf{p}_1, \mathbf{q}_1)] \tag{6.38}$$

There is no quantum structure in this statistical density, for the short-time limit can obviously not describe the dynamical periodicities of the system which have been seen to be responsible for quantization.

## VII. CONCLUDING REMARKS

It is thus possible to construct an internally consistent classical-limit quantum mechanics which in all respects parallels the ordinary quantum mechanical description of physical systems. The correspondence relations established in Section II are the definition of the classical-limit theory: the basic formalism is purely quantum mechanical—that is, the objects of the theory are probability amplitudes that transform by matrix multiplication (the quantum principle of superposition)—but all dynamical relations between coordinates and momenta are those of classical mechanics. The key (and only) mathematical approximation necessary to convert quantum mechanics into classical-limit quantum mechanics is the stationary phase approximation for evaluating integrals over oscillatory integrands whose phases are proportional to $\hbar^{-1}$; since this integral approximation is of an asymptotic nature, the classical limit of quantum mechanics is an asymptotic approximation to quantum mechanics, becoming exact as $\hbar \to 0$. Since all intrinsically quantum effects (interference, tunneling, and quantization) are a direct consequence of the superposition of probability amplitudes, these effects are all contained at least qualitatively in the classical-limit theory; for molecular systems, for which the potential energy function is "smooth," the description is also often quantitative.

Applied to a collision system, the theory leads to a prescription for constructing the classical limit of $S$-matrix elements, which describe transitions between specific quantum states of the collision partners, in terms of the classical trajectories of the system. Classically forbidden processes, that is, those for which no classical trajectories satisfy the appropriate initial and final boundary conditions, can be described by analytically continuing the classical $S$-matrix and finding complex-valued solutions to the classical equations of motion which do satisfy the correct boundary conditions. Furthermore, it is also possible to describe transitions between different adiabatic electronic states within the same formulation as any other classically forbidden process, the only difference being that the potential energy function, having come from the solution of an electronic secular equation, is multivalued so that different choices of the complex time path can cause the trajectory to emerge on different branches of the potential function, that is, in different electronic states.

Numerical applications to atom-diatom collision systems have shown that interference and tunneling effects are accurately described by classical

$S$-matrix theory. It has also been observed that the interference effects that are prominent in collinear $A + BC$ models are often quenched in the three-dimensional systems because of averages over variables such as impact parameter and $m$-components of rotational states. If interference effects in classically allowed processes are quenched, then the classical-limit treatment reduces effectively to the purely classical Monte Carlo trajectory approach.

It seems likely, therefore, that the ability to describe classically forbidden processes, for which purely classical methods are not applicable, may be the most important practical contribution of classical $S$-matrix theory. This is particularly true in light of the possibility of combining within a dynamically exact framework the semiclassical treatment of some degrees of freedom with a purely classical Monte Carlo treatment of other degrees of freedom. Vibrationally inelastic transitions at low collision energy, reactive tunneling in reactions with significant activation barriers, and electronic transitions between different potential energy surfaces are some of the important examples of classically forbidden phenomena.

The accuracy and usefulness of the semiclassical eigenvalue relations[52,53,55] have yet to be established. Since it is also possible to describe metastable states within this theory,[53] there may be the possibility of applying it to the theory of unimolecular reactions.

## Acknowledgements

This work was supported by the donors to the Petroleum Research Fund of the American Chemical Society and by the National Science Foundation under grant GP-34199X.

## References

1. K. W. Ford and J. A. Wheeler, *Ann. Phys. N.Y.*, **7,** 259, 287 (1959).
2. R. B. Bernstein, *Advan. Chem. Phys.*, **10,** 75 (1966).
3. M. V. Berry and K. E. Mount, *Rept. Progr. Phys.*, **35,** 315 (1972).
4 See, for example, R. G. Newton, *Scattering Theory of Waves and Particles*, McGraw-Hill, New York, 1966, pp. 491–492.
5. R. J. Cross, Jr., *J. Chem. Phys.*, **47,** 3724 (1967); **48,** 4838 (1968); **49,** 1753 (1968); **50,** 1036 (1969).
   M. A. Wartell and R. J. Cross, *Chem. Phys. Lett.*, **5,** 477 (1970); *J. Chem. Phys.*, **55,** 4983 (1971).
6. J. C. Y. Chen and K M Watson, *Phys. Rev.*, **174,** 152 (1968); **188,** 236 (1969).
7. R. J. Glauber, in *Lectures in Theoretical Physics*, Vol. I, W. E. Brittin and L. G. Dunham, Eds., Interscience, New York, 1959, p. 315. See also E. Gerjuoy, in *Physics of Electronic and Atomic Collisions*, VII ICPEAC, 1971, T. R. Govers and F. J. deHeer, Eds., North-Holland, 1972, p. 247.
8. (*a*) W. H. Miller, *J. Chem. Phys.*, **53,** 1949 (1970); (*b*) **53,** 3578 (1970); (*c*) *Chem. Phys. Lett.*, **7,** 431 (1970); (*d*) *J. Chem. Phys.*, **54,** 5386 (1971); (*e*) **55,** 3150 (1971); (*f*) T. F. George and W. H. Miller, *J. Chem. Phys.*, **56,** 5637 (1972); (*g*) **56,** 5668

(1972); (*h*) **56,** 5722 (1972); (*i*) **57,** 2458 (1972); (*j*) J. D. Doll and W. H. Miller, *J. Chem. Phys.*, **57,** 5019 (1972); (*k*) J. D. Doll, T. F. George, and W. H. Miller, *J. Chem. Phys.*, **58,** 1343 (1973); (*l*) W. H. Miller and A. W. Raczkowski, *Faraday Discussions Chem. Soc.*, **55,** 45 (1973).

9. (*a*) R. A. Marcus, *Chem. Phys. Lett.*, **7,** 525 (1970); (*b*) *J. Chem. Phys.*, **54,** 3965 (1971); (*c*) J. N. L. Connor and R. A. Marcus, *J. Chem. Phys.*, **55,** 5636 (1971); (*d*) W. H. Wong and R. A. Marcus, *J. Chem. Phys.*, **55,** 5663 (1971); (*e*) R. A. Marcus, *J. Chem. Phys.*, **56,** 311 (1972); (*f*) J. Stine and R. A. Marcus, *Chem. Phys. Lett.*, **15,** 536 (1972); (*g*) R. A. Marcus, *J. Chem. Phys.*, **57,** 4903 (1972).
10. Other studies similar in spirit to those in Refs. 8 and 9 are (*a*) P. Pechukas, *Phys. Rev.*, **181,** 166 (1969); **181,** 174 (1969); (*b*) I. L. Beigman, L. A. Vainshtein, and I. I. Sobel'man, *Zh. Eksperim. i Teor. Fiz.*, **57,** 1703 (1969) [*Soviet Phys. JETP*, **30,** 920 (1970)]; (*c*) I C. Percival and D. Richards, *J. Phys. B*, **3,** 315 (1970); **3,** 1035 (1970); (*d*) R. D. Levine and B. R. Johnson, *Chem. Phys. Lett.*, **4,** 404 (1970); **8,** 501 (1971); (*e*) M. D. Pattengill, C. F. Curtiss, and R. B. Bernstein, *J. Chem. Phys.*, **54,** 2197 (1971).
11. For a recent review of classical trajectory methods, see D. L. Bunker, *Methods Comp. Phys.*, **10,** 287 (1971).
12. For a discussion of many of the results obtained for classical trajectory studies, see J. C. Polanyi, *Acc. Chem. Res.*, **5,** 161, (1972).
13. W. H. Miller, *Advances in Molecular Beams*, K. P. Lawley, Ed., Wiley, to be published.
14. P. A. M. Dirac, *The Principles of Quantum Mechanics*, 4th Ed., Oxford, 1958.
15. H. Goldstein, *Classical Mechanics*, Addison-Wesley, Reading, Mass., 1950, pp. 237–314.
16. See, for example, A. Erdélyi, *Asymptotic Expansions*, Dover, New York, 1956, p. 51.
17. See, for example, L. I. Schiff, *Quantum Mechanics*, 3rd ed., McGraw-Hill, New York, 1968, pp. 268–279.
18. M. E. Rose, *Elementary Theory of Angular Momentum*, Wiley, New York, 1957, pp. 32–47.
19. A. R. Edmonds, *Angular Momentum in Quantum Mechanics*, Princeton University Press, Princeton, N.J., 1960, p. 122.
20. P. J. Brussard and H. A. Tolhoek, *Physica*, **23,** 955 (1957).
21. See, for example, (*a*) K. Gottfried, *Quantum Mechanics*, Benjamin, New York, 1966, pp. 70–74; (*b*) A. S. Davydov, *Quantum Mechanics*, Addison-Wesley, Reading, Mass., 1965, pp. 69–73; (*c*) A. Messiah, *Quantum Mechanics*, Wiley, New York, 1961, pp. 222–228.
22. R. P. Feynman and A. R. Hibbs, *Quantum Mechanics and Path Integrals*, McGraw-Hill, New York, 1965.
23. J. H. vanVleck, *Proc. Natl. Acad. Sci. U.S.*, **14,** 178 (1928).
24. C. Morette, *Phys. Rev.*, **81,** 848 (1951).
25. I. Fujiwara, *Prog. Theoret. Phys.* (*Kyoto*), **21,** 902 (1959).
26. R. Schiller, *Phys. Rev.*, **125,** 1109 (1962).
27. See, for example, Ref. 4, pp. 177–197.
28. J. L. Dunham, *Phys. Rev.*, **41,** 721 (1932).
29. D. Secrest and B. R. Johnson, *J. Chem. Phys.*, **45,** 4556 (1966).
30. P. M. Morse and H. Feshbach, *Methods of Theoretical Physics*, McGraw-Hill, New York, 1953, p. 467.
31. R. G. Newton, Ref. 4, pp. 478–482.

32. See, for example, F. T. Smith, *Phys. Rev.*, **118,** 349 (1960).
33. See, for example, Ref. 16, p. 39, or Ref. 30, pp. 437–441.
34. R. N. Porter and M. Karplus, *J. Chem. Phys.*, **40,** 1105 (1964).
35. D. J. Diestler, *J. Chem. Phys.*, **54,** 4547 (1971).
36 S.-F. Wu and R. D. Levine, *Mol. Phys.*, **22,** 881 (1972).
37. D. J. Diestler and M. Karplus, *J. Chem. Phys.*, **55,** 5832 (1971).
38. H. S. Johnston, *Gas Phase Reaction Rate Theory*, Ronald Press, New York, 1966, pp. 190–196.
39. M. Karplus, R. N. Porter, and R. D. Sharma, *J. Chem. Phys.*, **43,** 3259 (1965).
40. Ref. 22, pp. 68–71.
41. E. C. G. Stuckelberg, *Helv. Phys. Acta*, **5,** 369 (1932).
42. L. D. Landau and E. M. Lifschitz, *Quantum Mechanics*, Pergamon Press, New York, 1965, pp. 185–187.
43. See also A. M. Dykhne, *Soviet Phys. JETP*, **14,** 941 (1962);
A. M. Dykhne and A. V. Chaplik, *Soviet Phys. JETP*, **16,** 631 (1963);
G. V. Dubrovskii, *Soviet Phys. JETP*, **19,** 591 (1964).
44. L. D. Landau, *Physik. Z. Sowjetunion U.R.S.S.*, **2,** 46 (1932);
C. Zener, *Proc. Roy. Soc. (London) Ser. A*, **137,** 696 (1932).
45. See also the discussion in N. F. Mott and H. S. W. Massey, *The Theory of Atomic Collisions*, Oxford University Press, 1965, p. 353.
46. Yu. N. Demkov, *Soviet Phys. JETP*, **18,** 138 (1964).
47. R. E. Olson, and F. T. Smith, *Phys. Rev. A*, **7,** 1529 (1973).
48. E. E. Nikitin, *Opt. Spektrosk.*, **19,** 91 (1965); E. I. Dashevskaya, E. E Nikitin, and A. I. Reznikov, *J. Chem. Phys.*, **53,** 1175 (1970).
49. J. C. Tully and R. K. Preston, *J. Chem. Phys.*, **55,** 562 (1971).
50. P. Pechukas and J. P. Davis, *J. Chem. Phys.*, **56,** 4970 (1972).
51. W. H. Miller, *J. Chem. Phys.*, **55,** 3146 (1971); S. M. Hornstein and W. H. Miller, *Chem. Phys. Lett.*, **13,** 398 (1972); W. H. Miller, *J. Chem. Phys.*, **58,** 1664 (1973).
52. M. G. Gutzwiller, *J. Math. Phys.*, **8,** 1979 (1967); **10,** 1004 (1969); **11,** 1791 (1970); **12,** 343 (1971).
53. See also, W. H. Miller, *J. Chem. Phys.*, **56,** 38 (1972).
54. E. T. Whittaker, *A Treatise on The Analytical Dynamics of Particles and Rigid Bodies*, Cambridge University Press, London, 1965, pp. 395–396.
55. See, for example, the discussion by P. Pechukas, *J. Chem. Phys.*, **57,** 5577 (1972); and I. C. Percival, *J. Phys. B*, **6,** L229 (1973).

# SOURCES OF ERROR AND EXPECTED ACCURACY IN *AB INITIO* ONE-ELECTRON OPERATOR PROPERTIES: THE MOLECULAR DIPOLE MOMENT

SHELDON GREEN*

*Institute for Space Studies, New York, New York*

## CONTENTS



## I. INTRODUCTION

According to quantum theory all information about molecular properties is contained in the molecular wave function. The wave function, which effectively describes the motion of the electrons and nuclei, may be

* NSF predoctoral fellow 1966–1971; NRC-NASA Research Associate 1971–1973.

obtained by solving the eigenvalue equation (Schrödinger equation) for the Hamiltonian operator. Unfortunately, exact solutions to this equation have not been obtained for systems with more than one electron; however, a great deal of effort has been expended in finding approximate solutions for atomic and molecular systems.[1,2] In many cases, *ab initio* results are now accurate enough to provide chemically significant information. Most calculations to date have been concerned primarily with energy properties, and other information contained in the wave function has often been ignored. In this article attention is directed toward those properties that can be obtained as one-electron operator expectation values. All properties that depend on the charge density distribution, such as the molecular electric multipole moments and the field gradient at the nuclei, fall into this category. Other one-electron operator properties include spin hyperfine constants and the diamagnetic contributions to magnetic susceptibility and shielding at the nuclei. The property for which the most accurate and extensive experimental results are available is the electric dipole moment. A recent series of papers by the author[3–7] has examined methods for the accurate calculation of diatomic dipole moments. Attention was limited to diatomics because of the greater availability of calculations and accurate experimental values; however, similar results are expected for larger systems. Other one-electron operator properties should be amenable to the same techniques as the dipole moment but have been examined to a lesser extent because of the more limited experimental data. (See, however, a discussion of spin hyperfine constants in Ref. 7.)

Experimental techniques for determining dipole moments and their expected accuracy have been discussed elsewhere, and tabulations of experimental values are available.[8] The most accurate measurements rely on the Stark effect. If the species can be studied in a molecular beam spectrometer, accuracy of better than 0.001 D can be obtained and the dependence on vibrational and rotational level is generally determined. Even for unstable free radicals experimental accuracy on the order of 0.01 to 0.05 D is often obtained. The ability of quantum theory to reproduce observed dipole moments has obviously been important in understanding the nature of chemical bonding. However, in view of the availability of accurate experimental values, one might question the practical need for accurate theoretical dipole moments. Of course, the dipole was selected for the opposite reason: to use accurate experimental values to gage the reliability of *ab initio* techniques. Once the sources of error are well understood other properties, such as the molecular quadrupole moment which is not so easy to determine experimentally, can then be calculated with some degree of confidence. However, dipole calculations are not entirely unnecessary. Not all species are equally amenable to experimental techniques; short-lived states may be particularly troublesome. The CN radical is one example where the answer to several astrophysically interesting questions depended on an unknown dipole moment. (See discussion and references in Ref. 7.) For CN, an experimental value was eventually obtained; in the future, *ab initio* determinations may be cheaper, faster, and as reliable as experimental measurements for obtaining such information. Calculations have also been useful in choosing between disparate experimental determinations.[9] Finally, the dipole moment of a charged species is currently amenable only to *ab initio* techniques, and these moments are needed, for example, for an understanding of rotationally inelastic scattering of electrons by molecular ions.[10]

The purpose of this article is to compare a number of experimental and theoretical dipole moments in order to draw conclusions about the sources of error and the accuracy that can be expected from different levels of approximation. Attention has been limited to *ab initio* values obtained within the self-consistent field (SCF) and configuration interaction (CI) approximations. Also, only those SCF calculations that employ a large enough basis set to be near the Hartree-Fock (HF) limit have been included. A general discussion of sources of error and expected accuracy is given in Section II. By comparing a large number of theoretical and experimental values for diatomic molecules it will be shown that rather specific estimates may be given for the dipole moment error that can be expected in *ab initio* calculations. In particular, both the size and direction of the error in HF dipole moments can be predicted from simple, qualitative ideas of the molecular bonding. This capability is especially important because HF wave functions are now rather readily available, but the HF method can never yield the exact solution. The quantum chemist will now be able to determine the feasibility of obtaining an unknown dipole by *ab initio* methods based on knowledge of the level of approximation and hence the cost of the calculation necessary to obtain a desired accuracy. In a very real sense the computer should be considered as a tool for determining dipole moments which is competitive with other experimental techniques. If very high accuracy is necessary or if the system has very many electrons, the standard experimental techniques are likely to be preferred; however, for highly reactive and short-lived states, and especially for ionic species, *ab initio* methods will often be cheaper, faster, and more reliable than experiments. In Section III examples of SCF calculations for polyatomic molecules are presented which indicate that the HF errors for these are analogous to those found for diatomics. Several new calculations have been performed to increase the basis for comparison, and these are presented in Section IV. Section V contains a tabulation of the diatomic results upon which the conclusions have been based, and some concluding remarks are given in Section VI.

## II. ACCURACY OF *AB INITIO* DIPOLE MOMENTS

For *ab initio* dipole moments to be of maximum utility, it is imperative to understand the sources of error inherent in the various approximations which are used. Ideally, one should be able to estimate the probable errors for calculated values just as one establishes experimental error bounds.

### A. Sources of Error

The sources of error in current *ab initio* calculations may be conveniently divided into the following categories: (*1*) Born-Oppenheimer approximation and relativistic effects, (*2*) variation with internuclear distance, (*3*) finite expansion basis set, and (*4*) correlation and polarization effects.

#### 1. *Born-Oppenheimer and Relativistic Effects*

The Born-Oppenheimer approximation assumes that the nuclear and electronic motions are separable. If all terms in the Hamiltonian that couple the nuclear coordinates **R** and the electron coordinates **r** are ignored, the total wave function is the product of a nuclear, vibration-rotation wave function $\Psi_{v,J}(\mathbf{R})$ and an electronic wave function $\Psi(\mathbf{r};\mathbf{R})$ which depends parametrically on **R**. The SCF and CI methods give approximate solutions to the electronic Schrödinger equation with the nuclei fixed at $\mathbf{R}_0$.

$$\mathbf{H}(\mathbf{r};\mathbf{R}_0)\Psi(\mathbf{r};\mathbf{R}_0) = E(\mathbf{R}_0)\Psi(\mathbf{r};\mathbf{R}_0) \tag{1}$$

Then $E(\mathbf{R})$ provides the potential in which the nuclei move. Experimental estimates of the dipole moment error introduced by neglecting terms that couple nuclear and electronic motion are available. For example, the permanent dipole moment of HD ($\mu = 6 \times 10^{-4}$ D)[11] is caused entirely by non-Born-Oppenheimer effects. There is also data on Born-Oppenheimer breakdown in HCl and DCl which indicates an effect of about $10^{-3}$ D.[12]

Relativistic terms in the Hamiltonian are normally neglected in (1). For small atoms the most important such terms are spin-orbit interactions. The spin-orbit (and non-Born-Oppenheimer) terms mix adiabatic spectroscopic states, producing perturbations in the vibration-rotation energy levels. For such perturbations to be significant, the states involved must be of similar energy. The ground state is usually well separated from other states; however, there are often many excited states with similar energies, and such perturbations are common.[13] In CO $a^3\Pi$ $v = 4$ and 5 are strongly perturbed by the $a'^3\Sigma^+$ state, and the effect on the dipole moment has been studied experimentally.[14] (See also Section IV.E.)

### 2. *Nuclear Motion*

Because (in the Born-Oppenheimer approximation) the electronic wave function depends parametrically on the nuclear coordinates, the expectation value of a one-electron operator **M** also depends on the internuclear distance:

$$M(\mathbf{R}) = \int d\mathbf{r}\Psi^*(\mathbf{r};\,\mathbf{R})\mathbf{M}(\mathbf{r})\Psi(\mathbf{r};\,\mathbf{R}) \tag{2}$$

The effect of nuclear motion on **M** is then found by averaging $M(\mathbf{R})$ over the vibration-rotation wave function

$$M_{vJ} = \int d\mathbf{R}\Psi_{vJ}^*(\mathbf{R})M(\mathbf{R})\Psi_{vJ}(\mathbf{R}) \tag{3}$$

The dipole moment of diatomic molecules typically changes by 0.05 to 0.1 D with vibrational level. Rotational effects are generally two orders of magnitude smaller.

It has become common practice to compare experimental properties with calculations performed at the equilibrium internuclear distance $R_e$. $M(R_e)$ can correctly be compared with the experimental value extrapolated to the vibrationless, rotationless state $v = -\frac{1}{2}$, $J = 0$. However, it is undoubtedly better practice to compare vibrationally averaged properties [computed via (3)] with experimental data. There has also been some question, especially in SCF calculations, as to whether it is preferable to use the calculated or experimental $R_e$.[15,16] This is really the same question as whether a calculated $M(R)$ should be averaged over vibrational wave functions calculated from the *ab initio* or an experimental $E(R)$. If the calculation gives an accurate potential curve, of course, this is a moot point. However, many methods (including SCF) give better accuracy for $M(R)$ than for $E(R)$; then it is preferable to use experimental vibrational wave functions (obtained from $E(R)$ constructed, e.g., by the Rydberg-Klein-Rees procedure) to obtain $M_v$. For consistency it seems desirable to use the experimental $E(R)$ whenever available, and this practice has been followed in the comparisons of calculated and experimental dipole moments tabulated in Section V.

### 3. *Basis Set Error*

To make molecular calculations numerically tractable, the wave function is expanded in a finite basis set of one-electron functions, and it is crucial to select an adequate yet economical basis. For diatomic molecules, Slater-type orbitals are by far the most commonly used functions.[2] Other choices that have been employed are elliptical orbitals[17] and various

Gaussian-type functions.[2] Nearly all polyatomic calculations employ Gaussians.

The variation principle provides a means for comparing the accuracy of basis sets. Since any approximate wave function gives an upper bound to the true energy, the basis set that gives the lowest energy is the best. A basis set may be improved by optimizing parameters (e.g., orbital exponents), by increasing the basis, or both. In this way the basis set dependence of SCF functions has been extensively studied. (The infinite basis set limit of the SCF method is the Hartree-Fock function.) It has been economically unfeasible to study similarly the effect of basis set in going beyond the HF limit, that is, to consider the additional basis functions that might be necessary to describe correlation effects. This would entail a direct optimization of basis set in large-scale CI calculations. However, the effect on a CI calculation of a basis set that cannot reach the HF limit was examined in Ref. 6. It should be noted that the variation principle does *not* imply that properties other than the energy improve monotonically with improved basis set, but in all cases the dipole moment has been found to "converge" as the basis approaches the HF limit.

The following generalizations may be made about SCF basis sets. (*1*) It is more economical and generally better to use a larger basis set with little optimization than a smaller, highly optimized set. (*2*) A good initial basis for molecular calculations can be constructed from optimized atomic basis sets. Clementi[18] has given accurate atomic basis sets which may be used. He has also given double-zeta (DZ) bases which are somewhat less accurate; these contain two basis functions for each occupied atomic *nl* shell. Similarly optimized atomic basis sets have been given by Bagus, Gilbert, and Roothaan[19]; these authors provide "accurate" and "nominal" sets, the latter being comparable to the DZ set of Clementi. For highly ionic species one should start with atomic basis sets optimized for the ions. For Rydberg states, diffuse atomic orbitals must be included; initial values for the Rydberg orbital exponents can be obtained from Slater's rules.[20] (*3*) For dipole moment calculations it is imperative to add polarization functions to the basis. These are atomic orbitals with higher angular momentum which are not occupied in the separated atoms (e.g., $3d\sigma$, $3d\pi$, and $4f\sigma$) and which allow the atomic charge density to be polarized into the bonding region. Good results have been obtained by choosing polarization functions optimized in a smaller molecular calculation[21] or by interpolating from optimized calculations on other molecules.[16] (*4*) The minimum basis set that appears acceptable for dipole moment calculations is atomic double-zeta plus polarization (DZ + P). Systematic study by Yoshimine and McLean[21] indicates that DZ + P dipole moments are usually within 0.2 D of the HF limit. (*5*) Some properties depend strongly on basis functions to which the energy is not very sensitive, and it is better to saturate the basis with such functions than to optimize a single such function. For example, the field gradient at the nucleus is sensitive to *d* atomic orbitals; the inclusion of several 3*d* orbitals with varying exponents has been found to be much more reliable (for LiH) than attempting to optimize the exponent of one such function.[22] States with diffuse charge distributions (e.g., Rydberg states or weakly bound states with large equilibrium internuclear distance) may be very sensitive to small basis set changes. (*6*) Some basis set optimization or at least experimentation is very desirable. The dipole moment is most sensitive to polarization and valence functions, and stability of the SCF dipole with respect to optimization of these functions is a necessary although not a sufficient condition for being near the HF limit. (*7*) A deficiency in the SCF basis will not be corrected by a subsequent CI calculation, and the importance of limiting this source of error cannot be overemphasized. It is important to be near the HF limit *for the property of interest*. An indication of the nature of the basis set is given for each calculation tabulated in Section V.

### 4. *Correlation Effects*

Because the nonrelativistic, Born-Oppenheimer Hamiltonian contains terms involving the interelectronic coordinates, a single-configuration (i.e., HF) wave function cannot be an exact solution to (1). The difference between the HF approximation and the exact solution is defined to be the correlation error. Various symmetry restrictions are normally also imposed in the HF method, and these may introduce further error.[23,24] The error between a restricted HF function and the exact function can be separated into polarization and correlation effects.[24] Other authors have made somewhat different divisions into "internal" and "external" correlation.[25] These differences are not important in the following discussion. However, they do give insight into the difference between closed- and open-shell states, and the nature of the additional correction terms that are necessary for the latter.

Properties calculated from Hartree-Fock functions therefore contain an error due to lack of correlation. In some cases, that is, open-shell states, there is an additional error due to symmetry restrictions. Both of these errors can be eliminated by using CI wave functions. Correlation errors for diatomic dipole moments are generally 0.1 to 0.5 D; this is the size error that is found in the HF dipole moment of closed-shell molecules. Polarization errors vary greatly from system to system; also the distinction between polarization and correlation error is not unambiguous. It can be noted, however, that restricted HF dipole moments for open-shell diatomic molecules are found to have errors which vary from 0.1 to 1.0 D.

### B. Expected Accuracy for Diatomic Molecules

Since (1) can be solved exactly with an infinite complete CI calculation, it is instructive to consider this hypothetical CI and relate the error in real calculations to the terms which are omitted. Starting with a complete orthonormal set of one-electron functions (orbitals)—$\{\varphi_i\}$, $i = 1, \infty$—an $N$-electron reference function is chosen as the normalized, determinantal function

$$\Phi_0 = (N!)^{-1/2} \det \{\varphi_1(1)\varphi_2(2)\varphi_3(3) \cdots \varphi_N(N)\} \tag{4}$$

Then the set of all determinants formed by replacing $1, 2, \ldots, N$ of the $\varphi_i$, $i = 1, N$ with $\varphi_a$, $a = N + 1, \infty$ is a complete set of $N$-electron functions. These will be denoted as $\Phi_i^a$, $\Phi_{ij}^{ab}, \ldots,$ and referred to as single, double, . . . , excitations from $\Phi_0$. Rather than using single determinants, linear combinations may be taken to form symmetry adapted configurations, which are eigenfunctions of the appropriate spin and orbital angular momenta and spatial symmetries. All the arguments that follow are valid for such symmetry adapted configurations as well as for single determinant configurations. The complete CI can be written as

$$\Psi = c_0\Phi_0 + c_i^a\Phi_i^a + c_{ij}^{ab}\Phi_{ij}^{ab} + \cdots \tag{5}$$

Summation is implied over repeated indices in each term, and the coefficients are to be determined by minimizing the energy, that is, by diagonalizing the (infinite order) Hamiltonian matrix. This variational procedure can be replaced with the equivalent infinite order perturbation calculation, and it is useful to estimate the relative importance of various terms by low-order perturbation theory (see Ref. 4).

The HF method prescribes a choice of orbitals $\{\varphi_i\}$ such that $\Phi_0$ is the (energetically) best single determinantal wave function. For an unrestricted HF function and also for a closed-shell restricted HF function, it can be shown (Brillouin's theorem)[26] that

$$\langle \Phi_0 | H | \Phi_i^a \rangle = 0 \tag{6}$$

Then, through first order in perturbation theory, only double excitations, $\Phi_{ij}^{ab}$, contribute to the CI function because only these have nonvanishing Hamiltonian matrix elements with $\Phi_0$. Single, triple, and quadruple excitations enter in second order by coupling to the double excitations. If Brillouin's theorem does not apply, for example, in the case of open-shell restricted HF functions, then both single and double excitations contribute in first order. In actual calculations the largest coefficients of "first order" terms are about 0.1, and the largest coefficients of "second order" terms are about 0.01.

The exact one-electron operator expectation value can be expanded as

$$\langle\Psi| M |\Psi\rangle = c_0^2\langle\Phi_0| M |\Phi_0\rangle + 2c_0c_i^a\langle\Phi_0| M |\Phi^a\rangle + c_i^a c_j^b\langle\Phi_i^a| M |\Phi_j^b\rangle$$
$$+ 2c_i^a c_{jk}^{bc}\langle\Phi_i^a| M |\Phi_{jk}^{bc}\rangle + c_{ij}^{ab}c_{kl}^{cd}\langle\Phi_{ij}^{ab}| M |\Phi_{kl}^{cd}\rangle + \cdots \quad (7)$$

If Brillouin's theorem applies, the leading correction to the HF value is on the order of $(c_{ij}^{ab})^2$, which explains the usual statement that HF properties are correct to second order for closed-shell molecules. (See, however, Section II.B.1.) If Brillouin's theorem does not apply, the leading term is on the order of $c_0c_i^a$, and the correction to the HF value may be quite large.

A qualitative idea of the error introduced by truncating the CI expansion (i.e., using a finite calculation) can be seen as follows. If some configurations are excluded, those that are retained tend to be overemphasized. The amount of overemphasis depends, of course, on the importance of the excluded configurations. This is partially a simple renormalization effect. However, owing to the variational nature of the calculation, there is also selective emphasis (and deemphasis) in an attempt to describe the correlation effects of the excluded configurations in terms of those that are retained. The correction to the dipole moment is usually dominated by a few configurations. If these are included in the CI but other important terms are omitted, one might expect an "overcorrection" due to the resulting overemphasis of configurations which give the dipole correction. Indeed, this effect is generally obtained. It was shown in Ref. 4 that excluding all configurations involving excitations from the core orbitals (fixed-core approximation) led to overcorrection in going from the SCF to the CI dipole. (Note, however, that the error in this approximation was less than 0.05 D in LiH and CS.) It was found in Ref. 5 that exclusion of higher-than-double excitations when they were important also led to too large a CI correction. As a final example, it appears that exclusion of higher angular correlation (i.e., exclusion of delta orbitals) can also give a CI correction to the dipole which is too large. (See Section IV.C.)

The above discussion has assumed that $\Phi_0$ is a symmetry restricted HF function, that is, essentially a single determinant. The generalization in which $\Phi_0$ is a linear combination of a (small) finite number of configurations can also be considered. $\Phi_0$ may be chosen to dissociate correctly to atoms (OVC[27]), to include the "internal" correlation (NCMET[25]), to represent an "extended" or "projected" HF function,[28] etc. If $\Phi_0$ is determined by the variation principle, a "generalized Brillouin's theorem"[29] applies. An appropriate choice for $\Phi_0$ could ensure that the error in one-electron properties is "second-order" for open- as well as closed-shell systems. Such extended HF functions are not currently available for very many molecules, and they are not considered further here. (A. C. Wahl,

M. Krauss, and coworkers[62] have recently completed calculations which indicate that good dipole moments are obtained from the OVC scheme using small CI expansions where the orbital *and* CI coefficients are simultaneously varied to minimize the energy. The following OVC dipole moments may be compared with SCF, CI, and experimental values tabulated in Section V: for CH $X^2\Pi$, 10 configuration OVC $\mu = 1.53$ D; for OH $X^2\Pi$, 17 configuration OVC $\mu = 1.675$ D; for HF $X^1\Sigma^+$, 8 configuration OVC $\mu = 1.805$ D; and for CO $X^1\Sigma^+$, 11 configuration OVC $\mu = -0.164$ D.)

Although the dipole moment can, in principle, be obtained to any degree of accuracy by employing a large enough CI, in practice only severely truncated CI expansions can be handled. Even those calculations that have employed in excess of 10,000 configurations are actually quite limited.[30] Fortunately, it has been demonstrated that dipole moment errors for diatomics can be kept below about 0.05 D with relatively small CI expansions (a few hundred configurations). More extensive calculations (several thousand configurations) have reduced the error to 0.01 to 0.02 D. It now appears, however, that the CI method is not likely to produce dipole results much more reliable than this. For example, LiH with only four electrons has been the subject of many calculations. Recently an extensive CI function has been obtained[22] which recovers 97% of the correlation energy, but the dipole moment is still in error by 0.015 D (after proper vibrational averaging). Although this is an error of only two parts in $10^3$, the experimental accuracy is better than one part in $10^4$. The remaining error in LiH must be attributed to truncation of the $N$-electron space, primarily lack of higher excitations, and inability of the one-electron basis set to describe correlation effects, although it is certainly at the HF limit.

The framework is now laid for categorizing the dipole moment error found by systematically comparing a large number of calculated and experimental values. It then appears that rather specific predictions can be made concerning the size errors expected in calculations on species for which experimental values are not available. It is convenient to discuss closed- and open-shell states separately.

### 1. *Closed-Shell Ground States*

As mentioned previously, the error in HF dipole moments is found quite uniformly to be 0.1 to 0.5 D for closed-shell ground states of diatomic molecules, and CI calculations have generally reduced the error to 0.01 to 0.05 D. (See also Section V.) The most interesting result, because it had not been anticipated, was that the CI correction was produced almost entirely by single excitations.[31,32] This apparent contradiction of Brillouin's theorem is easily explained. The most important double excitations have coefficients $c_{ij}^{ab} \sim 0.1$, and the most important single excitations have coefficients $c_i^a \sim 0.01$. This is denoted formally by $c_{ij}^{ab} = O(\epsilon)$ and $c_i^a = O(\epsilon^2)$. Hence the dipole correction terms due to single excitations and those due to double excitations [see (7)] are both $O(\epsilon^2)$. Furthermore, the effect of double excitations can be shown to be nearly canceled by the renormalization, $(c_0)^2$, of the HF term. From the orthonormality of the total wave function and the individual configurations,

$$(c_0)^2 + \sum_{\substack{ij\\ab}} (c_{ij}^{ab})^2 = 1 + O(\epsilon^4) \tag{8}$$

Because $M$ is a one-electron operator

$$\langle \Phi_{ij}^{ab} | M | \Phi_{ij}^{ab} \rangle = \langle \Phi_0 | M | \Phi_0 \rangle + \langle \varphi_a | M | \varphi_a \rangle + \langle \varphi_b | M | \varphi_b \rangle - \langle \varphi_i | M | \varphi_i \rangle - \langle \varphi_j | M | \varphi_j \rangle \tag{9}$$

and

$$\langle \Phi_i^a | M | \Phi_0 \rangle = \langle \varphi_i | M | \varphi_a \rangle \tag{10}$$

Using (9) and (10), (7) can be rewritten as

$$\begin{aligned}\langle \Psi | M | \Psi \rangle = {} & c_0^2 \langle \Phi_0 | M | \Phi_0 \rangle + 2c_0 c_i^a \langle \varphi_i | M | \varphi_a \rangle \\ & + (c_{ij}^{ab})^2 \{ \langle \Phi_0 | M | \Phi_0 \rangle + \langle \varphi_a | M | \varphi_a \rangle + \langle \varphi_b | M | \varphi_b \rangle \\ & - \langle \varphi_i | M | \varphi_i \rangle - \langle \varphi_j | M | \varphi_j \rangle \} + O(\epsilon^3)\end{aligned} \tag{11}$$

Using (8) and collecting terms then gives

$$\begin{aligned}\langle \Psi | M | \Psi \rangle = {} & \langle \Phi_0 | M | \Phi_0 \rangle + 2c_0 c_i^a \langle \varphi_i | M | \varphi_a \rangle \\ & + (c_{ij}^{ab})^2 \{ \langle \varphi_a | M | \varphi_a \rangle + \langle \varphi_b | M | \varphi_b \rangle - \langle \varphi_i | M | \varphi_i \rangle \\ & - \langle \varphi_j | M | \varphi_j \rangle \} + O(\epsilon^3)\end{aligned} \tag{12}$$

Now the most important double excitations are those that most effectively correlate $\Phi_0$, and this implies[33] that $\varphi_a$ and $\varphi_b$ are localized in the same region of space as $\varphi_i$ and $\varphi_j$. Hence the term in brackets in (12) is expected to be very small; this completes the demonstration that double excitations effectively cancel the renormalization of the HF term. On the other hand, the most important single excitations are polarization effects, and these tend to transfer charge across the molecule. Indeed, CI calculations (see Section V for references) indicate that the bulk of the correction to the HF dipole is due to one single excitation from the highest occupied sigma (pi) orbital to the lowest unoccupied sigma (pi) orbital in LiH, HF, HCl, and ClF (CO and CS).

The choice of configurations necessary for an accurate CI dipole moment follows straightforwardly from the above discussion. Single and double excitations from the HF function must be included; the single excitations that correct the dipole moment are coupled to the HF function by the double excitations that provide the bulk of the correlation. The number of possible single excitations is comparatively small and all are usually included; however, polarization of the core orbitals may be omitted since it has been found to be unimportant. The number of possible double excitations, on the other hand, is quite large; it has proved effective to select from these on the basis of their contribution to the second-order perturbation energy.[34] Adequate convergence of the dipole moment has been found on increasing the number of double excitations.[3] An indication of the level of convergence can be seen by comparing medium- and large-scale CI calculations which have been reported for LiH and HF. Using 200 configurations, Grimaldi[34] obtained dipole moments of 5.814 D for LiH and 1.832 D for HF. Starting from similar SCF functions, Bender and Davidson[35] obtained 5.853 D for LiH using 939 configurations and 1.816 D for HF with 1517 configurations. Bender and Davidson also included delta and phi orbitals, but higher angular correlation is apparently unnecessary in these dipole moment calculations. The inclusion of triple excitations has also been shown to be unimportant for closed-shell dipole moment calculations.[3]

There is an additional potential source of error in HF calculations on closed-shell ground states which has not been mentioned because it has not yet caused trouble. It should be kept in mind, however, especially if HF values are used to predict properties of unobserved species. It is not necessarily true that the ground state of a species has a closed-shell electronic structure even if such a structure seems perfectly reasonable. For example, Hartree-Fock calculations on CaO indicate that the ground $^1\Sigma^+$ state may be an open-shell configuration[36]; preliminary results on SiC also indicate the strong possibility of an open-shell ground state.[37] In such cases, the HF method cannot be trusted, but a small CI that includes excitations among the valence orbitals should be reliable.

**a. Type of Bonding and Size of HF Error.** Two rather interesting trends are found in comparing the HF dipole errors for a variety of molecules. First, it is found that for molecules with *single* sigma bonds the error is only 0.1 to 0.2 D whereas for molecules with *multiple* or *pi* bonds the error is 0.4 to 0.6 D. The few cases that appear to violate this rule (NaBr, RbF, BH, and BF) are uncertain due to inadequate basis set, large experimental uncertainty, and/or unknown vibrational dependence. Because the CI correction to the HF dipole appears to be caused almost entirely by one singly excited configuration, it seems reasonable that some correlation can be found between the size error and the availability of a suitably low-lying charge transfer single excitation. Indeed the observed difference in dipole error between sigma- and pi-bonded systems is in accord with simple molecular orbital theories based on linear combinations of atomic orbitals which show the $\sigma$ to $\sigma^*$ energy splitting to be larger than the $\pi$ to $\pi^*$ splitting.[20] It should be noted that this picture may be somewhat oversimplified; using HF orbital energies, a few cases are found where the dipole error does not correlate with the relative $\pi$–$\pi^*$ and $\sigma$–$\sigma^*$ excitation energies. However, the distinction between sigma- and pi-bonded systems does appear to be quite generally valid and therefore useful in predicting the expected accuracy of a HF dipole moment calculation.

**b. Direction of HF Error.** The second interesting trend concerns the direction of the HF error. If the convention is adopted that a positive dipole implies the polarity $A^+B^-$, and if B is more electronegative[38] than A (which is the normal convention for writing AB), then the HF dipole is almost always algebraically larger than the experimental value. Apparent exceptions to this rule (see Section V) are LiBr, NaBr, AlF, BH, and BF; however, SCF values for the first two may not be near the HF limit, and the rest have large uncertainties in the experimental value. For CS the electronegativity difference is zero, but written in the usual order this rule is obeyed. Thus the HF dipole is such that the electropositive end of the molecule is too positive and the electronegative end is too negative. This is in keeping with the fact that the HF method tends to overestimate ionic contributions to the structure at the expense of covalent contributions.[1,39]

### 2. *Open-Shell States*

It is not possible to discuss open-shell states with as much precision as closed-shell states. This is due mainly to a lack of correspondingly accurate experimental data with which to compare the *ab initio* results. For example, CO $a^3\Pi$ is the only open-shell state for which the vibrational dependence of the dipole moment is known experimentally. There is, however, one obvious and important difference between closed- and open-shell HF results: the latter may contain substantially larger errors. For example the HF dipole moments of CN $X^2\Sigma^+$, CO $a^3\Pi$, and CS $A^1\Pi$ are in error by about 100% (0.6 to 1.0 D). As in closed-shell states, the size error in open-shell species is related to the availability of low-lying charge transfer excitations. In open-shell states, excitations among the occupied orbitals are possible, and these are usually much more favorable energetically than the excitations in closed-shell states, which must be from occupied to unoccupied orbitals. However, for hydrides the lowest available charge transfer is to hydrogen $2s$ or $2p$ orbitals which are at moderately high energies, and the HF dipole moment is fairly accurate; the restricted HF dipole for diatomic hydrides has been found consistently to be 0.1 to 0.2 D too large for both open- and closed-shell ground and excited states. For ionic species, such as LiO, the charge transfer excitations are again at moderately high energies, and the HF dipole moment error is expected to be small.

Because both single and double excitations may have coefficients greater than 0.1 in open-shell CI functions, higher excitations are expected to be significant. Triple excitations were indeed found to be important in CO $a^3\Pi$ and CS $A^1\Pi$, although not in CN $X^2\Sigma^+$. Therefore, higher-than-double excitations must be considered to obtain reliable open-shell CI dipole moments. It has proved effective to include a few triple and quadruple excitations selected by combining the most important single and double excitations. This technique can be justified by perturbation theory. As might be expected, the higher excitations found to be important are valence excitations and are among those necessary to correct for "internal" correlation.[25]

In order to achieve a dipole moment accuracy of 0.05 D with a CI limited to 200 configurations, it has been necessary to use the iterative natural orbital (INO) method to improve convergence.[33] The INO method partially obviates the necessity of starting with the HF orbitals; however, if available, they appear to be the best choice. For a large enough CI calculation, the INO method will not be necessary.[31,40]

Errors in open-shell calculations may be summarized as follows. If there are no available low-lying charge transfer states (as is true for hydrides and highly ionic species) the HF dipole moment is likely to be in error by less than 0.3 D. Otherwise, the error may be much larger (0.5 D or more); however, a small CI containing only valence excitations reduces this error to a few tenths of a debye. In any case, moderate-size CI calculations, as described in Ref. 5, can reduce the error to less than about 0.05 D.

### 3. *Higher Excited States of a Symmetry*

The usual HF method is applicable only to the lowest state of a symmetry; for higher excited states it must be modified to ensure orthogonality to the HF functions of all lower states. In a CI calculation, the $n$th lowest eigenvalue is an upper bound to the energy of the $n$th lowest level, and the eigenfunction is an approximation to the wave function. There have been very few dipole moment determinations for such states, either experimental or theoretical. (See, however, CN $B^2\Sigma^+$ in Ref. 7.) For one-electron operator properties, it is important to recall that in a CI, errors in lower roots propagate to higher roots. Therefore, if the $n$th root is of interest, it is imperative to consider simultaneously all lower roots to at least the same degree of accuracy as is desired in the *n*th root.

## III. EXTENSION TO POLYATOMIC MOLECULES

The results presented in Section II would be interesting but of relatively limited utility if they applied only to diatomics. Although all the empirical comparisons in the previous section were based on diatomic data, the theoretical treatment was quite general and there is evidence that the conclusions about HF errors for diatomics may be extrapolated to polyatomic molecules in straightforward ways. This capability is quite important since SCF results with adequate basis sets are rapidly becoming available for larger systems. To demonstrate the ways in which the diatomic results apply to polyatomic systems, a selection of SCF results employing DZ + P or better basis sets is presented here. It should be emphasized that these examples have been chosen more or less at random and are not intended to represent a complete survey of existing results.

Dipole moments for linear polyatomic molecules have been calculated by McLean and Yoshimine[21] using Slater-type basis sets of varying quality. In particular they report DZ + P and "accurate atom plus polarization" values for HCN, NNO, FCN, and HCCF. The "accurate" SCF (and experimental)[8] values, in debye, are for HCN, 3.29 (2.98); for NNO, 0.64 (0.17); for FCN, 2.28 (2.17 $\pm$ 0.1); and for HCCF, 0.85 (0.75 $\pm$ 0.05). In each case the DZ + P basis was within 0.1 D of the larger basis, which is in accord with the observation that DZ + P values are within a few tenths of a debye of the HF limit for diatomic molecules. Just as for diatomic molecules, the SCF value is found to be too large for polyatomic molecules. The four species reported above are closed-shell ground states and the SCF dipoles are between 0.1 and 0.5 D too large. Unlike the diatomic cases, however, the distinction between sigma and pi bonds is somewhat unclear here; further study of the correlation between type of bonding and dipole error in polyatomic molecules seems warranted. McLean and Yoshimine[21] also report DZ + P values (experimental value in parentheses) for SCO, 0.99 (0.71); for ClCN, 3.05 (2.82 $\pm$ 0.05); for HCCCl, 0.42 (0.44 $\pm$ 0.05); and for HCCCN, 4.13 (3.72 $\pm$ 0.04).

For nonlinear molecules almost all calculations employ Gaussian orbitals. Several calculations are available with DZ + P basis sets, but only a few have employed larger sets. Two systems that seem representative of sigma-bonded molecules are $H_2O$ and $H_3SiCH_3$. For water the calculated dipole is 2.04 D at the experimental $R_e$ and 2.00 D at the calculated $R_e$,[41] both of which are about 0.2 D larger than the experimental value of 1.85 D. Methylsilane is interesting because, although Si is more electropositive than C, Shoemaker and Flygare[42] have found that the polarity is

$Si^-C^+$. SCF calculations[43] also indicate this polarity, and the SCF dipole of −0.58 D is algebraically larger than the experimental value of −0.735 D by 0.15 D. A typical pi-bonded system for which good *ab initio* calculations are available is $H_2CO$.[44] One would expect the bonding in this system to be rather similar to that in CO, and indeed, the HF dipole moment of 2.83 D is about 0.5 D larger than the experimental value, an error that is very similar to that found in CO. The ozone molecule can be described as a resonance structure with one sigma and one pi bond. SCF calculations[45] give a dipole of 0.82 D; this is 0.3 D larger than the experimental value, an error halfway between that expected for sigma and that expected for pi systems. The $NO_2$ molecule is similar to $O_3$ with one electron removed giving an open-shell ground state. SCF calculations are available[45] which are comparable to those for $O_3$, but for $NO_2$ the SCF dipole of 0.79 D is 0.5 D larger than experiment. The fact that the SCF error is larger for $NO_2$ than for $O_3$ is apparently a consequence of the open-shell structure of the former. As a final example, a DZ + P SCF on $SO_2$[46] gave a dipole moment of 2.28 D, which is 0.65 D larger than experiment. It is not apparent why the SCF error here is so much larger than that found in the isovalent $O_3$. One possibility, however, is that the basis set for S may not be as good as that used for O; this would cause a basis set error in the same direction as the HF error and could explain the large observed SCF error in $SO_2$.

These examples indicate that the HF dipoles for closed-shell polyatomic molecules are 0.1 to 0.5 D *too large*, just as they are for closed-shell diatomic molecules. There is some indication that pi-bonded systems are subject to larger errors than are sigma-bonded systems; however, with several bonds, the correlation is not so clear-cut as in the diatomic case. Also, open-shell states appear to have larger HF errors than closed-shell states for polyatomic as well as diatomic molecules. This overall similarity at the HF level makes it reasonable to assume that CI dipole moment calculations for polyatomic molecules can be done successfully with the same techniques found to be effective for diatomic molecules.

## IV. RESULTS OF NEW CALCULATIONS

### A. BH $A^1\Pi$

SCF calculations were done using the basis set of Cade and Huo[47] which was optimized for the ground state. This procedure has been found to be effective for excited states of CH and NH.[48] Exponents for pi functions were taken from the sigma set. The calculation was done at $R_e(A^1\Pi) =$ 2.316 a.u. The SCF energy and dipole moment are included in Section V.

The SCF $X - A$ transition energy is $T_e = 20050$ cm$^{-1}$, which may be compared with the experimental value of 23105 cm$^{-1}$.

### B. SiS $X^1\Sigma^+$

The electronic structure of SiS is expected to be similar to that found in the isovalent species, CO, CS, and SiO. Both theoretical and experimental results are available for the latter molecules, and it seemed desirable to complete the series by performing calculations on SiS. Consequently, SCF calculations were performed at the experimental $R_e = 3.645$ a.u. A $(28\sigma, 16\pi)$ basis set was constructed from the Bagus, Gilbert, and Roothaan[19] "nominal" atomic bases plus polarization functions—Si $3d(2.4, 0.9)$, Si $4f(1.8)$, S $3d(2.6, 1.1)$, and S $4f(2.2)$. No optimization was performed, and an SCF energy of $-686.5106$ a.u. was obtained. The dissociation energy with respect to SCF atoms is $D_e^\circ = 4.13$ eV which may be compared with the experimental value of $D_0^\circ = 6.5$ eV. A selection of calculated properties is presented in Table I. The calculated dipole of 2.205 D is about 0.5 D larger than the experimental value; this is the expected error in a closed-shell pi-bonded system and is quite similar to the SCF errors found for CO, CS, and SiO. It should be noted that a

TABLE I

Molecular Properties of SiS $X^1\Sigma^+$ Calculated From an SCF Function at $R_e = 3.645$ a.u. (All Values are Given in Atomic Units)

| Property | Value |
|---|---|
| $\langle 1/r_{Si} \rangle$ | 53.6051 |
| $\langle 1/r_S \rangle$ | 63.0824 |
| $\langle (3\cos^2\theta_{Si} - 1)/r_{Si}^3 \rangle$ | 2.7242 |
| $q_{Si}$ [a] | $-2.0634$ |
| $\langle (3\cos^2\theta_S - 1)/r_S^3 \rangle$ | 1.1936 |
| $q_S$ [a] | $-0.6154$ |
| $\langle \cos\theta_{Si}/r_{Si}^2 \rangle$ | 1.1966 |
| $F_{Si}$ [b] | $-0.1080$ |
| $\langle \cos\theta_S/r_S^2 \rangle$ | $-1.0539$ |
| $F_S$ [b] | $-0.0026$ |
| $\langle r_{Si}^2 \rangle$ | 273.5466 |
| $\langle r_S^2 \rangle$ | 240.6493 |
| $\langle z_{Si}^2 \rangle$ | 235.1690 |
| $\langle z_S^2 \rangle$ | 202.2717 |
| $\langle z_S \rangle$ | 59.1876 |

[a] Field gradient at the nucleus.
[b] Hellmann-Feynman force on the nucleus.

## TABLE II

Summary of Calculations on ClF $X^1\Sigma^+$ at $R_e = 3.0766$ a.u. For Basis 1 the Two Values Quoted for Properties are SCF/CI; for Bases 2 and 3 Only SCF Values Were Calculated (Values in a.u. Unless Otherwise Noted)

| | Basis 1 ($23\sigma$, $13\pi$, $4\delta$) | Basis 2 ($27\sigma$, $15\pi$) | Basis 3 ($30\sigma$, $18\pi$) |
|---|---|---|---|
| Chlorine | | | |
| 1*s* | (18.0) | (17.36, 28.02) | (17.362, 28.02) |
| 2*s* | (6.78, 15.6) | (6.78, 15.6) | (6.78, 15.6) |
| 3*s* | (2.97, 1.86, 6.0) | (2.97, 1.86, 6.0) | (2.973, 1.859, 6.0) |
| 2*p* | (8.0) | (7.68, 14.195) | (8.455, 14.458, 5.298) |
| 3*p* | (2.62, 1.46, 5.92) | (2.62, 1.46, 5.92) | (2.623, 1.459, 5.922, 0.9) |
| 3*d* | (1.2, 2.0, 3.0) | (1.2,[a] 2.0, 3.0[a]) | (1.2, 2.0, 3.0) |
| 4*f* | (2.45[b]) | (2.45[a]) | (2.45) |
| Fluorine | | | |
| 1*s* | (9.0) | (8.86, 14.41) | (8.809, 14.413) |
| 2*s* | (3.27, 8.07, 2.0) | (3.27, 8.07, 2.0) | (3.259, 8.02, 1.996) |
| 2*p* | (2.36, 4.74, 1.32) | (2.36, 4.74, 1.32) | (2.03, 3.911, 1.265, 8.636) |
| 3*d* | (1.5) | (1.5,[a] 3.0[a]) | (1.5, 3.0) |
| 4*f* | (2.1[b]) | (2.1[a]) | (2.1) |
| Energy | −558.8704/−559.1366 | −558.91627 | −558.91793 |
| $\mu$ (D) | 1.096/0.839 | 1.098 | 1.101 |
| $\Theta_{zz}$ [c,d] | 1.18/1.10 | 1.16 | 1.16 |
| $eqQ$ [c] (MHz) | −148.9/−141.6 | −151.66 | −152.65 |
| $F_{Cl}$ [e] | −0.775/−0.759 | −0.269 | −0.269 |
| $F_F$ [e] | 2.345/2.331 | 0.129 | −0.127 |
| $\langle 1/r_{Cl} \rangle$ | 67.2115/67.2239 | 67.2461 | 67.2462 |
| $\langle 1/r_F \rangle$ | 32.0484/32.0131 | 32.0607 | 32.0618 |

[a] Value optimized in this basis set.

[b] Not included in $\delta$ orbital set.

[c] For $^{35}$ClF.

[d] In $10^{-26}$ esu cm$^2$.

[e] Force on the nucleus; positive direction from Cl to F.

basis set error of about 0.2 D might be expected in this calculation; however, results for the other members of this series indicate that an error of less than 0.1 D is likely here.

### C. ClF $X^1\Sigma^+$

A recent molecular Zeeman effect experiment[49] indicated a dipole direction of $Cl^-F^+$, which is contrary to chemical intuition. As discussed in Section II, the HF method is expected to give a dipole moment for this molecule which is 0.1 to 0.2 D too large, and such a calculation should unequivocally determine the dipole direction. Indeed, SCF calculations have been reported[50] which indicate that the experimental determination is incorrect. Employing a basis set chosen from optimized alkali halide calculations and including a large set of polarization functions (see Basis 2 in Table II) the SCF dipole is $+1.10$ D, which may be compared with the accurate molecular beam value of $\pm 0.888$ D. Because the calculated and experimental polarities differ, it seemed desirable to reconsider the possible sources of error in the calculation to see if some anomalously large effect might be present. Two new calculations have been done and they are reported here.

To check that the SCF dipole error is indeed not unusually large for ClF, a CI calculation was done. For reasons of economy the basis set size was reduced, mostly by eliminating functions which describe the core orbitals. (See Basis 1 in Table II.) Although such functions are energetically important, they do not significantly influence the one-electron operator properties of interest. (A similar conclusion was reached in the case of CS.[5]) The CI was performed in the usual manner for closed-shell states[3] including single and double excitations from the SCF function; a convergence level of a few hundredths of a debye was obtained as the number of double excitations was increased to a total of 200 configurations. The CI lowered the dipole moment by 0.26 D to a value of 0.84 D. Since the dipole derivative is positive,[50] vibrational averaging will increase this value somewhat for the $v = 0$ level giving excellent agreement with the experimental result. CI values for other properties are listed in Table II.

Evidence was obtained in these CI calculations that delta orbitals are significant in this molecule. In preliminary calculations, certain configurations involving delta functions were excluded. In particular, only $\sigma^2 \rightarrow \delta\bar{\delta}$ excitations were included. The 200-configuration result was $E = -559.1112$ a.u. and $\mu = 0.781$ D. From comparison with both the experimental value and the results in Table II in which all excitations to $\delta$ orbitals were included, the change here from SCF to CI dipole is apparently too large. If $\delta$ orbitals are important, excluding them would be expected to cause such an overcorrection. It thus appears that higher angular correlation may be important for molecules containing third-row atoms, although it was not important for molecules containing only first- and second-row atoms.[5] (From previously reported calculations[3,5] on CS $X^1\Sigma^+$ the CI dipole corrected for basis set error is 2.09 D, which is 0.12 D above the experimental value. Although not noted previously, it is reasonable to assume that lack of $\delta$ orbitals is responsible here also for too large a change in going from SCF to CI.)

Finally, to check for possible basis set inadequacies in the ClF calculation, a larger ($30\sigma$, $18\pi$) basis was constructed by adding both valence and polarization function to the Clementi[18] atomic bases. (See Basis 3 in Table II.) This gave an SCF energy 0.0017 a.u. lower than Basis 2, but the dipole moment changed by less than 0.01 D. It appears that the basis sets used are, conservatively, within 0.1 D of the HF limit and that the HF dipole is about 0.2 D too large, as anticipated. The uncertainty due to lack of vibrational averaging is an additional 0.1 D, and other sources of error (e.g., Born-Oppenheimer breakdown and relativistic effects) are expected to be negligible. Therefore the *ab initio* results are felt to be quite conclusive concerning the direction of the ClF dipole moment.

### D. SO $X^3\Sigma^-$, $a^1\Delta$, and $b^1\Sigma^+$

The electronic structure of SO is expected to be similar to $O_2$. Indeed, as in $O_2$, the ground state of SO is $X^3\Sigma^-$ corresponding to the open-shell

TABLE III

Basis Sets Used in Calculations on SO. The Smaller Basis was Used for SCF and CI Calculations on the Ground State. The Larger Basis was Used for SCF Calculations on the Ground and Valence Excited States

| Center *nl* | ($24\sigma$, $14\pi$) Basis | ($30\sigma$, $18\pi$) Basis |
|---|---|---|
| Sulfur | | |
| 1*s* | 17.6 | 15.488, 22.0 |
| 2*s* | 15.45, 5.362 | 14.069, 6.753 |
| 3*s* | 9.681, 2.587, 1.63 | 5.743, 2.611, 1.643 |
| 2*p* | 13.419, 7.752, 4.8 | 13.5, 7.176 |
| 3*p* | 2.324, 1.318 | 5.44, 2.328, 1.319, 0.8 |
| 3*d* | 1.2, 2.4 | 1.2, 2.4 |
| 4*f* | 2.1 | 2.2, 1.9 |
| Oxygen | | |
| 1*s* | 11.844, 7.311 | 13.224, 7.606 |
| 2*s* | 1.974 | 1.897, 3.144, 6.378 |
| 3*s* | 4.35 | — |
| 2*p* | 5.8, 2.87, 1.47 | 3.438, 1.796, 1.154, 7.907, 0.8 |
| 3*d* | 1.6, 2.8 | 1.6, 2.8 |
| 4*f* | 2.2 | 2.2 |

TABLE IV

Calculated Properties for SO $X^3\Sigma^-$ at $R_e = 2.7989$ a.u. The ($24\sigma$, $14\pi$) Basis set Listed in Table III was Used. The CI Values are From a 200 Configuration INO Wave Function. SCF Results From a Larger Basis Set are Given in Table V (All Values in Atomic Units, Unless Otherwise Noted)

| Property | SCF | CI |
|---|---|---|
| Energy | −472.40071 | −472.6361 |
| $\langle 1/r_S \rangle$ | 61.9718 | 61.9977 |
| $\langle 1/r_O \rangle$ | 28.0175 | 27.9918 |
| $\mu$ [a] | 2.081 | 1.274 |
| $\langle r_S^2 \rangle$ | 104.0927 | 103.4889 |
| $\langle r_O^2 \rangle$ | 162.1811 | 163.3546 |
| $\langle z_S^2 \rangle$ | 78.1736 | 77.2106 |
| $\langle z_O^2 \rangle$ | 136.2620 | 137.0763 |
| $\langle P_2(\cos\theta_S)/r_S^3 \rangle$ | 0.1059 | −0.0264 |
| $\langle P_2(\cos\theta_O)/r_O^3 \rangle$ | 0.3586 | 0.4485 |
| $\langle P_1(\cos\theta_S)/r_S^2 \rangle$ | 1.0158 | 1.0108 |
| $\langle P_1(\cos\theta_O)/r_O^2 \rangle$ | −2.0636 | −2.0586 |
| $F_S$ [b] | −0.0872 | −0.1669 |
| $F_O$ [b] | 0.1693 | 0.1298 |
| Spin density at S | 0.0 | 0.1105 |
| Spin density at O | 0.0 | 0.3017 |

[a] Dipole moment in debye. Experimental value is $1.55 \pm 0.02$.
[b] Force on the nucleus; positive value from S to O.

configuration $1\sigma^2 2\sigma^2 3\sigma^2 4\sigma^2 5\sigma^2 6\sigma^2 7\sigma^2 1\pi^4 2\pi^4 3\pi^2$. This configuration also gives rise to the $a^1\Delta$ and $b^1\Sigma^+$ states. The dipole moment has been measured in both the $X$ and the $a$ states. Calculations have not been performed previously for this species; however, the $\pi^2$ structure leads to several low-lying single excitations, and the HF method seemed likely to give dipole moments in error by half a debye or more. Calculations on these states were undertaken to determine the adequacy of the HF dipole for the $X$ and $a$ states, to predict a dipole for the $b$ state, and to calculate other properties for these states. In addition to SCF calculations for all three states, CI calculations were performed for the ground state.

An initial basis set was constructed by adding polarization functions to the Bagus, Gilbert, and Roothaan[19] "nominal" atomic set. With this ($24\sigma$, $14\pi$) basis, the SCF function was calculated for the $X$ state at $R_e(X) = 2.7989$ a.u. Starting with the SCF orbitals, CI calculations were performed in the usual manner,[5] including single, double, and a few triple excitations and performing four cycles of natural orbital iteration. The basis set is given in Table III and the SCF and CI results for the ground state are given in Table IV. The CI dipole moment for the $X$ state is seen to be about 0.3 D lower than the experimental value. This error is consistent with the uncertainties expected from basis set error (±0.2 D for "nominal" plus polarization) and lack of proper vibrational averaging (±0.1 D). It is probable that part of the error is also due to lack of delta functions in the CI calculation.

An interesting demonstration of the importance of basis set was discovered accidentally in the course of these calculations, and it seems instructive to describe this here. Owing to a keypunch error, the calculations described in the previous paragraph were performed initially with a basis set differing from that in Table III only by O $2p(1.1)$ in place of O $2p(1.47)$. The SCF (and CI) results were $E = -472.3767$ $(-472.6099)$ a.u. and $\mu = 2.716$ (1.873) D. These results may be compared with those in Table IV which were obtained by exactly repeating the calculation with the corrected orbital exponent. This comparison illustrates two points very nicely. First, the rather small basis set "error" introduces a rather large error in the SCF energy (0.024 a.u.) and dipole moment (0.6 D). Second, the basis set error is the same at the SCF and CI levels in agreement with the theory presented in Ref. 6.

TABLE V

Results of SCF Calculations for Valence States of SO. The ($30\sigma$, $18\pi$) Basis Set is Given in Table III (All Values are in Atomic Units Unless Otherwise Noted). Experimental Values are Given in Parentheses

| Property | $X^3\Sigma^-$ | $a^1\Delta$ | $b^1\Sigma^+$ |
|---|---|---|---|
| $R_e$ | (2.7989) | (2.8233) | (2.8354) |
| Energy | −472.40260 | −472.36228 | −472.32441 |
| $T_e$ [a] | — | 8848. | 17159. |
| | | (6350.) | (10510.) |
| $\mu$ [b] | 2.224 | 1.997 | 1.833 |
| | (1.55 ± 0.02) | (1.31 ± 0.04) | |
| $\langle 1/r_S \rangle$ | 61.9693 | 61.9430 | 61.9276 |
| $\langle 1/r_O \rangle$ | 28.0164 | 27.9494 | 27.9130 |
| $\langle r_S^2 \rangle$ | 104.2689 | 105.5045 | 106.2581 |
| $\langle r_O^2 \rangle$ | 162.0422 | 164.8350 | 166.4838 |
| $\langle z_S^2 \rangle$ | 78.3297 | 79.1947 | 79.6284 |
| $\langle z_O^2 \rangle$ | 136.1029 | 138.5252 | 139.8541 |
| $\langle P_2(\cos\theta_S)/r_S^3 \rangle$ | 0.1214 | −0.0007 | −0.0611 |
| $\langle P_2(\cos\theta_O)/r_O^3 \rangle$ | 0.3415 | 0.3857 | 0.4238 |
| $F_S$ [c] | −0.126 | −0.111 | −0.106 |
| $F_O$ [c] | −0.141 | −0.150 | −0.153 |

[a] Excitation energy in $cm^{-1}$.
[b] Dipole moment in debye.
[c] Force on the nucleus; positive direction is S to O.

The largest error in the calculations described above was expected to be from the limited basis set. Therefore SCF calculations were undertaken employing a much larger basis. The atomic functions of Clementi[18] were augmented with both additional valence and polarization functions. After some experimentation with these functions the ($30\sigma$, $18\pi$) set listed in Table III was adopted. Judging from the changes found in incrementally increasing the basis, this set is estimated to be within 0.003 a.u. and 0.05 D of the HF limit. SCF calculations were performed for the $X$, $a$, and $b$ states, and results are given in Table V. Using the theory of Ref. 6 to account for basis set error gives a "corrected CI" dipole moment of 1.42 D for the ground state. Again the remaining error is within the expected uncertainties due to basis set error and lack of vibrational averaging. However, it seems probable that lack of delta functions is responsible for much of the error; as discussed in Section II.B, lack of delta functions would lead to an overcorrection from the SCF to the CI dipole as is found here. It did not seem worth the additional effort necessary to pin down these remaining errors since the calculations are adequate to answer the question of primary interest, which is the size error in the SCF dipole. As suspected, the SCF error is more than half a debye; in fact, in both the $X$ and $a$ states the SCF dipole is 0.7 D too large. It seems likely that the same trend will be found for the $b$ state leading to a predicted dipole of $1.15 \pm 0.1$ D for this state.

### E. CO $a'^3\Sigma^+$

The dipole moment of CO $a'^3\Sigma^+$ has been determined recently in an elegant molecular beam experiment.[14] This state is interesting as a case in which spin-orbit (relativistic) interactions are not small. The $a'$ state interacts strongly with the $a^3\Pi$ state and, indeed, the $^3\Sigma^+$ dipole moment was obtained by analyzing perturbations in the $^3\Pi$ data. The actual levels are of course mixtures of both states. If enough data are available they can be analyzed in terms of the "pure" states, and "deperturbed" molecular constants may be obtained for both such states.[51] The correct *ab initio* approach is the reverse of this: the relativistic and/or Born-Oppenheimer violating terms that cause the perturbation must be included in the Hamiltonian to obtain mixing of states. In this way properties can be calculated that correspond directly to the experimental observations. It is important to note that, although often small in magnitude, such perturbations are universally present, and this causes some ambiguity in the comparison of theoretical and experimental values. Either the experimental data must be deperturbed or else the perturbations must be included in the *ab initio* calculation to a degree consistent with the level of accuracy expected in the comparison. Although this is generally not a problem with the current level of *ab initio* accuracy, it is an effect that should be considered as a possible source of discrepancy between theoretical and experimental results.[9]

TABLE VI
Results of SCF Calculations on CO $a'^3\Sigma^+$ at $R_e(a') = 2.5547$ a.u. (All Values are in Atomic Units Unless Otherwise Noted)

| Property | Value |
|---|---|
| Energy | −112.58006 |
| $\mu$ (debye) | −0.664 |
| $\langle 1/r_C \rangle$ | 17.7834 |
| $\langle 1/r_O \rangle$ | 24.5656 |
| $\langle r_C^2 \rangle$ | 77.3184 |
| $\langle r_O^2 \rangle$ | 65.6020 |
| $\langle z_C^2 \rangle$ | 61.4926 |
| $\langle z_O^2 \rangle$ | 49.7762 |
| $\langle P_2(\cos \theta_C)/r_C^3 \rangle$ | 0.6677 |
| $\langle P_2(\cos \theta_O)/r_O^3 \rangle$ | 0.5465 |
| $F_C$ [a] | 0.0992 |
| $F_O$ [a] | −0.1216 |

[a] Force on the nucleus; positive direction from C to O.

Calculations on CO $a^3\Pi$ provided the first indication that HF dipole moments might not be adequate for open-shell states.[52] It therefore seemed interesting to see how adequate the HF dipole might be for the $a'^3\Sigma^+$ state. Accordingly, SCF calculations were performed by augmenting Clementi's[18] atomic bases with C $3d(1.8, 2.8)$, C $4f(2.0)$, O $3d(1.4, 2.38)$, and O $4f(2.0)$ to give a $(24\sigma, 14\pi)$ basis set. Results are presented in Table VI. The computed dipole moment is −0.66 D, which is only about 0.4 D larger (algebraically) than the experimental value. The expansion error is estimated from previous experience with CO $a^3\Pi$ basis sets to be less than 0.1 D. Thus the HF dipole error is substantially less for the $a'$ than for the $a$ state. It is not obvious why this is true since the $\pi^3\pi$ configuration of the $a'$ state should give rise to many low-lying single excitations. It is possible, however, that excitations out of both open pi shells give large, but opposing, correction terms. CI calculations should be able to clarify this point.

## V. TABULATION OF DIATOMIC RESULTS

The conclusions in this paper about *ab initio* accuracy for dipole moments were based on comparison of the calculated and experimental values which are presented in Tables VII and VIII. This is believed to be a complete compilation of available results for diatomic molecules. Closed-shell ground states and open-shell states are listed separately, in each case according to increasing number of electrons. The diatomic species AB is listed such that B is more electronegative than A. A positive dipole moment implies $A^+B^-$; the polarity is rarely determined experimentally and exceptions are noted. Experimental error estimates are taken from the original reference; if none are given, the moment is accurate to all figures quoted. The symbol ($v$) following an experimental value indicates that vibrational data have been extrapolated to $v = -\frac{1}{2}$. Calculations were done at the experimental $R_e$ unless otherwise noted. It should be recalled that the uncertainty from lack of correct vibrational averaging is about 0.05 to 0.1 D.

Because of the influence of basis set on *ab initio* accuracy, an indication of the basis set quality is given. The following abbreviations are used: "opt" and "part opt" indicate that the functions that follow have been optimized or partially optimized in the *molecular* calculation. Most of these calculations started with atomic bases, referred to as "acc" for accurate (from Ref. 18 or 19), DZ for double zeta (from Ref. 18), or "nom" for nominal (from Ref. 19). All calculations included polarization (P) functions. Unless further optimization has been performed, "DZ + P" or "nom + P" functions may have basis set errors of about 0.2 D. The symbol "→HF" is used in some cases to indicate that the basis is believed to be near (within 0.03 D) the HF limit. CI results are listed on the same line as SCF results with the same or a comparable basis set. When possible, the SCF energy is given to compare with other (future) calculations for basis set accuracy. The term "corr CI" implies that the theory of Ref. 6 has been used to correct a CI dipole moment for a known basis set error.

All dipole moments are in debye. All energies and internuclear distances are given in atomic units.

## VI. CONCLUDING REMARKS

In the past few years there has been a great deal of progress in the ability of quantum mechanics to produce chemically useful results. Unfortunately many experimentalists have a great distrust of such theoretical results and this has severely limited their application. Much of the fault lies with the theoreticians who often present (possibly) erroneous results with no indication of the sources of error or the expected accuracy. The purpose of this paper has been to demonstrate that these errors (at least for diatomic dipole moments and probably in general) are neither unexpected nor unpredictable. It is hoped that this will serve several

TABLE VII

Diatomic Results for Closed-Shell Ground States

| Molecule | SCF | CI | Experimental results | Notes |
|---|---|---|---|---|
| LiH $X^1\Sigma^+$ | 5.990[53] | 5.853[53] | 5.828 ($v$)[54] | $E_{SCF} = -7.9870$, $E_{CI} = -8.0600$ |
| | 6.002[55] | 5.843[56] | | Opt ($16\sigma$) $\rightarrow$ HF; $E_{SCF} = -7.9873$, $E_{CI} = -8.0591$ |
| | 6.001[22] | 5.814[22] | | $\rightarrow$ HF; $E_{CI} = -8.0682$ |
| BH $X^1\Sigma^+$ | −1.733[55] | −1.470[53] | 1.27 ± 0.21[57] | $R_e = 2.336$; $\rightarrow$ HF; $E_{SCF} = -25.1314$ |
| HF $X^1\Sigma^+$ | 1.934[58] | 1.816[53] | 1.826[59] | Expt $v = 0$; opt ($16\sigma, 8\pi$) $\rightarrow$ HF; $E_{SCF} = -100.0705$ |
| LiF $X^1\Sigma^+$ | 6.453[21] | | 6.284 ($v$)[60] | Part opt acc + P $\rightarrow$ HF; calculation of Ref. 21 repeated at $R_e = 2.9553$, $E_{SCF} = -106.99180$ |
| NaLi $X^1\Sigma^+$ | 0.94[3] | 0.99[3] | 0.463[61] | Expt $v = 0$; calc at $R = 5.5$; ($19\sigma, 9\pi$), acc + P; $E_{SCF} = -169.2928$ |
| | 0.70[62] | | | $\rightarrow$ HF; $R_e = 5.33$ |
| BF $X^1\Sigma^+$ | −0.88[21] | | 0.5 ± 0.2[63] | ($22\sigma, 12\pi$) $\rightarrow$ HF; $E_{SCF} = -124.1671$ |
| CO $X^1\Sigma^+$ | 0.280[21] | −0.121[3] | −0.112 ± 0.005[64] | ($22\sigma, 12\pi$), acc + opt P $\rightarrow$ HF; $E_{SCF} = -112.7891$ |
| HCl $X^1\Sigma^+$ | 1.215[58] | 1.043[65] | 1.093 ($v$)[12] | ($23\sigma, 13\pi$) $\rightarrow$ HF; $E_{SCF} = -460.11185$ |
| LiCl $X^1\Sigma^+$ | 7.204[16a] | | 7.075 ($v$)[66] | $\mu$ interpolated to $R_e = 3.819$; ($22\sigma, 10\pi$) acc + P |
| NaF $X^1\Sigma^+$ | 8.354[16b] | | 8.124 ($v$)[67] | $\mu$ interpolated to $R_e = 3.639$; ($23\sigma, 11\pi$) acc + P |
| AlF $X^1\Sigma^+$ | 1.34[21] | | 1.53 ± 0.10[68] | ($25\sigma, 13\pi$) acc + P; $E_{SCF} = -341.4832$ |
| CS $X^1\Sigma^+$ | −1.56[69] | −2.03[3] | −1.97 ($v$)[70] | Ref. 75 implies $\mu < 0$; $E_{SCF} = -435.3317$ |
| | −1.63[5] | | | ($30\sigma, 17\pi$); $E_{SCF} = -435.3601$; corr CI $\mu = -2.09$ |
| PN $X^1\Sigma^+$ | 3.24[21] | | 2.751 ($v$)[71] | ($25\sigma, 13\pi$) acc + opt P; $E_{SCF} = -395.1848$ |
| SiO $X^1\Sigma^+$ | 3.68[21] | | 3.088 ($v$)[72] | ($25\sigma, 13\pi$) acc + opt P; $E_{SCF} = -363.8517$ |
| ClF $X^1\Sigma^+$ | 1.101 | | 0.888[73] | Ref 49 implies $\mu < 0$; expt $v = 0$; ($30\sigma, 18\pi$) acc + opt P; $E_{SCF} = -558.91793$ |
| | 1.096 | 0.839 | | ($23\sigma, 13\pi$); $E_{SCF} = -558.8704$ |
| KF $X^1\Sigma^+$ | 8.689[16d] | | 8.558 ($v$)[74] | ($27\sigma, 13\pi$) part opt nom + P; $E_{SCF} = -698.6850$ |
| SiS $X^1\Sigma^+$ | 2.21 | | 1.74 ± 0.07[75] | Expt $v = 0$; ($28\sigma, 16\pi$) nom + P; $E_{SCF} = -686.5106$ |
| KCl $X^1\Sigma^+$ | 10.456[16e] | | 10.238 ($v$)[76] | ($27\sigma, 13\pi$) DZ + P; $E_{SCF} = -1058.7583$ |
| LiBr $X^1\Sigma^+$ | 7.054[16e] | | 7.226 ($v$)[77] | $\mu$ interpolated to $R_e = 4.102$; ($25\sigma, 13\pi, 2\delta$) DZ + P |
| NaCl $X^1\Sigma^+$ | 9.10[16c] | | 8.973 ($v$)[78] | $\mu$ interpolated to $R_e = 4.4601$; ($27\sigma, 13\pi$) acc + P |
| NaBr $X^1\Sigma^+$ | 8.72[16f] | | 9.092 ($v$)[76] | ($29\sigma, 15\pi, 2\delta$) DZ + P; $E_{SCF} = -2734.2876$ |
| RbF $X^1\Sigma^+$ | 8.76[16f] | | 8.513 ($v$)[76] | ($29\sigma, 15\pi, 2\delta$) DZ + P; $E_{SCF} = -3037.7727$ |

## TABLE VIII

Diatomic Results for Open-Shell States

| Molecule | SCF | CI | Experimental results | Notes |
|---|---|---|---|---|
| BH $A^1\Pi$ | −0.590 | | 0.58 ± 0.04[79] | $(16\sigma, 8\pi)$ opt for $X$ state; $E_{SCF} = -25.0400$ |
| HC $X^2\Pi$ | 1.57[55] | 1.43[53] | 1.46 ± 0.06[80] | Opt $(16\sigma, 8\pi)$; $E_{SCF} = -38.2794$ |
| | 1.619[81] | 1.45[81] | | → HF; calc $\mu_{CI}$ $(v = 0) = 1.41$ |
| HN $a^1\Delta$ | 1.64[48] | | 1.49 ± 0.06[82] | Opt $(16\sigma, 8\pi)$; $E_{SCF} = -54.91095$ |
| $A^3\Pi$ | 1.34[48] | | 1.31 ± 0.03[82] | Opt $(16\sigma, 8\pi)$; $E_{SCF} = -54.83697$ |
| $c^1\Pi$ | 1.85[48] | | 1.70 ± 0.07[82] | Opt $(16\sigma, 8\pi)$; $E_{SCF} = -54.75710$ |
| HO $X^2\Pi$ | 1.78[55] | 1.63[53] | 1.66 ± 0.01[83] | Opt $(16\sigma, 8\pi)$; $E_{SCF} = -54.9781$ |
| HO $A^2\Sigma^+$ | 1.959[9] | | 1.97 ± 0.08[84] | Expt $v = 0$; $(27\sigma, 15\pi)$ → HF; $E_{SCF} = -75.26595$ |
| DO $A^2\Sigma^+$ | 1.989[9] | 1.834[9] | 2.16 ± 0.08[84] | Expt $v = 0$; part opt $(18\sigma, 10\pi)$; $E_{SCF} = -75.26586$ |
| | | | 1.72 ± 0.10[85] | Ref. 9 gives corr CI $\mu$ (OD, $v = 0$) = 1.84 ± 0.06 |
| LiO $X^2\Pi$ | 6.87[86] | 6.76[86] | 6.84 ± 0.03[87] | Expt $v = 0$; calc at $R_e$ (CI) = 3.20; $(36\sigma, 22\pi)$ → HF; $E_{SCF} = -82.3115$, calc $\mu_{CI}$ $(v = 0) = 6.80$ |
| CN $X^2\Sigma^+$ | 2.301[7] | 1.465[7] | 1.45 ± 0.08[88] | $(22\sigma, 14\pi)$ acc + opt P; $E_{SCF} = -92.2232$; $R_e = 2.2143$ |
| $B^2\Sigma^+$ | — | −0.958[7] | −1.15 ± 0.08[88] | Calc at $R_e(X)$; $R_e(B) = 2.1742$; only relative $X - B$ polarity known experimentally |
| CO $a^3\Pi$ | 2.46[52] | 1.55[5] | 1.374 $(v)$[89] | Opt $(16\sigma, 8\pi)$; $E_{SCF} = -112.5742$ |
| | 2.34[6] | 1.43[6] | | Part opt $(26\sigma, 14\pi)$; $E_{SCF} = -112.5822$ |
| | 2.30[5] | | | $(36\sigma, 22\pi)$ → HF; $E_{SCF} = -112.5832$; corr CI $\mu = 1.39$ |
| $a'^3\Sigma^+$ | −0.66 | | −1.06 ± 0.2[14] | $(24\sigma, 14\pi)$ acc + P; $E_{SCF} = -112.5801$ |
| CF $X^2\Pi$ | −0.42[90] | | 0.65 ± 0.05[91] | $(21\sigma, 11\pi)$ acc + P; $E_{SCF} = -137.2259$ |
| NO $X^2\Pi$ | 0.260[92] | −0.25[92] | 0.159[93] | Expt $v = 0$; $(28\sigma, 16\pi)$; $E_{SCF} = -129.2953$ |
| $A^2\Sigma^+$ | 0.62[92] | 0.4[92] | 1.12 ± 0.10[94] | Expt $v = 3$; $(28\sigma, 16\pi)$ opt Rydberg; $E_{SCF} = -129.10365$ |
| HS $X^2\Pi$ | 0.86[55] | | 0.62 ± 0.01[91] | Opt $(16\sigma, 8\pi)$; $E_{SCF} = -398.1015$ |
| CS $A^1\Pi$ | 0.09[5] | | −0.63 ± 0.04[95] | $(30\sigma, 17\pi)$; $E_{SCF} = -435.1833$; $R_e = 2.9754$ |
| | 0.02[5] | −0.63[5] | | Calc at $R_e(X) = 2.89964$; $(19\sigma, 9\pi)$; $E_{SCF} = -435.1475$ |
| SN $X^2\Pi$ | 1.73[96] | | 1.86 ± 0.03[91] | $(23\sigma, 12\pi)$ acc + P; $E_{SCF} = -451.9329$ |
| SO $X^3\Sigma^-$ | 2.08 | 1.27 | 1.55 ± 0.02[97] | $(24\sigma, 14\pi)$ nom + P; $E_{SCF} = -472.4007$ |
| | 2.22 | | | $(30\sigma, 18\pi)$ acc + P; $E_{SCF} = -472.4026$; corr CI $\mu = 1.42$ |
| $a^1\Delta$ | 2.00 | | 1.31 ± 0.04[91] | $(30\sigma, 18\pi)$ acc + P; $E_{SCF} = -472.3623$ |
| SF $X^2\Pi$ | 1.40[98] | | 0.87 ± 0.05[91] | $(22\sigma, 11\pi)$ nom S/acc F + P, possible basis imbalance? $E_{SCF} = -496.9670$ |
| ClO $X^2\Pi$ | 0.81[99] | | 1.24 ± 0.01[100] | $(23\sigma, 12\pi)$ nom Cl/acc O + P, possible basis imbalance? $E_{SCF} = -534.2929$ |
| SeF $X^2\Pi$ | 2.21[98] | | 1.52 ± 0.05[91] | $(30\sigma, 17\pi, 6\delta)$ nom Se/acc F + P, possible basis imbalance? $E_{SCF} = -2499.3205$ |

purposes. First, for the theoretician who wishes to calculate an unknown dipole moment, an indication has been given of the level of approximation necessary for a desired degree of accuracy. Unfortunately, calculations are still being done where the level of approximation is determined by the amount of available computer time rather than the ability, or inability, of such a calculation to produce meaningful results. Second, this analysis should be a useful guide for those wishing to examine the errors in other calculated properties. Third, it is hoped that the information presented here will help spur theoreticians to report expected error limits along with computed properties. Only in this way will quantum chemists produce results that are generally useful to others. Finally, it is hoped that this paper will help the nontheoretician in assessing the reliability of published *ab initio* results. In particular, an attempt has been made to specify the sources of error and the expected size of each so that an experimentalist will know what questions to ask about a calculation.

## Acknowledgments

I would like to thank Prof. William Klemperer for directing my interest toward the calculation of molecular properties and Dr. Winifred Huo for providing early instruction in the art of molecular computation. The assistance of Mr. John Yeanacopolis in quotidian matters is gratefully acknowledged.

## References

1. J. C. Slater, *Quantum Theory of Molecules and Solids*, McGraw-Hill, New York, 1963.
2. H. F. Schaefer III, *The Electronic Structure of Atoms and Molecules: A Survey of Rigorous Quantum Mechanical Results*, Addison-Wesley, Reading, Mass., 1972; R. K. Nesbet, *Advan. Quantum Chem.*, **3,** 1 (1966).
3. S. Green, *J. Chem. Phys.*, **54,** 827 (1971).
4. S. Green, *J. Chem. Phys.*, **54,** 3051 (1971).
5. S. Green, *J. Chem. Phys.*, **56,** 739 (1972).
6. S. Green, *J. Chem. Phys.*, **57,** 2830 (1972).
7. S. Green, *J. Chem. Phys.*, **57,** 4694 (1972).
8. W. Gordy and R. L. Cook, *Microwave Molecular Spectra*, Wiley, New York, 1970; A. L. McClellan, *Tables of Experimental Dipole Moments*, Freeman, San Francisco, 1963; R. D. Nelson, D. R. Lide, Jr., and A. A. Maryott, "Selected Values of Electric Dipole Moments for Molecules in the Gas Phase," *Natl. Bur. Std.* (*U.S.*) *NSRDS-NBS10*, U.S. Government Printing Office, Washington, D.C., 1967.
9. S. Green, *J. Chem. Phys.*, **58,** 4327 (1973).
10. A. Dalgarno and R. A. McCray, *Ann. Rev. Astron. Astrophys.*, **10,** 375 (1972); A. Dalgarno, private communication.
11. M. Trefler and H. P. Gush, *Phys. Rev. Letters*, **20,** 703 (1968).
12. E. W. Kaiser, *J. Chem. Phys.*, **53,** 1686 (1970).

13. J. H. Van Vleck, *Phys. Rev.*, **33,** 467 (1929).
14. B. G. Wicke, R. W. Field, and W. Klemperer, *J. Chem. Phys.*, **56,** 5758 (1972).
15. W. M. Huo, *J. Chem. Phys.*, **43,** 624 (1965).
16. R. L. Matcha, *J. Chem. Phys.*, (*a*) **47,** 4595 (1967); (*b*) **47,** 5295 (1967); (*c*) **48,** 335 (1968); (*d*) **49,** 1264 (1968); (*e*) **53,** 485 (1970); (*f*) **53,** 4490 (1970).
17. D. D. Ebbings, *J. Chem. Phys.*, **36,** 1361 (1962).
18. E. Clementi, "Tables of Atomic Functions," *IBM J. Res. Develop. Suppl.*, **9,** 2 (1965); *J. Chem. Phys.*, **40,** 1944 (1964).
19. P. S. Bagus, T. L Gilbert, and C. C. J. Roothaan, *J. Chem. Phys.*, **56,** 5195 (1972).
20. M. Karplus and R. N. Porter, *Atoms and Molecules: An Introduction for Students of Physical Chemistry*, Benjamin, New York, 1970.
21. M. Yoshimine and A. D. McLean, *Intern. J. Quantum Chem.*, **1S,** 313 (1967).
22. M. Yoshimine, A. D. McLean, and H. J. Preston, private communication.
23. R. K. Nesbet, *Proc. Roy. Soc.* (*London*) *Ser. A*, **230,** 312 (1955).
24. S. Green, *J. Chem. Phys.*, **52,** 3100 (1970).
25. H. J. Silverstone and O. Sinanoglu, *J. Chem. Phys.*, **44,** 1899, 3608 (1966); for a somewhat different separation into "near-degeneracy" and "true correlation," see E. Clementi and A. Veillard, *J. Chem. Phys.*, **44,** 3050 (1966).
26. M. L. Brillouin, *Actual. Sci. Ind.*, **71** (1933); **159** (1934).
27. G. Das and A. C. Wahl, *J. Chem Phys.*, **47,** 2934 (1967).
28. U. Kaldor, *J. Chem. Phys.*, **48,** 835 (1968).
29. B. Levy and G. Berthier, *Intern. J. Quantum Chem.*, **2,** 307 (1968).
30. I. Shavitt, *J. Comput. Phys.*, **6,** 124 (1970).
31. C. F. Bender and E. R. Davidson, *J. Chem. Phys.*, **49,** 4222 (1968).
32. F. Grimaldi, A. Lecourt, and C. Moser, *Intern. J. Quantum Chem.*, **1S,** 153 (1967).
33. C. F. Bender and E. R. Davidson, *J. Phys. Chem.*, **70,** 2675 (1966).
34. F. Grimaldi, *Advan. Chem. Phys.*, **14,** 341 (1969).
35. C. F. Bender and E. R. Davidson, *Phys. Rev.*, **183,** 23 (1969).
36. K. D. Carlson, K. Kaiser, C. Moser, and A. C. Wahl, *J. Chem. Phys.*, **52,** 4678 (1970).
37. B. Lutz and J. Ryan, private communication.
38. L. Pauling, *The Nature of the Chemical Bond*, 3rd ed., Cornell University Press, Ithaca, N.Y., 1960.
39. S. Fraga and B. J. Ransil, *J. Chem. Phys.*, **36,** 1127 (1962).
40. S. Green, P. S. Bagus, B. Liu, A. D. McLean, and M. Yoshimine, *Phys. Rev. A*, **5,** 1614 (1972).
41. W. C. Ermler and C. W. Kern, *J. Chem. Phys.*, **55,** 4851 (1971).
42. R. L. Shoemaker and W. H. Flygare, *J. Am. Chem. Soc.*, **94,** 684 (1972).
43. D. H. Liskow and H. F. Schaefer, *J. Am. Chem Soc.*, **94,** 664 (1972).
44. D. B. Neumann and J. W. Moskowitz, *J. Chem. Phys.*, **50,** 2216 (1969).
45. S. Rothenberg and H. F. Schaefer, *Mol. Phys.*, **21,** 317 (1971).
46. S. Rothenberg and H. F. Schaefer, *J. Chem. Phys.*, **53,** 3014 (1970).
47. P. E. Cade and W. M. Huo, *J. Chem. Phys.*, **47,** 614 (1967).
48. W. M. Huo, *J. Chem. Phys.*, **49,** 1482 (1968).
49. J. J. Ewing, H. L. Tigelaar, and W. H. Flygare, *J. Chem. Phys.*, **56,** 1957 (1972).
50. S. Green, *J. Chem. Phys.*, **58,** 3117 (1973).
51. R. W. Field, S. G. Tilford, R. A. Howard, and J. D. Simmons, *J. Mol. Spectry.*, **44,** 347 (1972); R. W. Field, B. G. Wicke, J. D. Simmons, and S. G. Tilford, *J. Mol. Spectry.*, **44,** 383 (1972).
52. W. M. Huo, *J. Chem. Phys.*, **45,** 1554 (1966).

53. C. F. Bender and E. R. Davidson, *Phys. Rev.*, **183,** 23 (1969).
54. L. Wharton, L. P. Gold, and W. Klemperer, *Phys. Rev.*, **133,** B270 (1964).
55. P. E. Cade and W. M. Huo, *J. Chem. Phys.*, **45,** 1063 (1966).
56. S. Green, *Phys. Rev. A*, **4,** 251 (1971).
57. R. Thompson and F. W. Dalby, *Can. J. Phys.*, **47,** 1155 (1969).
58. A. D. McLean and M. Yoshimine, *J. Chem. Phys.*, **47,** 3256 (1967).
59. J. S. Muenter and W. Klemperer, *J. Chem. Phys.*, **52,** 6033 (1970).
60. L. Wharton, W. Klemperer, L. P. Gold, R. Strauch, J. J. Gallagher, and V. E. Deer, *J. Chem. Phys.*, **38,** 1203 (1963).
61. P. J. Dagdigian, J. Graff, and L. Wharton, *J. Chem. Phys.*, **55,** 4980 (1971).
62. A. C. Wahl, private communication.
63. F. J. Lovas and D. R. Johnson, *J. Chem. Phys.*, **55,** 41 (1971).
64. C. A. Burrus, *J. Chem. Phys.*, **28,** 427 (1958).
65. F. Grimaldi, A. Lecourt, and C. Moser, *Symp. Faraday Soc.*, **2,** p. 59.
66. D. R. Lide, P. Cahill, and L. P. Gold, *J. Chem. Phys.*, **40,** 156 (1964).
67. C. D. Hollowell, A. J. Hebert, and K. Street, *J. Chem. Phys.*, **41,** 3540 (1964).
68. D. R. Lide, *J. Chem. Phys.*, **42,** 1013 (1965).
69. W. G. Richards, *Trans. Faraday Soc.*, **63,** 257 (1967).
70. G. Winnewisser and R. L. Cook, *J. Mol. Spectry.*, **28,** 266 (1967).
71. J. Raymonda and W. Klemperer, *J. Chem. Phys.*, **55,** 232 (1971).
72. J. W. Raymonda, J. S. Muenter, and W. A. Klemperer, *J. Chem. Phys.*, **52,** 3458 (1970).
73. R. E. Davis and J. S. Muenter, *J. Chem. Phys.*, **57,** 2836 (1972).
74. R. van Wachem, F. H. de Leeuw, and A. Dymanus, *J. Chem. Phys.*, **47,** 2256 (1967).
75. A. N. Murty and R. F. Curl, *J. Mol. Spectry.*, **30,** 102 (1969).
76. A. J. Hebert, F. J. Lovas, C. A. Melendres, C. D. Hollowell, T. L. Story, and K. Street, *J. Chem. Phys.*, **48,** 2824 (1968).
77. A. J. Hebert, F. W. Breivogel, and K. Street, *J. Chem. Phys.*, **41,** 2368 (1964).
78. L. P. Gold, Ph.D. Thesis, Harvard University, 1962 (unpublished).
79. R. Thomson and F. W. Dalby, *Can. J. Phys.*, **47,** 1155 (1969).
80. D. H. Phelps and F. W. Dalby, *Phys. Rev. Letters*, **16,** 3 (1966).
81. G. Lie, B. Liu, and J. Hinze, *J. Chem. Phys.*, **59,** 1887 (1973).
82. T. A. R. Irwin and F. W. Dalby, *Can. J. Phys.*, **43,** 1766 (1965).
83. F. X. Powell and D. R. Lide, *J. Chem. Phys.*, **42,** 4201 (1965).
84. E. A. Scarl and F. W. Dalby, *Can. J. Phys.*, **49,** 2825 (1971).
85. E. Weinstock and R. N. Zare, *J. Chem. Phys.*, **58,** 4319 (1973).
86. M. Yoshimine, *J. Chem. Phys.*, **57,** 1108 (1972).
87. S. M. Freund, E. Herbst, R. Mariella, and W. Klemperer, *J. Chem. Phys.*, **56,** 1467 (1972).
88. R. Thomson and F. W. Dalby, *Can. J. Phys.*, **46,** 2815 (1968).
89. R. C. Stern, R. H. Gammon, M. E. Lesk, R. S. Freund, and W. Klemperer, *J. Chem. Phys.*, **52,** 3467 (1970).
90. P. A. G. O'Hare and A. C. Wahl, *J. Chem. Phys.*, **55,** 666 (1971).
91. C. R. Byfleet, A. Carrington, and D. K. Russell, *Mol. Phys.*, **20,** 271 (1971).
92. S. Green, *Chem. Phys. Lett.*, **13,** 552 (1972); S. Green, *Chem. Phys. Lett.*, to be published.
93. R. M. Neumann, *Astrophys. J.*, **161,** 779 (1970).
94. T. H. Bergeman and R. N. Zare, private communication.
95. R. W. Field and T. H. Bergeman, *J. Chem. Phys.*, **54,** 2936 (1971).

96. P. A. G. O'Hare, *J. Chem. Phys.*, **52,** 2992 (1970).
97. F. X. Powell and D. R. Lide, *J. Chem. Phys.*, **41,** 1413 (1964).
98. P. A. G. O'Hare and A. C. Wahl, *J. Chem. Phys.*, **53,** 2834 (1970).
99. P. A. G. O'Hare and A. C. Wahl, *J. Chem. Phys.*, **54,** 3770 (1971).
100. T. Amano, S. Saito, E. Hirota, Y. Morino, D. R. Johnson, and F. X. Powell, *J. Mol. Spectry.*, **30,** 275 (1969).

# ALGEBRAIC VARIATIONAL METHODS IN SCATTERING THEORY*

DONALD G. TRUHLAR, JOSEPH ABDALLAH, JR., AND RICHARD L. SMITH

*Department of Chemistry, University of Minnesota, Minneapolis, Minnesota*

## CONTENTS



* Supported in part by the National Science Foundation (grant no. GP-28684) and the Graduate School of the University of Minnesota.

# I. INTRODUCTION

The variational methods commonly used in scattering calculations differ in one very important respect from those commonly used in bound-state calculations. Calculations of approximate bound-state energies and the associated square-integrable wave functions are generally performed using a variational principle for the energy which provides an upper bound to the exact energy. For collision processes, however, the energy is given and the scattering parameters are to be obtained from the asymptotic part of the nonsquare-integrable wave function. The variational methods commonly used do not give rigorous upper or lower bounds to these scattering parameters. Instead the scattering variational principles are of the form that if a trial wave function is determined, then the scattering parameters can be variationally corrected to eliminate errors of first order in the inaccuracies of the trial wave function.

Algebraic variational methods are methods in which the wave function is expressed in terms of a function containing parameters and the problem is solved by obtaining the best values for the parameters. Often the function is a linear combination of predetermined basis functions. Functional variational methods are methods in which the wave function is determined without such an expansion, usually by numerically integrating the differential equations. In multidimensional problems a combination of these two techniques, such as the close-coupling method, is often used. This article is concerned with algebraic variational methods for scattering problems, but it also considers some algebraic and basis function methods which are not variationally derived.

The scattering parameters for the scattering of a particle by a central potential can all be expressed in terms of the phase shifts. These can be obtained exactly by numerical integration of a series of ordinary differential equations. This problem is called single-channel scattering. Algebraic variational methods allow the determination of scattering parameters by performing a number of one-dimensional integrals and solving a set of coupled algebraic equations. In Section II we discuss this scattering problem in detail because it affords the simplest illustrations of the techniques used in the variational methods of scattering and because it illustrates the historical development of the variational methods. It is interesting to note that the same dichotomy of numerical integration methods[1] versus algebraic methods[2] exists for the solution of one-dimensional bound-state problems.

For multichannel problems (scattering of a particle by a noncentral potential or scattering of composite particles), the exact solution is much more difficult. The most commonly used approximate techniques have been perturbation theory and the close-coupling method. Generally the latter is a more accurate procedure. It involves numerical integration of a set of coupled differential or integrodifferential equations. Algebraic variational methods for this problem involve solving a set of coupled algebraic equations. In order to obtain these equations a number of multidimensional integrals must be performed. It is interesting that the same dichotomy of types of methods exists for multidimensional bound-state problems. For example, for the problem of electronic structure of atoms, the numerical Hartree-Fock method[3] is an analogue of the close-coupling method and the matrix Hartree-Fock method[4] is an algebraic variational method. As the complexities of bound-state problems increase (e.g., in going from atomic to molecular problems), the algebraic methods seem to become more and more useful as compared to numerical integration methods. It is anticipated that as algebraic variational techniques for scattering problems continue to improve, the same trend will be manifest in scattering problems also.

The oldest algebraic variational method for scattering is that developed for single-channel scattering by Hulthén[5] in 1944. A disadvantage of the Hulthén method is that a quadratic equation must be solved. This means that, depending on the problem and the energy, either two phase shifts or nonphysical complex phase shifts may be obtained. In 1948, Kohn[6] and Hulthén[7] introduced new methods that do not involve nonlinear equations. In addition, Kohn explicitly showed how to apply his method to multichannel scattering. The first application of a variational method to a problem in chemical physics was made by Huang[8] in 1949. The method used was an extension of the work of Tamm.[9] This method, which has been of less historical importance than the work of Hulthén and Kohn, involves the minimization of an Euler-type integral.

The methods of Hulthén and Kohn prompted a number of theoretical investigations introducing new variational methods along similar lines, clarifying the relationships between various methods, and testing the applicability of variational methods for electron-atom scattering. The most significant study was the work of Schwartz[10,11] in 1961. Using the Kohn procedure he found that the calculations of the phase shifts are plagued by nonphysical singularities. These singularities, which are just artifacts of the Kohn method, show the phase shift increases (or decreases) rapidly by $\pi$ radians as a function of energy. In other words, the Kohn method predicts a resonance or antiresonance where no physical structure should exist. Schwartz also found that as the basis set is increased in size the number of singularities increases, but the widths of the spurious structures become narrower. This suggests that as the trial wave function approaches the exact wave function these singularities become undetectable. Nevertheless, they provide serious problems in practical applications of the Kohn method. Thus the work of Schwartz uncovered a disappointing feature of the use of the Kohn method.

Harris[12] stimulated renewed interest in algebraic scattering calculations by proposing in 1967 an algebraic method that could easily be used for single-channel scattering calculations. The Harris method can be used only at certain energies determined by the basis set, but the basis set can be adjusted to make one of these energies equal to the energy of interest. Nesbet,[13] in a critical evaluation of the Hulthén, Kohn, and Harris methods, presented the first correct analysis of the source of spurious singularities in Kohn method calculations, suggested a method (the anomaly-free method) for avoiding these spurious singularities, and clarified the relation between the Harris method and the variational methods. In another important contribution, Nesbet[14] clearly and explicitly extended the Kohn method and the anomaly-free method to multichannel scattering.

Harris and Michels[15,16] modified the Harris method for single-channel scattering so it is variational and can be used at any energy. They also extended the modified method (called the minimum-norm method) to inelastic scattering and reviewed[16] all previous work. References to the work done in the period from 1967 to 1971 also have been summarized elsewhere.[17–19]

Recently there have been many new important ideas and these (along with a fuller discussion of the material briefly reviewed above) are discussed in subsequent sections. The present outlook for algebraic variational methods in scattering is very promising but actual applications of the ideas that have been developed have still barely scratched the surface. Further, applications in chemical physics have been restricted almost entirely to electron-atom collisions.

Since scattering problems not involving explicitly time-dependent fields can always be solved by time-independent quantum mechanics, and since doing so is usually more convenient than using time-dependent quantum mechanics, we restrict our attention to time-independent formulations in this article.

Section II involves the application of algebraic variational methods to single-channel scattering problems. We use the $s$-wave scattering of a particle with the mass of an electron off an attractive exponential as an illustrative example. Another example of single-channel scattering is the scattering of electrons by atoms in the static approximation.[20,21] In the static approximation, one approximates the wave function as a ground state atomic wave function times a nonsquare-integrable function of the scattering electron's coordinates. In this way the problem reduces to a particle scattering off a central potential. The inclusion of electron exchange into the calculation by antisymmetrizing the wave function in the coordinates of all the electrons results in an additional potential term which is nonlocal. This is called the static exchange approximation (or often just the exchange approximation).[21,22] We defer discussion of the exchange approximation and other treatments of electron-atom scattering which go beyond the static approximation to Section III.

Since the phase shifts in the single-channel case can be obtained to any desired accuracy by numerical integration of the ordinary differential equation (not a difficult computational task), the motivation of considering algebraic variational methods in that case is to learn more about them so they can be applied most advantageously to multichannel cases. Multichannel cases are considered explicitly in Section III.

## II. SINGLE-CHANNEL SCATTERING

### A. Wave Functions

Consider the scattering of a particle off a potential $V(r)$, where $\mathbf{r}$ is the coordinate of the particle with respect to the origin. The wave function $\psi(\mathbf{r})$ must satisfy the Schrödinger equation

$$(H - E)\psi(\mathbf{r}) = 0 \tag{1}$$

where $H$ is the Hamiltonian operator, and $E$ is the total energy of the system. For collision problems the total energy is given, and the asymptotic part of the wave function is to be determined. If the wave function is expanded in partial waves

$$\psi(\mathbf{r}) = \sum_l \frac{X_l(r)}{r} P_l(\cos\theta) \tag{2}$$

where $P_l(\cos\theta)$ is the Legendre polynomial of order $l$; the function $X_l(r)$ is the solution to [23]

$$L_l X_l = 0 \tag{3}$$

where the operator $L_l$ is defined as

$$L_l = -\frac{\hbar^2}{2m}\frac{d^2}{dr^2} + \frac{l(l+1)\hbar^2}{2mr^2} + V(r) - E \tag{4}$$

with the boundary condition

$$X_l(0) = 0 \tag{5}$$

The solutions of (4) with $V(r) = 0$ are $r$ times the spherical Bessel ($j_l$) and Neumann ($-\xi_l$) functions. They are characterized by the boundary conditions[24]

$$krj_l(kr) \underset{r\to\infty}{\sim} \sin(kr - \tfrac{1}{2}l\pi) \tag{6a}$$

$$kr\xi_l(kr) \underset{r\to\infty}{\sim} \cos(kr - \tfrac{1}{2}l\pi) \tag{6b}$$

$$j_l(kr) \underset{r\to 0}{\sim} \frac{(kr)^l}{(2l+1)!!} \tag{6c}$$

and

$$\xi_l(kr) \underset{r\to 0}{\sim} \frac{(2l-1)!!}{(kr)^{l+1}} \tag{6d}$$

The wave number vector $\mathbf{k}$ is related to the energy by

$$E = \frac{\hbar^2 k^2}{(2m)} \tag{7}$$

For potentials of shorter range than a coulomb potential the solution of (3) is asymptotically equal to some linear combination of $rj_l(kr)$ and $r\xi_l(kr)$. If we force the coefficient of $\sin(kr - \frac{1}{2}l\pi)$ to be unity, we have

$$X_l(r) \underset{r\to\infty}{\sim} a_0^{-1/2} kr[j_l(kr) + t_l \xi_l(kr)] \tag{8}$$

where $a_0$ is a unit of length (taken to be the bohr in numerical examples) and $t_l$ is the tangent of the partial wave phase shift $\eta_l$. The problem is the calculation of $\eta_l$ and the solution of (3) subject to the boundary conditions (5) and (8). The partial cross-sections $\sigma_l$ are related to the partial wave phase shifts by

$$\sigma_l = \left(\frac{4\pi}{k^2}\right)(2l+1)\sin^2\eta_l \tag{9}$$

The total cross-section is

$$Q = \sum_l \sigma_l \tag{10}$$

The total cross-section may also be written

$$Q = 2\pi \int d\theta \, |f(\theta)|^2 \tag{11}$$

in terms of the scattering amplitude. The scattering amplitude is related to the phase shifts by

$$f(\theta) = \frac{1}{2ik} \sum_l (2l+1)(e^{2i\eta_l} - 1)P_l(\cos\theta) \tag{12}$$

We shall consider the problem of determining $X_l(r)$ and $t_l$. Methods exist for the determination of $\psi(\mathbf{r})$ and $f(\theta)$ without making the expansion

(2) but they are generally less useful for accurate work and we do not consider them here.

### B. The Variational Expressions

Consider the functional[25,26]

$$I_l(X_l^0) = \int_0^\infty X_l^0(r) L_l X_l^0(r)\, dr \tag{13}$$

where $X_l^0$ is a trial radial wave function satisfying

$$X_l^0(0) = 0 \tag{14}$$

$$X_l^0(r) \underset{r\to\infty}{\sim} a_0^{-1/2} kr[j_l(kr) + t_l^0 \xi_l(kr)] \tag{15}$$

where $t_l^0 = \tan \eta_l^0$. The trial radial wave function can be written

$$X_l^0(r) = X_l(r) + \delta X_l(r) \tag{16}$$

where $X_l(r)$ is the solution of (3). Then

$$\delta X_l(0) = 0 \tag{17}$$

$$\delta X_l(r) \underset{r\to\infty}{\sim} a_0^{-1/2} kr \xi_l(kr)\, \delta t_l \tag{18}$$

where

$$\delta t_l = t_l^0 - t_l \tag{19}$$

We assume $V(r)$ is real so that $X_l^0(r)$, $X_l(r)$, and $\delta X_l(r)$ are real. We can express the error in $I_l$ as

$$\delta I_l = I_l(X_l^0) - I_l(X_l) \tag{20}$$

where the last term is zero. Using (16) this becomes

$$\delta I_l \cong \int_0^\infty dr X_l(r) L_l\, \delta X_l(r) \tag{21}$$

where we have neglected terms that are of order $(\delta X_l)^2$. Using (3) and (4), (21) becomes

$$\delta I_l \cong -\left(\frac{\hbar^2}{2m}\right) \int_0^\infty dr \left(X_l \frac{d^2}{dr^2}\, \delta X_l - \delta X_l \frac{d^2}{dr^2} X_l\right) \tag{22a}$$

$$\cong -\frac{\hbar^2}{2m}\left[X_l \frac{d}{dr}\, \delta X_l - \delta X_l \frac{d}{dr} X_l\right]_0^\infty \tag{22b}$$

Using (5), (6), (8), (17), and (18), (22b) becomes

$$\delta I_l \cong \left(\frac{\hbar^2}{2ma_0}\right) k\, \delta t_l \tag{23}$$

Thus the expression

$$t_l^K = t_l^0 - \left(\frac{2ma_0}{\hbar^2 k}\right) I_l(X_l^0) \tag{24}$$

is a stationary expression for the tangent of the phase shift; that is, the first-order correction to the variationally corrected quantity $t_l^K$ vanishes by (23). An expression equivalent to (24) was first derived by Hulthén[5] and (24) may be called Hulthén's variational expression. Equation (24) is also used in Kohn's variational method[6] and because of this it will be called Kohn's variational expression.[26] To complete the specification of the variational methods[26] requires giving procedures to determine $t_l^0$ and $X_l^0(r)$, which are needed to evaluate the right-hand side of (24).

Hulthén's second method[7] can be reformulated[27] using a variational expression for the cotangent of the phase shift. This expression may be derived as above except using the normalizations

$$X_l(r) \underset{r\to\infty}{\sim} a_0^{-1/2} kr[t_l^{-1} j_l(kr) + \xi_l(kr)] \tag{25}$$

$$X_l^0(r) \underset{r\to\infty}{\sim} a_0^{-1/2} kr[(t_l^0)^{-1} j_l(kr) + \xi_l(kr)] \tag{26}$$

Then we obtain the following stationary expression for the cotangent of the phase shift

$$(t_l^R)^{-1} = (t_l^0)^{-1} + \left(\frac{2ma_0}{\hbar^2 k}\right) I_l(X_l^0) \tag{27}$$

Note that $X_l^0(r)$ in (27) is normalized according to (26). If we wish to use $X_l^0(r)$ normalized according to (15) we must use

$$(t_l^R)^{-1} = (t_l^0)^{-1} + \left[\frac{2ma_0}{\hbar^2 k (t_l^0)^2}\right] I(X_l^0) \tag{28}$$

(27) or (28) can be called the variational expression of the second Hulthén method or the inverse Kohn variational expression. Moiseiwitsch and Stacey[9,47] and Williams[28] have shown that Hulthén's second method is equivalent to the method derived later by Rubinow.[29,30] To avoid ambiguity we use the symbol $R$ (for Rubinow) for this variational expression.

A more general expression was derived by Kato for $s$-waves.[31] Generalizing to arbitrary $l$ we consider

$$X_l(r) \underset{r\to\infty}{\sim} a_0^{-1/2} kr[\lambda_{\theta,l} j_l(kr + \theta) + \xi_l(kr + \theta)] \tag{29}$$

$$X_l^0(r) \underset{r\to\infty}{\sim} a_0^{-1/2} kr[\lambda_{\theta,l}^0 j_l(kr + \theta) + \xi_l(kr + \theta)] \tag{30}$$

where $\theta$ is an arbitrary constant. Note

$$\lambda_{\theta,l} = [\tan(\eta_l - \theta)]^{-1} \tag{31}$$

Then we obtain the following stationary expression for $\lambda_{\theta,l}$

$$\lambda_{\theta,l}^{\text{Kato}} = \lambda_{\theta,l}^{0} + \frac{2ma_0}{\hbar^2 k} I(X_l^0) \tag{32}$$

Note that $X_l^0(r)$ in (32) is normalized as in (30). Kato[31] also derived an explicit expression for the error in (32). This result may be stated in the form of the following equation for the exact $\lambda_{\theta,l}$:

$$\lambda_{\theta,l} = \lambda_{\theta,l}^{0} + \frac{2ma_0}{\hbar^2 k} [I(X_l^0) - I(\delta X_l)] \tag{33}$$

which is called the Kato identity.

Percival[32] suggested using an alternative form of (24). He used the normalizations

$$X_l^0(r) \underset{r\to\infty}{\sim} kr[\alpha_{0l}^0 j_l(kr) + \alpha_{1l}^0 \xi_l(kr)] \tag{34}$$

and

$$\begin{aligned}\delta X_l(r) \underset{r\to\infty}{\sim} & [Q\cos(\eta_l^0 + \delta\eta_l) - \alpha_{0l}^0]krj_l(kr) \\ & + [Q\sin(\eta_l^0 + \delta\eta_l) - \alpha_{1l}^0]kr\xi_l(kr)\end{aligned} \tag{35}$$

where

$$\alpha_{0l}^0 = Q\cos\eta_l^0 \tag{36a}$$

and

$$\alpha_{1l}^0 = Q\sin\eta_l^0 \tag{36b}$$

to obtain the stationary expression for the phase shift given by

$$\eta_l^P = \eta_l^0 - \frac{2m}{\hbar^2 k\,|Q|^2} I(X_l^0) \tag{37}$$

Note that $X_l^0(r)$ in (37) is normalized according to (34) and (36). If we wish to use $X_l^0(r)$ normalized according to (15) we must use

$$\eta_l^P = \eta_l^0 - \frac{2ma_0}{\hbar^2 k}\cos^2\eta_l^0 I(X_l^0) \tag{38}$$

Seaton[33] showed that (37) can also be obtained starting from

$$\delta X_l(r) \underset{r\to\infty}{\sim} [-\alpha_{1l}^0 j_l(kr) + \alpha_{0l}^0 \xi_l(kr)]kr\,\delta\eta_l \tag{39}$$

(35) and (39) are identical through first order in $\delta\eta_l$.

Above we have given variationally correct expressions for $t_l$, $t_l^{-1}$, $[\tan(\eta_l - \theta)]^{-1}$, and $\eta_l$. Variationally correct expressions may be derived for other functions of $\eta_l$ but they have been of less interest.

### C. The Hulthén, Kohn, Rubinow, and Percival Methods

For basis set calculations, the trial radial wave function $X_l^0(r)$ is written as the following linear combination

$$X_l^0(r) = S_l(r) + t_l^0 C_l(r) + \sum_{a=1}^{n} c_a^l \eta_a^l(r) \tag{40}$$

where

$$S_l(0) = 0 \tag{41}$$

$$C_l(0) = 0 \tag{42}$$

$$S_l(r) \underset{r\to\infty}{\sim} a_0^{-1/2} kr j_l(kr) \tag{43}$$

$$C_l(r) \underset{r\to\infty}{\sim} a_0^{-1/2} kr \xi_l(kr) \tag{44}$$

the $\eta_a^l(r)$ are a set of square-integrable functions, and the $c_a^l$ and $t_l^0$ are $(n + 1)$ coefficients to be determined. To satisfy (5) we require

$$\eta_a^l(0) = 0 \tag{45}$$

Note that (8) is automatically satisfied by $X_l^0(r)$ of (40).

The Kohn variational method[6,26] for obtaining the $(n + 1)$ coefficients in the trial function consists in solving the $(n + 1)$ equations

$$\frac{\partial t_l^K}{\partial c_a^l} = 0 \qquad a = 1, 2, \ldots, n \tag{46}$$

$$\frac{\partial t_l^K}{\partial t_l^0} = 0 \tag{47}$$

Equation (46) may be written

$$\int_0^\infty dr \eta_a^l(r) L_l X_l^0(r) = 0 \qquad a = 1, 2, \ldots, n \tag{48}$$

and (47) may be written

$$\int_0^\infty dr C_l(r) L_l X_l^0(r) = 0 \tag{49}$$

[The latter is derived by using (70).] The trial $t_l^0$ obtained by this procedure will be called the Kohn zero-order result $t_K^0$, where we have dropped the subscript $l$.

The Hulthén variational method[5] for the $(n + 1)$ coefficients in the trial function consists in solving (46) [or the equivalent (48)] and

$$I(X_l^0) = 0 \tag{50a}$$

Equation (50a) has two solutions (discussed later). One of these is called the Hulthén zero-order result $t_H^0$. Equation (50a) may be written [using (48)]

$$\int_0^\infty dr[S_l(r) + t_l^0 C_l(r)]L_l X_l^0(r) = 0 \tag{50b}$$

We have already mentioned that the second Hulthén[7] and Rubinow methods[29,30] are identical. They yield a $t_l^0$ which will be called the Rubinow zero-order result $t_R^0$. To obtain the $t_R^0$, we use the trial function

$$X_l^0(r) = (t_l^0)^{-1}S_l(r) + C_l(r) + \sum_{a=1}^{n} c_a^l \eta_a^l(r) \tag{51}$$

and solve the equations

$$\frac{\partial(t_l^R)^{-1}}{\partial c_a^l} = 0 \qquad a = 1, 2, \ldots, n \tag{52}$$

$$\frac{\partial(t_l^R)^{-1}}{\partial(t_l^0)^{-1}} = 0 \tag{53}$$

for the $(n + 1)$ coefficients in (51). Equation (52) may be written as (48), and (53) may be written

$$\int_0^\infty dr S_l(r) L_l X_l^0(r) = 0 \tag{54}$$

These procedures yield variationally uncorrected tangents of the phase shift which may be improved (or corrected through first order, i.e., corrected to second order[26]) by using (24), (28), or (38). Usually $t_K^0$ is corrected to give $t_K^K$ by (24) and $t_R^0$ is corrected to give $t_R^R$ by (28) because[13,16] the corrected tangents are then stationary with respect to the zero-order tangents, that is,

$$\left.\frac{\partial t_l^K}{\partial t_l^0}\right|_{t_l^0 = t_K^0} = 0 \tag{55}$$

and

$$\left.\frac{\partial t_l^R}{\partial t_l^0}\right|_{t_l^0 = t_R^0} = 0 \tag{56}$$

The properties (55) and (56) are discussed later in this section. Kohn zero-order tangents may be corrected by the Rubinow variational expression to give nonstationary $t_K^R$ or by the Percival expression to give $t_K^P$, or the Rubinow zero-order tangents may be corrected by the Kohn variational expression to give the nonstationary $t_R^K$. The nonstationary quantity $t_R^P$ may also be obtained, but it is not considered in this article.

Hulthén's variational method obviously yields a trial wave function for which the variational correction computed by (24), (28), or (38) vanishes. Thus

$$t_H{}^0 = t_H{}^K = t_H{}^R = t_H{}^P \tag{57}$$

and this quantity may simply be called the Hulthén tangent $t_H$.

Percival's variational expression[32] uses the trial function

$$X_l{}^0(r) = a_0^{-1/2}[\cos \eta_l{}^0 S_l(r) + \sin \eta_l{}^0 C_l(r)] + \sum_{a=1}^{n} c_a{}^l \eta_a{}^l \tag{58}$$

The $(n + 1)$ unknowns in (58) are obtained by solving the $(n + 1)$ equations

$$\frac{\partial \eta_l{}^P}{\partial c_a{}^l} = 0 \qquad a = 1, 2, \ldots, n \tag{59}$$

$$\frac{\partial \eta_l{}^P}{\partial \eta_l{}^0} = 0 \tag{60}$$

Equation (59) may be written as (48), and (60) is a transcendental equation for $\eta_P{}^0$ which can be rewritten as a quadratic equation for tan $2\eta_P{}^0$. The zero-order Percival result $\eta_P{}^0$ obtained this way is variationally corrected using (38) to yield $\eta_P{}^P$.

The methods explained above can be rederived in a fashion that yields equations more suitable for subsequent discussion. For this purpose we present the following analysis due to Nesbet.[13] We expand the radial wave function in a form slightly different from (40), namely,

$$X_l{}^0 = \phi_0{}^l(r) + S_l(r) + t_l{}^0[\phi_1{}^l(r) + C_l(r)] \tag{61}$$

where

$$\phi_0{}^l = \sum_{a=1}^{n} c_a{}^{Sl} \eta_a{}^l(r) \tag{62}$$

and

$$\phi_1{}^l = \sum_{a=1}^{n} c_a{}^{Cl} \eta_a{}^l(r) \tag{63}$$

Equations (48) are replaced by the equations

$$\int_0^\infty dr \eta_a{}^l(r) L_l \phi_0{}^l(r) = -\int_0^\infty dr \eta_a{}^l(r) L_l S_l(r) \qquad a = 1, 2, \ldots, n \tag{64}$$

and

$$\int_0^\infty dr \eta_a{}^l(r) L_l \phi_1{}^l = -\int_0^\infty dr \eta_a{}^l(r) L_l C_l(r) \qquad a = 1, 2, \ldots, n \tag{65}$$

From (13), (61), (64), and (65) we obtain

$$I(X_l{}^0) = M^{00} + (M^{01} + M^{10}) t_l{}^0 + (M^{11})(t_l{}^0)^2 \tag{66}$$

where

$$M^{\alpha\beta} = \int_0^\infty dr A_{l\alpha}(r) L_l[\phi_\beta{}^l(r) + A_{l\beta}(r)] \qquad \alpha, \beta = 0, 1 \tag{67}$$

and we have the following new labels

$$A_{l0}(r) = S_l(r) \tag{68}$$

$$A_{l1}(r) = C_l(r) \tag{69}$$

where we have suppressed the pair of subscripts $l$ which could have been added to the $M^{\alpha\beta}$'s. Note an important property of these matrix elements:

$$M^{01} - M^{10} = \frac{\hbar^2 k}{2ma_0} \tag{70}$$

which can be derived by integration-by-parts and evaluation of the resulting surface integral.

Now equations (50), (47), (53), and (60) can be written in the present notation as

$$M^{00} + (M^{01} + M^{10})t_H + M^{11}t_H{}^2 = 0 \tag{71}$$

$$M^{10} + M^{11}t_K{}^0 = 0 \tag{72}$$

$$M^{00} + M^{01}t_R{}^0 = 0 \tag{73}$$

and

$$\frac{\hbar^2 k}{2ma_0} - (M^{11} - M^{00}) \sin 2\eta_P{}^0 - (M^{01} + M^{10}) \cos 2\eta_P{}^0 = 0 \tag{74}$$

respectively.

The Hulthén tangent of the phase shift is given by the solution of the quadratic equation (71). At one time the fact that two phase shifts are obtained this way was considered a problem. It has been shown, however, that the physical solution, if any, is obtained with the positive sign[34–36] of the discriminant, giving

$$t_H = -\frac{M^{10}}{M^{11}} + \frac{\hbar^2 k}{4ma_0 M^{11}} \left\{ \left[ -16 \left( \frac{ma_0}{\hbar^2 k} \right)^2 \text{DET}\, M + 1 \right]^{1/2} - 1 \right\} \tag{75}$$

where we have used capital letters for a determinant of a matrix arranged by its superscripts, that is,

$$\text{DET}\, M = M^{00}M^{11} - M^{10}M^{01} \tag{76}$$

From (75) it is easily seen that $t_H$ is complex if

$$\text{DET}\, M > \frac{(\hbar^2 k/ma_0)^2}{16} \tag{77}$$

In this case there is no physical solution of (71).

The zero-order Kohn and Rubinow phase shifts given by (72) and (73), that is,

$$\eta_K{}^0 = \arctan\left(\frac{-M^{10}}{M^{11}}\right) \tag{78}$$

$$\eta_R{}^0 = \arctan\left(\frac{-M^{00}}{M^{01}}\right) \tag{79}$$

can be corrected through first order using the Kohn variational expression (24), the Rubinow variational expression (28), or the Percival variational expression (38). Using (70) these procedures yield

$$t_K{}^K = -\frac{M^{10}}{M^{11}} - \frac{2ma_0}{\hbar^2 k}\frac{\text{DET } M}{M^{11}} \tag{80a}$$

$$= -\frac{2ma_0}{\hbar^2 k}\left[M^{00} - \frac{(M^{10})^2}{M^{11}}\right] \tag{80b}$$

$$t_R{}^K = -\frac{M^{00}}{M^{01}} - \frac{2ma_0}{\hbar^2 k}\frac{M^{00}\text{ DET } M}{(M^{01})^2} \tag{81}$$

$$(t_K{}^R)^{-1} = -\frac{M^{11}}{M^{10}} + \frac{2ma_0}{\hbar^2 k}\frac{M^{11}\text{ DET } M}{(M^{10})^2} \tag{82}$$

$$(t_R{}^R)^{-1} = -\frac{M^{01}}{M^{00}} + \frac{2ma_0}{\hbar^2 k}\frac{\text{DET } M}{M^{00}} \tag{83}$$

and

$$t_K{}^P = \tan\left[-\arctan\frac{M^{10}}{M^{11}} - \frac{2ma_0}{\hbar^2 k}\cos^2\eta_K{}^0\frac{\text{DET } M}{M^{11}}\right] \tag{84}$$

We now consider the properties (55) and (56). From (24) and (66) we have

$$t^K = t^0 - \frac{2ma_0}{\hbar^2 k}[M^{00} + (M^{01} + M^{10})t^0 + M^{11}(t^0)^2] \tag{85}$$

Differentiating

$$\frac{\partial t^K}{\partial t^0} = 1 - \frac{2ma_0}{h^2 k}[M^{01} + M^{10} + 2M^{11}t^0] \tag{86}$$

Substituting $t^0 = t_K{}^0$ [from (78)] and using (70) yields (55). That this must occur was ensured by using (47) to define $t_K{}^0$. Thus $t_K{}^K$ is not only correct through first order but is also stationary with respect to variations of the zero-order tangent of the phase shift. A similar argument shows $(\partial t^K/\partial t^0)|_{t^0=t_R{}^0}$ is not zero. Thus the Kohn variational expression is stationary with respect to $t^0$ only if the Kohn method is used to obtain $t^0$. As a consequence, although $t_R{}^K$ is correct through first order, it is not stationary with respect to variations in the zero-order tangent of the phase shift.

Similarly we can show that $t_R{}^R$ and $t_P{}^P$ are stationary but $t_K{}^R$ and $t_K{}^P$ are not. The variationally correct and stationary results $t_K{}^K$ and $t_R{}^R$ are, as mentioned previously, the Kohn and Rubinow first-order results, respectively. The methods yielding $t_P{}^P$ and $t_K{}^P$ were used by Percival,[32] but the other two methods (yielding $t_R{}^K$ and $t_K{}^R$) do not appear to have been used in the literature.

Many examples of potential scattering calculations using the Hulthén, Kohn, and Rubinow methods may be found in the literature.[5–7,13,29,34–35,37–47] Applications to more complicated problems are discussed in Section III.

Notice from (75), (76), and (78) to (84) that when DET $M$ is 0 we have

$$t_K{}^0 = t_R{}^0 = t_H = t_K{}^K = t_R{}^R = t_R{}^K = t_K{}^R = t_K{}^P \tag{87}$$

For single-channel scattering the reactance matrix is a $1 \times 1$ matrix whose element is $t_l$. Thus we have been considering special cases of variational methods for the reactance matrix. Analogous variational methods of the types considered here and below may be written down for the scattering matrix.[6] For single-channel problems the scattering matrix is a $1 \times 1$ matrix whose element is $e^{2i\eta_l}$. Thus these methods may yield complex $\eta_l$. Further, they are often less convenient since they require use of complex asymptotic functions in place of $S_l$ and $C_l$. Thus we do not consider them explicitly.

### D. Integrals and Computational Procedures

Note that the calculations discussed in the last section reduce to the computation of the various $M^{\alpha\beta}$. Nesbet[13] has given expressions for the $M^{\alpha\beta}$ in terms of the set of basis functions

$$\tilde{\eta}_\nu{}^l = \sum_{a=1}^{N} c_a{}^{l\nu} \eta_a{}^l(r) \tag{88}$$

where the $c_a{}^{l\nu}$ are obtained by solving the standard eigenvalue problem

$$\sum_b H_{ab}{}^l c_b{}^{l\nu} = E_\nu{}^l c_a{}^{l\nu}, \qquad a = 1, 2, \ldots, n \quad \nu = 1, 2, \ldots, n \tag{89}$$

where

$$H_{ab}{}^l = \int_0^\infty dr \eta_a{}^l(r)(L_l + E)\eta_b{}^l(r) \tag{90}$$

These expressions are

$$M^{00} = M_{SS} + \sum_\nu M_{S\nu}(E - E_\nu)^{-1} M_{\nu S} \tag{91}$$

$$M^{01} = M_{SC} + \sum_\nu M_{S\nu}(E - E_\nu)^{-1} M_{\nu C} \tag{92}$$

$$M^{10} = M_{CS} + \sum_\nu M_{C\nu}(E - E_\nu)^{-1} M_{\nu S} \tag{93}$$

and

$$M^{11} = M_{CC} + \sum_{\nu} M_{C\nu}(E - E_\nu)^{-1} M_{\nu C} \tag{94}$$

where

$$M_{SS} = \int_0^\infty dr S_l(r) L_l S_l(r) \tag{95}$$

$$M_{SC} = \int_0^\infty dr S_l(r) L_l C_l(r) \tag{96}$$

$$M_{CS} = \int_0^\infty dr C_l(r) L_l S_l(r) \tag{97}$$

$$M_{CC} = \int_0^\infty dr C_l(r) L_l C_l(r) \tag{98}$$

and

$$M_{\nu C} = M_{C\nu} = \int_0^\infty dr C_l(r) L_l \tilde{\eta}_\nu^{\,l}(r) \tag{99}$$

$$M_{\nu S} = M_{S\nu} = \int_0^\infty dr S_l(r) L_l \tilde{\eta}_\nu^{\,l}(r) \tag{100}$$

Alternative expressions for the $M^{\alpha\beta}$ can be obtained which do not require the solution of (89). For these expressions we assume

$$\int_0^\infty dr \eta_a^{\,l}(r) \eta_b^{\,l}(r) = \delta_{ab} \tag{101}$$

and

$$\int_0^\infty dr A_{l\beta}(r) \eta_a^{\,l}(r) = 0 \tag{102}$$

Then[16]

$$M^{\alpha\beta} = (\mathbf{M}^{AA} - \mathbf{M}^{A\eta}\mathbf{M}^{\eta\eta-1}\mathbf{M}^{\eta A})_{\alpha\beta} \tag{103}$$

where

$$M_{ab}^{\ \eta\eta} = \int_0^\infty dr \eta_a^{\,l}(r) L_l \eta_b^{\,l}(r) \tag{104}$$

$$M_{\alpha\beta}^{\ AA} = \int_0^\infty dr A_{l\alpha}(r) L_l A_{l\beta}(r) \qquad \alpha, \beta = 0, 1 \tag{105}$$

$$M_{a\beta}^{\ \eta A} = \int_0^\infty dr \eta_a^{\,l}(r) L_l A_{l\beta}(r) \qquad \beta = 0, 1 \tag{106}$$

$$M_{\beta a}^{\ A\eta} = \int_0^\infty dr A_{l\beta}(r) L_l \eta_a^{\,l}(r) \qquad \beta = 0, 1 \tag{107}$$

Note that (101) and (102) are not really needed to derive (103), as was assumed by previous workers. This is discussed in Appendix 1.

The integrals (90) and (104) are called bound-bound integrals, the integrals (99), (100), (106), and (107) are called bound-free integrals, and the integrals (95) to (98) and (105) are called free-free integrals.

Equations (91) to (94) have the advantage over equation (103) that if (89) is once solved, scattering calculations may be performed at many energies $E$ by simply doing the new free-free integrals (95) to (98) and bound-free integrals (99) and (100). Using (103), however, we must invert $M^{\eta\eta}$ at each new energy.

### E. The Harris Method (for Energies Equal to Eigenvalues of the Bound-Bound Matrix)

Harris[12] pointed out a simple method for calculating the tangent of the phase shift in the case where the energy is equal to one of the eigenvalues $E_\mu{}^l$ of the bound-bound matrix. In this case, if we write the trial function as

$$X_l{}^0(r) = S_l(r) + t_l{}^0 C_l(r) + \sum_\nu d_\nu \tilde{\eta}_\nu{}^l(r) \tag{108}$$

involving the expansion coefficients $d_\nu$ and require $L_l X_l{}^0(r)$ to have no component in the space spanned by the $\eta_\alpha{}^l$ we obtain

$$t_{\text{Harris}} = \frac{-\displaystyle\int_0^\infty dr \tilde{\eta}_\mu{}^l(r) L_l S_l(r)}{\displaystyle\int_0^\infty dr \tilde{\eta}_\mu{}^l(r) L_l C_l(r)} \tag{109}$$

Nesbet[13] showed that when $E$ equals one of the eigenvalues $E_\mu{}^l$

$$t_{\text{Harris}} = t_R{}^0 = t_K{}^0 = t_H \tag{110}$$

However, the Harris method has the advantage over the previous methods that no free-free integrals need to be evaluated.

A possible difficulty with the Harris method is that the basis set has to be adjusted to perform a calculation at a preselected energy. If such basis set adjustment were required to calculate the phase shift as a function of energy in some energy region, spurious energy dependences of the phase shifts could be obtained. This is not expected to be a problem if large enough basis sets are used. Another disadvantage[18] however is that, because the eigenvalue $E_\mu{}^l$ depends on the trial function, the extrapolation technique as basis functions approach a complete set cannot be used.

Formally a more serious difficulty with the Harris method is that the tangent of the phase shift is not variationally corrected. When the minimum-norm method of Harris and Michels (discussed in the next section) is applied at an energy equal to one of the eigenvalues $E_\mu{}^l$, it reduces to the Harris method but with the addition of a variational correction. For this reason $t_{\text{Harris}}$ may be called $t_{MN}{}^0$.

Some applications of the original Harris method to potential scattering may be found in the literature.[12,13,44,45]

### F. The Minimum-Norm Method

Harris and Michels[15] derived an extension of the Harris method called the minimum-norm method which can be applied at any incident energy with any basis set. The radial wave function is expanded in the form

$$X_l{}^0(r) = \sum_{\beta=0}^{1} \alpha_{\beta l}{}^0 A_{l\beta}(r) + \sum_{a=1}^{n} c_a{}^l \eta_a{}^l(r) \tag{111}$$

where the $\alpha_{\beta l}{}^0$ and $c_a{}^l$ are coefficients. We must first determine $t_{MN}{}^0$, the zero-order minimum-norm tangent of the phase shift, given by

$$\tan \eta_l{}^0 = \frac{\alpha_{1l}{}^0}{\alpha_{0l}{}^0} \tag{112}$$

We consider a region defined as $0 \leqslant r \leqslant a$ where $a$ is large enough so that the integrals (101), (102), and (104) to (107) would all be approximately unchanged if the integration were carried out only over $0 \leqslant r \leqslant a$ instead of over $0 \leqslant r \leqslant \infty$. In the region $0 \leqslant r \leqslant a$, we expand

$$L_l X_l{}^0(r) = \sum_{\beta=0}^{1} g_{\beta l} A_{l\beta}(r) + \sum_{a=1}^{n} h_a{}^l \eta_a{}^l(r) \tag{113}$$

involving the coefficients $g_{\beta l}$ and $h_a{}^l$. Substituting (111) into (113) and requiring (48) or substituting (111) into (48) yields

$$\mathbf{M}^{\eta A}\boldsymbol{\alpha}^0 + \mathbf{M}^{\eta\eta}\mathbf{c}^l = 0 \tag{114}$$

where the column vectors $\boldsymbol{\alpha}$ and $\mathbf{c}^l$ are defined by

$$(\boldsymbol{\alpha}^0)_\beta = \alpha_{\beta l} \tag{115}$$

and

$$(\mathbf{c}^l)_a = c_a{}^l \tag{116}$$

Assume the $A_{l\beta}(r)$ are orthogonal to the $\eta_a{}^l(r)$. Then we obtain from (6a), (6b), (43), (44), (68), (69), (111), and (113) for very large $a$

$$\mathbf{M}^{AA}\boldsymbol{\alpha}^0 + \mathbf{M}^{A\eta}\mathbf{c}^l = \left(\frac{a}{2a_0}\right)\mathbf{g} \tag{117}$$

where

$$(\mathbf{g})_\beta = g_{\beta l} \tag{118}$$

and $g_{\beta l}$ is proportional to $a^{-1}$. We show in Appendix 1 that this orthogonality assumption is not necessary. Comparison of (114) and (117) shows we can require $\mathbf{g}$ to equal the null vector only if

$$\begin{vmatrix} \mathbf{M}^{AA} & \mathbf{M}^{A\eta} \\ \mathbf{M}^{\eta A} & \mathbf{M}^{\eta\eta} \end{vmatrix} = 0 \tag{119}$$

where the determinant is of order $n + 2$. In general the determinant does not vanish, and we cannot take $\mathbf{g}$ to be the null vector. Instead we minimize its norm $|\mathbf{g}|$ as follows. If $M^{\eta\eta}$ is nonsingular, we eliminate $\mathbf{c}^l$ from (114) and (117) to obtain

$$(\mathbf{M}^{AA} - \mathbf{M}^{A\eta}\mathbf{M}^{\eta\eta-1}\mathbf{M}^{\eta A})\boldsymbol{\alpha}^0 = \left(\frac{a}{2a_0}\right)\mathbf{g} \tag{120}$$

Comparing (120) to (103) we see that the $2 \times 2$ matrix in (120) is the matrix whose components are given in (67); thus, calling this matrix $\mathbf{M}$, we can write (120) as

$$\mathbf{M}\boldsymbol{\alpha}^0 = \left(\frac{a}{2a_0}\right)\mathbf{g} \tag{121}$$

We see that $\mathbf{g}$ can be taken as equal to the null vector only if DET $M = 0$. Multiplying each side of (121) by its adjoint yields

$$\boldsymbol{\alpha}^{0\dagger}\mathbf{Q}\boldsymbol{\alpha}^0 = \left(\frac{a}{2a_0}\right)^2 |\mathbf{g}|^2 \tag{122}$$

where

$$\mathbf{Q} = \mathbf{M}^\dagger\mathbf{M} \tag{123}$$

Now $\boldsymbol{\alpha}^0$ is chosen to minimize $|\mathbf{g}|$. This means the $\boldsymbol{\alpha}^0$ used in (112) must be the eigenvector of the Hermitian matrix $\mathbf{Q}$ corresponding to its lowest eigenvalue. Let the other eigenvector of $\mathbf{Q}$ be $\boldsymbol{\beta}$.

If $\boldsymbol{\alpha}^0$ and $\boldsymbol{\beta}$ are normalized they can be used to construct the $2 \times 2$ orthogonal transformation matrix

$$\boldsymbol{\mu} = (\boldsymbol{\alpha}^0\boldsymbol{\beta}) \tag{124}$$

Although $\boldsymbol{\mu}^+\mathbf{Q}\boldsymbol{\mu}$ is diagonal, it is interesting to point out[48,49] that the matrix $\mathbf{M}'$ defined as

$$\mathbf{M}' = \boldsymbol{\mu}^+\mathbf{M}\boldsymbol{\mu} \tag{125}$$

is not diagonal. The explicit result obtained from $\boldsymbol{\alpha}^0$ and (112) is[16]

$$t_{MN}{}^0 = \frac{(M^{11})^2 + (M^{01})^2 - (M^{10})^2 - (M^{00})^2 - \{[M^{00})^2 + (M^{01})^2 + (M^{10})^2 + (M^{11})^2]^2 - 4(\text{DET}\, M)^2\}^{1/2}}{2(M^{00}M^{01} + M^{11}M^{10})} \tag{126}$$

Note that we have corrected the typographical error in Ref. 16.

If $\mathbf{M}^{\eta\eta}$ is singular, that is, if $E = E_\mu{}^l$, then the $n$ linear equations (114) fully determine $\boldsymbol{\alpha}$ and the two linear equations (117) are not needed for this purpose. If we define the column vector obtained from (89) by

$$(\mathbf{c}^{l\mu})_a = c_a{}^{l\mu} \tag{127}$$

then multiplying (114) by the transpose conjugate of $\mathbf{c}_\mu{}^{l\mu}$ yields the single linear equation

$$[\mathbf{c}^{l\mu}]^\dagger \mathbf{M}^{\eta A} \boldsymbol{\alpha}^0 = 0 \tag{128}$$

which [with (112)] is identical to (109) of the original Harris method.

The zero-order result obtained from (126) or (128) may be corrected variationally by either the Kohn or the Rubinow variational expression to yield

$$t_{MN}{}^K = t_{MN}{}^0 - \frac{2ma_0}{\hbar^2 k}[M^{00} + (M^{01} + M^{10})t_{MN}{}^0 + M^{11}(t_{MN}{}^0)^2] \tag{129}$$

or

$$(t_{MN}{}^R)^{-1} = (t_{MN}{}^0)^{-1} + \frac{2ma_0}{\hbar^2 k}[(t_{MN}{}^0)^{-2}M^{00} + (t_{MN}{}^0)^{-1}(M^{01} + M^{10}) + M^{11}] \tag{130}$$

Harris has pointed out that these expressions are not stationary (see discussion in Section II.C above) and has suggested the use of a different variational expression.[50] This expression will be called the minimum-norm variational expression and it is

$$(t^{MN} + A)^{-1} = (t^0 + A)^{-1} + \left(\frac{2ma_0}{\hbar^2 k}\right)(t^0 + A)^{-2} I(X_l{}^0) \tag{131}$$

where

$$A = \frac{M^{00} + t^0 M^{01}}{M^{10} + t^0 M^{11}} \tag{132}$$

It can be verified that $t_K{}^{MN} = t_K{}^K$ and that $t_R{}^{MN} = t_R{}^R$. Applying (131) and (132) to $t_{MN}{}^0$ yields the variationally corrected minimum-norm result $t_{MN}{}^{MN}$. Notice that if DET $M = 0$

$$t_{MN}{}^0 = t_{MN}{}^K = t_{MN}{}^R = t_{MN}{}^{MN} \tag{133}$$

Nesbet and Oberoi[48,49] have also considered the problem of the non-stationary nature of the Kohn variational expression when the minimum-norm method is used for $t^0$. They have developed a method they call the optimized minimum-norm method. As the final result of the optimized minimum-norm method one calculates the quantity

$$t_{MN}^{OMN} = \frac{\sin \eta_{MN}{}^0 + t' \cos \eta_{MN}{}^0}{\cos \eta_{MN}{}^0 - t' \sin \eta_{MN}{}^0} \tag{134}$$

where

$$t' = -\frac{2ma_0}{\hbar^2 k}\left[M_{11}' - \frac{(M_{21}')^2}{M_{22}'}\right] \tag{135}$$

### G. Methods Involving No Free-Free Integrals

Of all the zero-order and variationally corrected calculational methods discussed above, the original Harris method is the only one that can be carried out without evaluating any free-free integrals. In most cases, the free-free integrals are the most difficult integrals to evaluate. In this section we discuss two other methods in which no free-free integrals need to be evaluated. Unlike the Harris method, these two methods can be applied at any energy for any basis set.

Euler integral methods have been used by Hulthén,[7] Huang,[8] Malik,[51] and Moiseiwitsch.[47] The method to be considered here is a more systematic version of the method used by Huang.[8] It differs from Huang's and Malik's approaches using Euler integrals in that the wave function is expressed as

$$X_l^0(r) = S_l(r) + t_l^0 C_l(r) + v_l(r) \tag{136}$$

where $v_l(r)$ is a square-integrable function with the boundary condition

$$v_l(0) = 0 \tag{137}$$

The function $v_l(r)$ is eventually determined by expanding it as

$$v_l(r) = \sum_{a=1}^{n} c_a^l \eta_a^l(r) \tag{138}$$

where the $\eta_a^l(r)$ are square-integrable functions satisfying (45) and the $c_a^l$ are coefficients. Substituting (136) into (3) yields

$$-\frac{\hbar^2}{2m}\frac{d^2 v_l}{dr^2} + \left[\frac{l(l+1)\hbar^2}{2mr^2} + V(r) - E\right]v_l(r) + L_l S_l(r) + t_l^0 L_l C_l(r) = 0 \tag{139}$$

This is a nonhomogeneous differential equation for $v_l(r)$ with homogeneous boundary conditions. It replaces the homogeneous differential equation (3) which had to be solved with nonhomogeneous boundary conditions at large $r$.

Now we seek a functional $F(r, v_l, v_l')$ where $v_l' = dv_l/dr$. It will be required that the Euler equation

$$\frac{d}{dr}\left(\frac{\partial F}{\partial v_l'}\right) - \frac{\partial F}{\partial v_l} = 0 \tag{140}$$

of this functional be equivalent to (139). It can be seen that the functional we seek is

$$F(r, v_l, v_l') = -\frac{\hbar^2}{4m}[v_l']^2 - \frac{1}{2}\left[\frac{l(l+1)\hbar^2}{2mr^2} + V(r) - E\right]v_l^2 - v_l[L_l S_l(r) + t_l^0 L_l C_l(r)] \quad (141)$$

Therefore, by Euler's theorem, the solution of (139) may be obtained by minimizing the Euler integral

$$\Gamma_l = \int_0^\infty dr F(r, v_l, v_l') \quad (142)$$

Substituting (138) and (141) into (142) yields

$$\Gamma_l(c_1^l, c_2^l, \ldots, c_n^l, t_l^0) = \tfrac{1}{2}\sum_{ab} c_a^l c_b^l A_{ab}^l - \sum_a c_a^l(M_{a0}^{\eta A} + t_l^0 M_{a1}^{\eta A}) \quad (143)$$

where

$$A_{ab}^l = \int_0^\infty dr\left\{-\frac{\hbar^2}{2m}\frac{d\eta_a^l}{dr}\frac{d\eta_b^l}{dr} - \left[\frac{\hbar^2 l(l+1)}{2mr^2} + V(r) - E\right]\eta_a^l(r)\eta_b^l(r)\right\} \quad (144)$$

To minimize (143) we require

$$\frac{\partial \Gamma_l}{\partial c_b^l} = 0 \qquad b = 1, 2, \ldots, n \quad (145)$$

$$\frac{\partial \Gamma_l}{\partial t_l^0} = 0 \quad (146)$$

This yields the $(n + 1)$ nonhomogeneous linear equations

$$\sum_{a=1}^n c_a^l A_{ab}^l - t_l^0 M_{b1}^{\eta A} = M_{b0}^{\eta A} \quad (147)$$

and

$$\sum_{a=1}^n c_a^l M_{a1}^{\eta A} = 0 \quad (148)$$

for the $n + 1$ unknowns. The $t_l^0$ determined this way is called $t_{EI}^0$. Extending a suggestion of Hulthén and Olsson[52] we may variationally correct $t_{EI}^0$ through first order using the Kohn variational expression, yielding $t_{EI}^K$. Such a correction involves the calculation of free-free integrals.

Ladányi, Lengyel, and Szondy[53] have proposed using the method of moments for scattering problems. In this case we use the $n + 1$ equations

$$\int_0^\infty dr w_a(r) L_l X_l^0(r) = 0 \quad (149)$$

where the $w_a(r)$ are "weight functions" to determine the $n + 1$ parameters in $X_l^0(r)$. For certain choices of trial function and weight functions, this method may be made to yield the zero-order Kohn or zero-order Rubinow method. Ladányi et al., however, recommend that all the weight functions be square-integrable. This eliminates all free-free integrals.

As an example of the method of moments we consider the trial function (40) and the solution of equations (48) and

$$\int_0^\infty dr \eta_{n+1}^{l*}(r) L_l X_l^0(r) = 0 \tag{150}$$

for the unknown coefficients $c_a^l$ and $t_l^0$. Equation (150) was obtained by using the basis functions [see (40)], including one more basis function than is used in $X_l^0(r)$, as weight functions in (149). The $t_l^0$ obtained by this method is called $t_{MM}^0$. We suggest that improved results can be obtained using the Kohn, Rubinow, Percival, or minimum-norm variational expression to correct $t_{MM}^0$. In this case, however, free-free integrals must be calculated.

### H. Discussion of Above Methods and the Anomaly-Free Method

In this section results are presented for a model problem, namely, the scattering of a particle with the mass of the electron and orbital angular momentum $l$ equal to zero off the potential

$$V = -2\mathscr{R}e^{-(r/a_0)} \tag{151}$$

where $\mathscr{R}$ is the Rydberg energy and $a_0$ is the bohr. The exact solution[54] to this problem exhibits no resonances. The trial function is

$$X_0^0(r) = a_0^{-1/2}\left[\sin kr + t_0^0(1 - e^{-r}) \cos kr + \sum_{a=1}^{n} c_a^0 \left(\frac{r}{a_0}\right)^a e^{-2.5r/a_0}\right] \tag{152}$$

where $n = 6$ for all results except some of those in Table IV. The phase shift is always given mod $\pi$ in discussing the results. For $n = 6$, the lowest two eigenvalues $E_\mu^0$ of $\mathbf{M}^{nn}$ correspond using

$$E_\mu^l = \frac{\hbar^2 k_\mu^2}{2m} \tag{153}$$

to $k_1 = 0.264362a_0^{-1}$ and $k_2 = 1.042019a_0^{-1}$.

We have done calculations for $k$ in the range 0.1 to $1.0a_0^{-1}$ designed to illustrate the various facets of the methods. Similar calculations were presented by Nesbet.[13] First we discuss the spurious singularities which are of great historical importance. Our explanations follow closely the work of

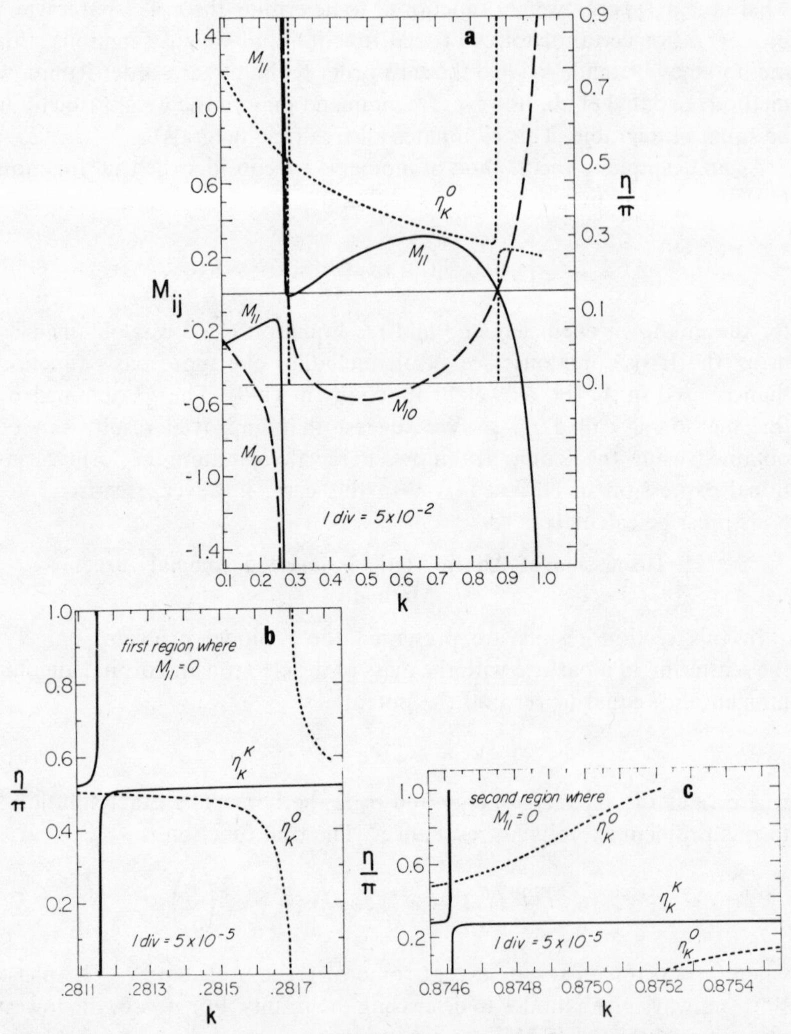


**Fig. 1.** (*a*) The quantities $M^{11}$ and $M^{10}$ (in Ry) and the zero-order Kohn phase shift $\eta_K{}^0$ (in radians) are shown as functions of momentum $k$ (in atomic units) for scattering of a particle with the mass of an electron from an attractive exponential potential. The scale on the left applies to the $M^{\alpha\beta}$ and the scale on the right to $\eta_K{}^0$. On the scale of (*a*), $\eta_K{}^0$ and $\eta_K{}^K$ could not be distinguished except near the regions where $M^{11}$ is not about equal to zero. (*b*) and (*c*) With a more expanded abscissa these are the only two regions where variationally corrected Kohn phase shift $\eta_K{}^K$ differs appreciably from $\eta_K{}^0$. The $M^{\alpha\beta}$ are labeled $M_{\alpha\beta}$ as in Ref. 13.

Nesbet's *Analysis of the Harris Variational Method in Scattering Theory.*[13] In this work, however, Nesbet missed a pseudoresonance in his Rubinow method calculations because it was too narrow to be seen in his Table V. This makes Sections VI and VII of his paper a little confusing. The mistake was corrected by Harris and Michels.[16] The correct conclusions of Nesbet's work are as follows: $t_K{}^K$ and $t_R{}^R$ show irregular behavior (i.e., the computed values show a wild fluctuation) when $E$ is near $E_\mu{}^l$. In addition, $t_K{}^K$ and $t_R{}^R$ have poles at the zeroes, respectively, of $M^{11}$ and $M^{00}$. For numerical accuracy, $t_K{}^K$ or $t_H$ should be used when

$$R = \frac{|M^{00}|}{|M^{11}|} \tag{154}$$

is less than unity and $t_R{}^R$ or $t_H$ should be used when $R$ is greater than unity. The following is an elaboration and continuation of this analysis.

Fig. 1*a* shows the zero-order Kohn phase shift and $M^{11}$ and $M^{10}$ for $k = 0.1$–$1.0a_0^{-1}$ and $n = 6$. The zero-order Kohn phase shift shows in one energy region a rapid increase by $\pi$ (a resonance) and in another energy region a rapid decrease by $\pi$ (an antiresonance). Since the exact solution[54] has no resonances the resonance is spurious and may be called a pseudoresonance. Further, an antiresonance as sharp as the one shown is forbidden by causality[55] and thus the antiresonance is spurious. It may be called a pseudoantiresonance. These spurious results may both be called spurious singularities since a change by $\pi$ of the phase shift in an energy interval causes both the tangent and the cotangent of the phase shift to become zero (singular) within that interval. The plot shows that the spurious singularities both occur near energies at which the matrix elements $M^{11}$ and $M^{10}$ become zero. Since $\eta_K{}^0$ converges to the exact result as $n$ is increased, (78) shows that the zeroes of $M^{11}$ and $M^{10}$ coincide in this limit except in regions where the phase shift goes smoothly through $\pi/2$ or 0. For finite $n$ these zeroes are shifted slightly relative to one another. It is this shift that causes the spurious zeroes. Fig. 1*b* is a blowup plot of the pseudoantiresonance shown in Fig. 1*a*, and Fig. 1*c* is a blowup of the pseudoresonance shown in Fig. 1*a*. From Fig. 1*b* we see that even though the zero-order Kohn phase shift exhibits pseudoantiresonance structure the variationally corrected Kohn phase shift shows pseudoresonance structure. Also note from Figs. 1*b* and 1*c* that the spurious zeroes of $\eta_K{}^K$ and $\eta_K{}^0$ are not centered about the same $k$.

The nature of the spurious results (increases and decreases of $\eta$ by $\pi$ as opposed to $\pi/2$, $2\pi$, or some other values) and their general location are explained as follows. Nesbet[13] has proved that when $E$ equals $E_\mu{}^l$ (i.e., when $\mathbf{M}^{\eta\eta}$ has a zero eigenvalue), DET $M$ and all four $M^{\alpha\beta}$ ($\alpha, \beta = 0, 1$) have

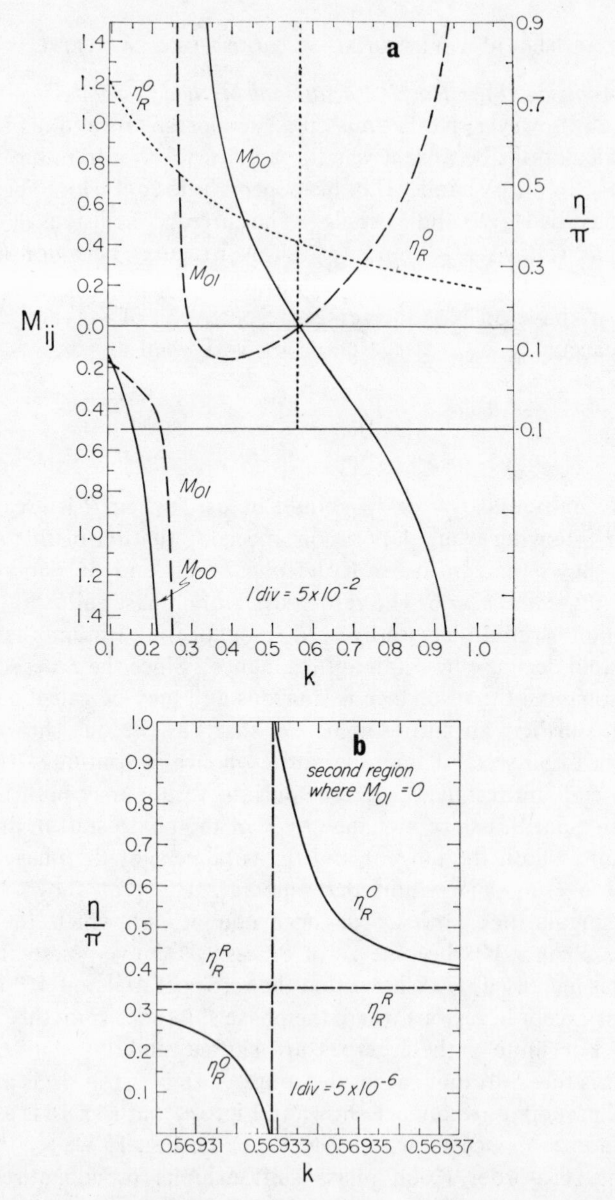


**Fig. 2.** (*a*) The integrals $M^{00}$ and $M^{01}$ (in Ry) and the zero-order Rubinow phase shift $\eta_R{}^0$ (in radians) are shown as functions of momentum $k$ (in atomic units) for scattering of a particle with the mass of an electron from an attractive exponential potential. The scale on the left applies to $M^{\alpha\beta}$ and the scale on the right to $\eta_K{}^0$. On the scale of part $a$, $\eta_R{}^0$ and $\eta_R{}^R$ could not be distinguished except near the region where $M^{00}$ is equal to zero. (*b*) With a more expanded abscissa this is the only region where the variationally corrected Rubinow phase shift $\eta_R{}^R$ differs appreciably from $\eta_R{}^0$. The $M^{\alpha\beta}$ are labeled $M_{\alpha\beta}$ as in Ref. 13.

poles of odd order. This behavior is illustrated graphically in Figs. 1*a*, 2*a*, and 3. Because of these odd-ordered poles, DET $M$ and each $M^{\alpha\beta}$ tend to pass through every value in the range $-\infty$ to $+\infty$ in the vicinity of each zero eigenvalue of $\mathbf{M}^{nn}$. Thus, for example, $M^{10}$ and $M^{11}$ each tend to pass through zero near each zero eigenvalue of $\mathbf{M}^{nn}$. By (78), this means $\eta_K{}^0$ tends to pass through both 0 and $\pi/2$ near each zero eigenvalue of $\mathbf{M}^{nn}$. Depending on the order in which $M^{10}$ and $M^{11}$ reach zero, this will be manifest as a resonancelike or antiresonancelike behavior. As explained above, in the limit where the basis set is large enough to represent the exact solution, if the structure is spurious the zeroes of $M^{10}$ and $M^{11}$ coincide and the structure disappears. Thus as the basis set is made more accurate, the zeroes of $M^{10}$ and $M^{11}$ move closer (except near real resonances) and thus the energies at which $\eta_K{}^0$ equals 0 and $\pi/2$ move closer; that is, the width of the pseudoresonance or pseudoantiresonance becomes zero.

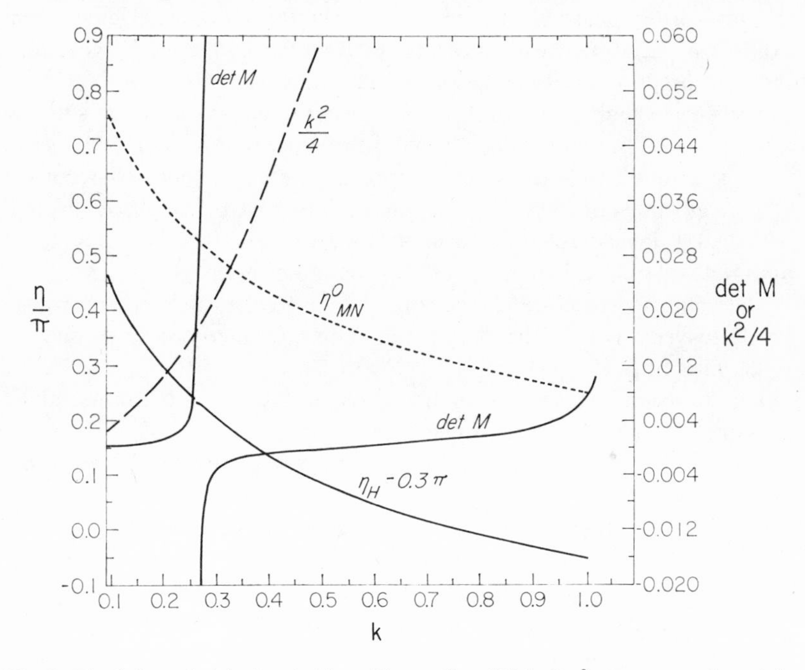


**Fig. 3.** The left- and right-hand sides of inequality (77) in $\text{Ry}^2$, the zero-order minimum-norm phase shift $\eta_{MN}{}^0$ (in radians), and the Hulthén phase shift $\eta_H$ (in radians) as functions of momentum $k$ (in atomic units) for scattering of a particle with the mass of an electron from an attractive exponential potential. The curve labeled det $M$ (following Ref. 13) is $\text{Ry}^{-2}$ DET $M$ and has been extended beyond $k = 1.0\, a_0{}^{-1}$ to indicate more clearly where its pole is.

Comparison of (78) and (80a) shows $\eta_K{}^0$ and $\eta_K{}^K$ are $\pi/2$ at the same place, namely, where $M^{11}$ vanishes (see Figs. 1*b* and 1*c* for examples). Since both $M^{10}$ and $(-2ma_0 \text{ DET } M/\hbar^2 k)$ tend to take on every value from $-\infty$ to $+\infty$ near a zero eigenvalue of $\mathbf{M}^{\eta\eta}$, they tend to cross; thus $t_K{}^K$ given by (80a) tends to have a zero near every zero eigenvalue of $\mathbf{M}^{\eta\eta}$. (Alternatively,[56] (80b) shows that $t_K{}^K$, like $t_K{}^0$, cannot vary smoothly near a zero of $M^{11}$ unless $M^{10}$ also vanishes.) Thus $\eta_K{}^K$, like $\eta_K{}^0$, tends to pass through both 0 and $\pi/2$ near each zero eigenvalue of $\mathbf{M}^{\eta\eta}$. This is the reason $\eta_K{}^K$ shows resonance, pseudoresonance, or pseudoantiresonance behavior near each zero eigenvalue of $\mathbf{M}^{\eta\eta}$. Figs. 1*b* and 1*c* illustrate the important fact that the width of the spurious structure in $\eta_K{}^K$ tends to be narrower than the width of the spurious structure in $\eta_K{}^0$.

Similar spurious singularities should occur in $\eta_R{}^0$ and $\eta_R{}^R$ for the same kinds of reasons. Fig. 2*a* is a graph similar to Fig. 1*a* except that quantities pertinent to the Rubinow method have been plotted. Fig. 2*a* shows that the zero-order phase shift varies smoothly through the eigenvalue but exhibits a pseudoantiresonance around $k$ equal to $0.55a_0{}^{-1}$. Note again that the pseudoantiresonance occurs in the region where both $M^{00}$ and $M^{01}$ vanish. The first zero of $M^{01}$ just causes the phase shift to go smoothly through $\pi/2$. Thus a zero of $M^{01}$ that is not close to a zero of $M^{00}$ does not cause spurious structure. Thus the spurious structure tends to occur near zero eigenvalues of $\mathbf{M}^{\eta\eta}$. This tendency is destroyed if the "background" $M^{01}$ or $M^{10}$ is near zero; however, using (70) we see the background $M^{01}$ and $M^{10}$ will not both be near zero for the same energy. Thus the tendency of the spurious structure to occur near a zero eigenvalue of $\mathbf{M}^{\eta\eta}$ will not be destroyed in both the Kohn and Rubinow methods near one zero eigenvalue of $\mathbf{M}^{\eta\eta}$.

Fig. 2*b* shows that the zeroes of $\eta_R{}^0$ and $\eta_R{}^R$ occur at the same value of $k$, namely, where $M^{00}$ is zero. This is explained by (79) and (83). Further the rapid changes by $\pi$ of $\eta_R{}^0$ and $\eta_R{}^R$ are centered about the same $k$ in this case.

The fact that the zeroes of $M^{00}$ and $M^{01}$ (which when they occur near each other cause structure in the Rubinow method) and the zeroes of $M^{10}$ and $M^{11}$ (which when they occur near each other cause structure in the Kohn method) do not generally occur in the same energy regions (the exception being when there really is a resonance and the basis set is accurate enough that both the Kohn and Rubinow methods show the resonance) forms the basis of Nesbet's original method, also called the anomaly-free method.[13,14] Nesbet suggested calculating the ratio $R$ [see (154)] and using the Kohn method when $R$ is less than unity and the Rubinow method otherwise. One possible difficulty with using this procedure can be seen as follows. Fig. 1 shows $M^{11}$ may be very large at energies very near to the spurious singularities identified by its zeroes. If this occurs where the accurate phase shift is passing smoothly through 0(mod $\pi$), then $M^{00}$ will be small but the Rubinow method may have no spurious structure at this or nearby energies.[57] In this case $R$ will be less than unity but the Rubinow method would be preferred. An alternative method to avoid spurious singularities is to calculate $\eta_K{}^K$, $\eta_R{}^R$, $\eta_H$, and one of the minimum-norm phase shifts. If three out of four of these quantities agree fairly well, they are probably free of spurious singularities. If no three out of four agree, the basis set should be improved.

Fig. 3 shows a plot of ($\mathscr{R}^{-2}$ DET $M$) against $(\hbar^2 k/4m\ \mathscr{R})^2$. Since DET $M$ tends to zero as the basis set becomes more accurate, it is generally small for a basis set that is fairly accurate. However, due to the odd-ordered pole of DET $M$ near each zero eigenvalue of $\mathbf{M}^{\prime\prime\prime}$, there will be a region near each zero eigenvalue of $\mathbf{M}^{\prime\prime\prime}$ in which (77) is true. This region will be bounded on one side by the zero eigenvalue of $\mathbf{M}^{\prime\prime\prime}$ and on the other by the energy where (77) with an equal sign holds. In this region then $t_H$ will be complex. Fig. 3 shows a plot of $\eta_H$ which has a narrow break in the energy region just below the first zero eigenvalue of $\mathbf{M}^{\prime\prime\prime}$. In this region, (77) is true. We have observed that the result $\eta_H$ often shows a change in slope very close to the break.

Fig. 3 also shows that $\eta_{MN}{}^0$ varies smoothly as one passes through the zero eigenvalue of $\mathbf{M}^{\prime\prime\prime}$. The denominator of (126) vanishes once in the energy region shown in Fig. 3; namely, it vanishes where $\eta_{MN}{}^0$ passes smoothly through $\pi/2$.

Fig. 4 is a plot of the variational corrections

$$\Delta_I{}^J = \eta_I{}^J - \eta_I{}^0 \tag{155}$$

These corrections are generally small but become large in the regions of spurious structure in the results. The spurious structure in $\eta_K{}^K$ and $\eta_R{}^R$ seen in Figs. 1 and 2 is seen again here in $\Delta_K{}^K$ and $\Delta_R{}^R$. In the energy region shown, $\Delta_{MN}{}^R$ is always small, indicating there is no spurious structure in $\eta_{MN}{}^R$. However, the figure shows there is one pseudo-resonance in $\eta_{MN}{}^K$. Thus the use of the Kohn variational expression to correct the singularity-free $\eta_{MN}{}^0$ has resulted in introducing a pseudo-resonance. In other cases (no examples are shown), the Rubinow variational expression introduces spurious singularities into the minimum-norm method. The figure shows, however, that $\eta_{MN}{}^{MN}$ is free of spurious singularities. Nesbet and Oberoi[48] have pointed out that $t_{MN}^{OMN}$ is not free of the spurious singularities.

Fig. 4 shows that $\Delta_R{}^K$ and $\Delta_K{}^R$ have two regions each of spurious

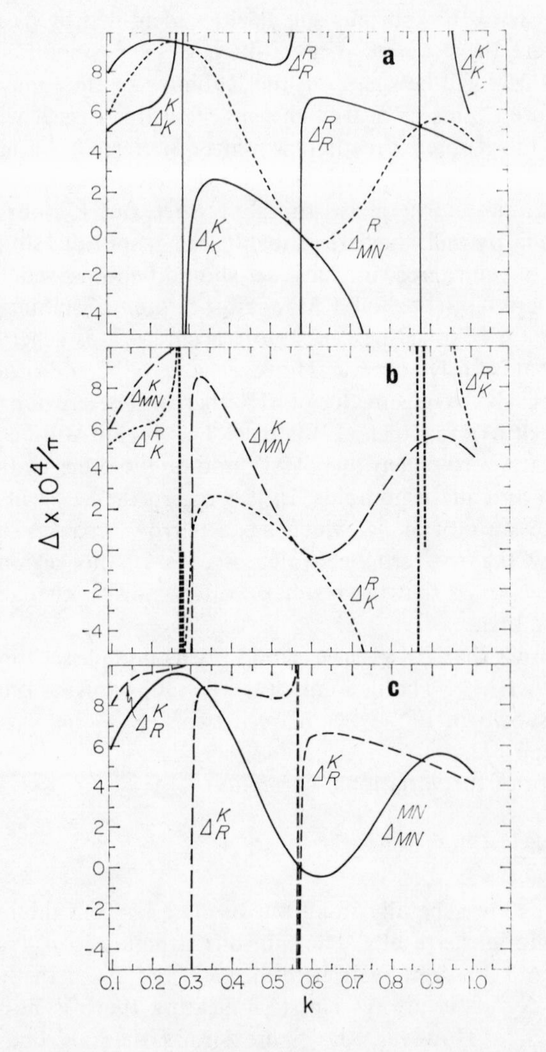


**Fig. 4.** The first-order corrections $\Delta_A{}^B$ (in radians) to the zero-order phase shift (mod $\pi$) in method $A$ as corrected using variational expression $B$ as functions of $k$ (in atomic units) for scattering of a particle with the mass of an electron from an attractive exponential potential. The abbreviations for the methods are $K$ = Kohn, $R$ = Rubinow, and $MN$ = minimum-norm.

structure. It can be shown that

$$\Delta_R{}^K = -\left(\frac{2m}{\hbar^2 k}\right) t_R{}^0 \frac{\text{DET } M}{M^{01}} \tag{156}$$

and thus $\Delta_R{}^K$ has a pole if and only if $M^{01} = 0$. This happens twice in the energy region shown, once because of a pseudoantiresonance in $t_R{}^0$ and once because $\eta_R{}^0$ passes smoothly through $\pi/2$. Further analysis shows that the Kohn variational correction does not make the region of spurious structure in $\eta_R{}^0$ narrower. In comparison, Fig. 2*b* shows that although the Rubinow variational correction does not remove the spurious structure it does make it narrower. Further the Kohn variational correction introduces a new region of spurious structure not found in $\eta_R{}^0$. The Rubinow variational correction does not introduce new regions of spurious structure not found in $\eta_R{}^0$. We conclude $\eta_R{}^R$ is to be preferred to $\eta_R{}^K$. A similar analysis would show $\eta_K{}^K$ is to be preferred to $\eta_K{}^R$.

Figs. 5*a* and 5*b* show with expanded abscissas two regions where the nonstationary methods lead to unusual spurious structure. These figures show spurious structures in $\eta_R{}^K$ and $\eta_K{}^R$ which are wider than those in $\eta_K{}^K$, $\eta_R{}^R$, $\eta_K{}^0$, and $\eta_R{}^0$. To analyze these structures we write (81) and (82) as

$$t_R{}^K = \frac{-M^{00}[M^{01} + (2ma_0/\hbar^2 k) \text{ DET } M]}{(M^{01})^2} \tag{81a}$$

and

$$t_K{}^R = \frac{(M^{10})^2}{M^{11}[-M^{10} + (2ma_0/\hbar^2 k) \text{ DET } M]} \tag{82a}$$

respectively. Equation (81a) and Figs. 1 through 3 can be used to show that $\eta_R{}^K$ must pass successively through $0 (\text{mod } \pi)$ when $M^{00}$ is zero, through $(\pi/2)(\text{mod } \pi)$ when $M^{01}$ is zero, and through $0(\text{mod } \pi)$ when $[M^{01} + (2ma_0/\hbar^2 k) \text{ DET } M]$ is zero and that these three locations are all very near to $k = 0.57 a_0{}^{-1}$. Notice that at the zeroes of $M^{01}$ and $M^{00}$, $\eta_R{}^K$ equals $\eta_R{}^0$ but $\eta_R{}^K$ has additional spurious behavior due to an extra zero of the denominator. This is not accidental; we can expect the zeroes of the three factors in $t_R{}^K$ will often occur near each other near a zero eigenvalue of $\mathbf{M}^{\eta\eta}$. A similar analysis holds for Fig. 5*a* and $t_K{}^R$.

Note that Figs. 4*c* and 4*b* show that $\eta_R{}^K$ and $\eta_{MN}{}^K$ exhibit spurious structure near $k = 0.30 a_0{}^{-1}$. These are attributed to unphysical zeroes of the numerators in the formulas for $t_R{}^K$ and $t_{MN}{}^K$. The former has a zero numerator near $k = 0.299245 a_0{}^{-1}$, and the latter has a zero numerator near $k = 0.299270 a_0{}^{-1}$. In this region $(0.299245 a_0{}^{-1} \leqslant k \leqslant 0.299270 a_0{}^{-1})$ the background phase shift has a value of $0.5018\pi$, and passes smoothly through $\pi/2$ near $k = 0.30160 a_0{}^{-1}$. The zero numerators mentioned above

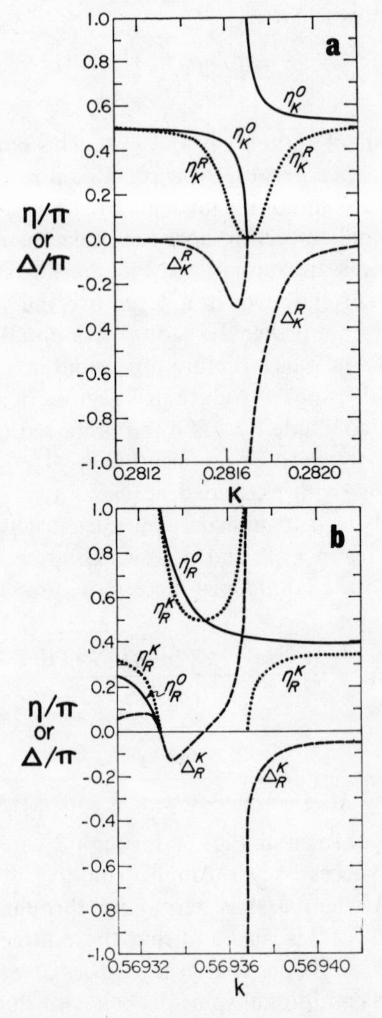


**Fig. 5.** (*a*) Phase shifts $\eta_K{}^0$ and $\eta_K{}^R$ (in radians) and first-order correction $\Delta_K{}^R$ (in radians) as functions of wave number $k$ (in atomic units) for the potential scattering problem of Section II. The range of the abscissa is chosen to more clearly illustrate spurious structure shown in Fig. 4*b*. (*b*) Phase shifts $\eta_R{}^0$ and $\eta_R{}^K$ (in radians) and first-order correction $\Delta_R{}^K$ (in radians) as functions of wave number $k$ (in atomic units) for the potential scattering problem of Section II. The range of the abscissa is chosen to more clearly illustrate spurious structure shown in Fig. 4*c*.

cause $\eta_R{}^K$ and $\eta_{MN}{}^K$ to increase to a value of $\pi(\text{mod }\pi)$ near $k = 0.299245a_0{}^{-1}$ and $k = 0.299270a_0{}^{-1}$, respectively. Then $\eta_{MN}{}^K$ and $\eta_R{}^K$ increase to the background phase shift (mod $\pi$) for $k$ a little larger than $0.30160a_0{}^{-1}$. A plot of $\Delta_R{}^K$ or $\Delta_{MN}{}^K$ with a more expanded ordinate would show a discontinuity where $\Delta$ changes from $0.4982\pi$ to $-0.5018\pi$. It is interesting to note that these variationally corrected phase shifts have spurious structures but their zero-order counterparts remain smooth throughout the region. Also note that $\eta_R{}^K$ and $\eta_{MN}{}^K$ do not attain the value of $\pi/2$ in this region.

The quantity $t_{OAF}$ appearing in Tables I–IV is discussed in Section II.I.

Table I illustrates the behavior of several of the computed tangents of the phase shifts near $E_1{}^0$. We see that in all cases the results vary smoothly. This smooth behavior is a result of the fact, which has been discussed, that the pseudoresonances and pseudoantiresonances do not occur at zero eigenvalues of $\mathbf{M}^{\eta\eta}$ but rather "nearby" where $M^{\alpha\beta}$ elements have correlated

TABLE I

The Values of det $M^{\eta\eta}$, DET $M$, $M^{00}$, $M^{01}$, $M^{11}$, and the Tangents of the Phase Shifts as Calculated by Various Methods Discussed in the Text for the Potential Scattering Problem of Section II. Values of $k$ Were Picked so That the Incident Energies Bracket an Eigenvalue of the Bound-Bound Matrix $\mathbf{H}^{\eta\eta}$. Energies are in Hartrees and $k$ is in $a_0{}^{-1}$

| $k$ | 0.26436165 | 0.26436170 | 0.26436175 | 0.26436180 |
|---|---|---|---|---|
| det $M^{\eta\eta}$ | 8.0189211(−23)[a] | 3.2692534(−23) | −1.4803981(−23) | −6.2300675(−23) |
| DET $M$ | 2.3254257(+2) | 5.7038538(+2) | −1.2596558(+3) | −2.9932065(+2) |
| $M^{00}$ | −5.7978167(+5) | −1.4221058(+6) | 3.1405237(+6) | 7.4625612(+5) |
| $M^{01}$ | −5.5541916(+4) | −1.3623467(+5) | 3.0085495(+5) | 7.1489513(+4) |
| $M^{11}$ | −5.3208168(+3) | −1.3051001(+4) | 2.8821199(+4) | 6.8485076(+3) |
| $t_K{}^0$ | −10.438632 | −10.438648 | −10.438664 | −10.438680 |
| $t_K{}^K$ | −10.107992 | −10.108008 | −10.108012 | −10.108028 |
| $t_K{}^R$ | −10.118143 | −10.118159 | −10.118164 | −10.118180 |
| $t_K{}^P$ | −10.118053 | −10.118069 | −10.118074 | −10.118089 |
| $t_R{}^0$ | −10.438633 | −10.438648 | −10.438664 | −10.438680 |
| $t_R{}^R$ | −10.118143 | −10.118159 | −10.118164 | −10.118180 |
| $t_R{}^K$ | −10.107992 | −10.108008 | −10.108012 | −10.108028 |
| $t_H$ | b | b | −10.437435 | −10.436164 |
| $t_P{}^0$ | −10.438633 | −10.438648 | −10.438664 | −10.438680 |
| $t_P{}^P$ | −10.118053 | −10.118069 | −10.118074 | −10.118089 |
| $t_{MN}{}^0$ | −10.438633 | −10.438648 | −10.438664 | −10.438680 |
| $t_{MN}{}^K$ | −10.107992 | −10.108008 | −10.010801 | −10.108028 |
| $t_{MN}{}^R$ | −10.118143 | −10.118159 | −10.118164 | −10.118180 |
| $t_{MN}{}^{MN}$ | −10.118054 | −10.118069 | −10.118075 | −10.118090 |
| $t_{MN}^{OMN}$ | −10.118054 | −10.118069 | −10.118075 | −10.118090 |
| $t_{EI}{}^0$ | −10.438624 | −10.438645 | −10.438666 | −10.438686 |
| $t_{EI}{}^K$ | −10.107985 | −10.108001 | −10.108017 | −10.108033 |
| $t_{MM}{}^0$ | −10.438633 | −10.438648 | −10.438664 | −10.438680 |
| $t_{MM}{}^K$ | −10.107988 | −10.108002 | −10.108017 | −10.108031 |
| $t_{AF}{}^{AF}$ | −10.118143 | −10.118159 | −10.118164 | −10.118180 |
| $t_{OAF}$ | −10.118054 | −10.118070 | −10.118074 | −10.118090 |
| $t_{\text{exact}}$ | −10.116471 | −10.116486 | −10.116500 | −10.116515 |

[a] Numbers in parentheses are multiplicative powers of 10.
[b] Complex (nonphysical) tangent.

TABLE II

The values of det $M^{\eta\eta}$, DET $M$, $M^{00}$, $M^{01}$, $M^{11}$, and Tangents of the Phase Shift as Calculated by Various Methods Discussed in the Text for the Potential Scattering Problem of Section II. Values of $k$ were Picked Such That DET $M$ Passes Through Zero as a Function of Momentum. Energies are in Hartrees and $k$ is in $a_0^{-1}$

| $k$ | 0.5675 | 0.5700 | 0.5725 | 0.5750 |
|---|---|---|---|---|
| det $M^{\eta\eta}$ | −3.0274153(−16)[a] | −3.0459831(−16) | −3.0643175(−16) | −3.0824155(−16) |
| DET $M$ | −6.4422255(−6) | −3.2612590(−6) | −6.5297276(−8) | 3.1459858(−6) |
| $M^{00}$ | 2.7988344(−3) | −1.0242731(−3) | −4.8290998(−3) | −8.6162553(−3) |
| $M^{01}$ | −1.3655824(−3) | 4.8366802(−4) | 2.3512792(−3) | 4.2372438(−3) |
| $M^{11}$ | 1.3680930(−1) | 1.3753433(−1) | 1.3824325(−1) | 1.3893592(−1) |
| $t_K{}^0$ | 2.0495537 | 2.1177192 | 2.0538181 | 2.0334575 |
| $t_K{}^K$ | 2.0842025 | 2.0687764 | 2.0536189 | 2.0387227 |
| $t_K{}^R$ | 2.0842025 | 2.0687764 | 2.0536189 | 2.0397227 |
| $t_K{}^P$ | 2.0842025 | 2.0687764 | 2.0536189 | 2.0387227 |
| $t_R{}^0$ | 2.0495537 | 2.1177192 | 2.0538181 | 2.0334575 |
| $t_R{}^R$ | 2.0842053 | 2.0687745 | 2.0536189 | 2.0387225 |
| $t_R{}^K$ | 2.0836292 | 2.0676165 | 2.0536189 | 2.0387089 |
| $t_H$ | 2.0842024 | 2.0687764 | 2.0536189 | 2.0387227 |
| $t_P{}^0$ | 2.0833340 | 2.0683340 | 2.0536100 | 2.0391558 |
| $t_P{}^P$ | 2.0842025 | 2.0687764 | 2.0536189 | 2.0387227 |
| $t_{MN}{}^0$ | 2.0840331 | 2.0686938 | 2.0536173 | 2.0387965 |
| $t_{MN}{}^K$ | 2.0842025 | 2.0687764 | 2.0536189 | 2.0387227 |
| $t_{MN}{}^R$ | 2.0842025 | 2.0687764 | 2.0536189 | 2.0387227 |
| $t_{MN}{}^{MN}$ | 2.0842025 | 2.0687764 | 2.0536189 | 2.0387227 |
| $t_{MN}^{OMN}$ | 2.0842025 | 2.0687764 | 2.0536189 | 2.0387227 |
| $t_{EI}{}^0$ | 2.0621128 | 2.0444639 | 2.0271283 | 2.0100965 |
| $t_{EI}{}^K$ | 2.0839707 | 2.0684931 | 2.0532800 | 2.0383245 |
| $t_{MM}{}^0$ | 2.0777692 | 2.0626769 | 2.0478431 | 2.0332608 |
| $t_{MM}{}^K$ | 2.0841835 | 2.0687589 | 2.0536028 | 2.0387078 |
| $t_{AF}{}^{AF}$ | 2.0842025 | 2.0687764 | 2.0536189 | 2.0334575 |
| $t_{OAF}$ | 2.0842025 | 2.0687764 | 2.0536189 | 2.0387227 |
| $t_{\text{exact}}$ | 2.0842069 | 2.0687798 | 2.0536219 | 2.0387254 |

[a] Numbers in parentheses are multiplicative powers of 10.

zeroes. It is especially interesting that $t_{MN}{}^0$, which at the eigenvalue is $t_{\text{Harris}}$, varies smoothly in this energy region. It is also interesting to note the much greater accuracy of the variationally corrected results when compared to the zero-order results.

Table I also illustrates (110). In addition it shows that $t_P{}^0$ may be added to the list in that equation.

Table II illustrates many of the computed tangents of the phase shifts near a zero of DET $M$. This provides an illustration of (87) and (133). In addition it shows that $t_P{}^P$, $t_{OAF}$, and $t_{MN}^{OMN}$ may be added to the lists in those equations.

Table III further illustrates the possible accuracies attainable by the various methods. Note that much more accurate values are obtained from variationally corrected results than from zero-order results. In particular, if the Kohn, Rubinow, Percival, minimum-norm, or Euler integral methods or the method of moments are used without a variational

TABLE III

The Values of det $M^{\eta\eta}$, DET $M$, $M^{C0}$, $M^{01}$, $M^{11}$, and the Tangents of the Phase Shifts Calculated by Various Methods Discussed in the Text for the Potential Scattering Problem of Section II for Various Values of $k$. Energies are in Hartrees and $k$ is in $a_0^{-1}$

| $k$ | 0.15 | 0.25 | 0.35 | 0.45 | 0.55 |
|---|---|---|---|---|---|
| det $M^{\eta\eta}$ | 9.1036780(−17)[a] | 1.3413482(−17) | −8.7504072(−17) | −1.9487821(−16) | −2.8911621(−16) |
| DET $M$ | 6.6835540(−5) | 1.2576473(−3) | −3.1940408(−4) | −1.4703525(−4) | −2.8317410(−5) |
| $M^{00}$ | −2.3153279(−1) | −3.2721365 | 6.2225763(−1) | 2.2012822(−1) | 3.0126828(−2) |
| $M^{01}$ | −1.3179002(−1) | −4.5252914(−1) | −7.0271727(−2) | −6.6204382(−2) | −1.3794608(−2) |
| $M^{11}$ | −1.1799494(−1) | −8.0255335(−2) | 2.7185306(−2) | 8.6912848(−2) | 1.3129464(−1) |
| $t_K{}^0$ | −1.7525330 | −7.1961464 | 9.0222171 | 3.3505332 | 2.1995918 |
| $t_K{}^K$ | −1.7449806 | −7.0707818 | 9.0893550 | 3.3580521 | 2.2003761 |
| $t_K{}^R$ | −1.7450130 | −7.0729283 | 9.0898584 | 3.3580690 | 2.2003763 |
| $t_K{}^P$ | −1.7450051 | −7.0728881 | 9.0898523 | 3.3580677 | 2.2003763 |
| $t_R{}^0$ | −1.7568310 | −7.2307752 | 8.8550212 | 3.3249797 | 2.1839568 |
| $t_R{}^R$ | −1.7450314 | −7.0735084 | 9.0811453 | 3.3581271 | 2.2003820 |
| $t_R{}^K$ | −1.7449516 | −7.0700118 | 9.0850125 | 3.3577999 | 2.2002594 |
| $t_H$ | −1.7448887 | −7.0586424 | 9.0886691 | 3.3580304 | 2.2003758 |
| $t_P{}^0$ | −1.7553792 | −7.2299461 | 8.8620146 | 3.3322018 | 2.1966600 |
| $t_P{}^P$ | −1.7450143 | −7.0734440 | 9.0910691 | 3.3581058 | 2.2003772 |
| $t_{MN}{}^0$ | −1.7549206 | −7.2297241 | 8.8771385 | 3.3411921 | 2.1994221 |
| $t_{MN}{}^K$ | −1.7449717 | −7.0700579 | 9.0860854 | 3.3580184 | 2.2003761 |
| $t_{MN}{}^R$ | −1.7450278 | −7.0735079 | 9.0811221 | 3.3581036 | 2.2003765 |
| $t_{MN}{}^{MN}$ | −1.7450089 | −7.0734274 | 9.0909030 | 3.3580794 | 2.2003761 |
| $t_{MN}^{OMN}$ | −1.7450143 | −7.0734445 | 9.0810686 | 3.3581058 | 2.2003772 |
| $t_{EI}{}^0$ | −1.6579729 | −6.7482295 | 15.5206952 | 3.5847924 | 2.1953115 |
| $t_{EI}{}^K$ | −1.7309131 | −6.9419690 | 2.5291187 | 3.3368541 | 2.2003673 |
| $t_{MM}{}^0$ | −1.7461007 | −7.2140997 | 8.7608185 | 3.3158178 | 2.1913031 |
| $t_{MM}{}^K$ | −1.7449155 | −7.0705748 | 9.0787405 | 3.3575866 | 2.2003433 |
| $t_{AF}{}^{AF}$ | −1.7450314 | −7.0735084 | 9.0911453 | 3.3581271 | 2.2003761 |
| $t_{OAF}$ | −1.7450143 | −7.0734445 | 9.0910685 | 3.3581058 | 2.2003772 |
| $t_{\text{exact}}$ | −1.7449393 | −7.0726191 | 9.0918095 | 3.3581425 | 2.2003827 |

[a] Numbers in parentheses are multiplicative powers of 10.

TABLE IV

The Tangents of the Phase Shifts for the Various Methods Discussed in the Text for the Potential Scattering Problem of Section II at $k = 0.55a_0^{-1}$ as a Function of the Number of Basis Functions used. At this Value of $k$ the Exact Tangent of the Phase Shift is Given by $t_{\text{exact}} = 2.2003827$. Energies are in Hartrees

| $n$ | 0 | 2 | 4 | 6 | 8 | 10 | 12 |
|---|---|---|---|---|---|---|---|
| det $M^{\eta\eta}$ | 0.0 | 1.2842504(−4) | −3.7528351(−11) | −2.8911621(−16) | −5.9559695(−21) | −4.65214224(−24) | 6.5801930(−24) |
| DET $M$ | 1.0728098(−3)[a] | 4.9480633(−3) | −5.5208351(−3) | −2.8317410(−5) | 2.9448730(−5) | 1.9167492(−5) | −7.4574873(−7) |
| $M^{00}$ | −2.7375566(−1) | −3.8358337(−1) | 1.4233691 | 3.0126829(−2) | −1.8205262(−1) | −6.8379396(−1) | 3.2180758(−1) |
| $M^{01}$ | 1.3169744(−1) | 1.9288660(−1) | −6.6702822(−1) | −1.3794608(−2) | 8.2843926(−2) | 3.1083106(−1) | −1.4625344(−1) |
| $M^{11}$ | 6.5020649(−2) | 2.8391510(−2) | 4.3758051(−1) | 1.3129464(−1) | 8.7279793(−2) | −1.6315694(−2) | 1.9144677(−1) |
| $t_K{}^0$ | 2.2039546 | 2.8921814 | 2.1528112 | 2.1995918 | 2.2016101 | 2.1961097 | 2.2003685 |
| $t_K{}^K$ | 2.1439563 | 2.2584372 | 2.1986902 | 2.2003761 | 2.2003832 | 2.2003817 | 2.2003827 |
| $t_K{}^R$ | 2.1455464 | 2.03723453 | 2.1996892 | 2.2003763 | 2.2003839 | 2.2003900 | 2.2003827 |
| $t_K{}^P$ | 2.1452790 | 2.3614951 | 2.1995098 | 2.2003763 | 2.2003838 | 2.2003886 | 2.2003827 |
| $t_R{}^0$ | 2.0786710 | 1.9886470 | 2.1338963 | 2.1839568 | 2.1975373 | 2.1998898 | 2.2003419 |
| $t_R{}^R$ | 2.1421248 | 2.1932379 | 2.2001139 | 2.2003820 | 2.2003816 | 2.2003827 | 2.2003827 |
| $t_R{}^K$ | 2.1402452 | 2.1741532 | 2.1981209 | 2.2002594 | 2.2003779 | 2.2003826 | 2.2003827 |
| $t_H$ | 2.1430802 | 2.2104555 | 2.1957556 | 2.2003758 | 2.2003827 | 2.2003828 | 2.2003827 |
| $t_P{}^0$ | 2.0386340 | 1.9133457 | 2.1383369 | 2.1966500 | 2.1937742 | 2.1998012 | 2.2003518 |
| $t_P{}^P$ | 2.1415288 | 2.1874769 | 2.1997742 | 2.2003772 | 2.2003801 | 2.2003826 | 2.2003827 |
| $t_{MN}{}^0$ | 2.1041835 | 2.0145743 | 2.1396162 | 2.1994221 | 2.1996814 | 2.1998789 | 2.2003587 |
| $t_{MN}{}^K$ | 2.1416028 | 2.1789209 | 2.1984131 | 2.2003761 | 2.2003820 | 2.2003825 | 2.2003827 |
| $t_{MN}{}^R$ | 2.1422803 | 2.1935190 | 2.2000746 | 2.2003765 | 2.2003822 | 2.2003827 | 2.2003827 |
| $t_{MN}{}^{MN}$ | 2.1425068 | 2.1952952 | 2.1996741 | 2.2003761 | 2.2003824 | 2.2003827 | 2.2003827 |
| $t_{MN}{}^{OMN}$ | 2.1415407 | 2.1875236 | 2.1997719 | 2.2003772 | 2.2003801 | 2.2003826 | 2.2003827 |
| $t_{EI}{}^0$ | b | 1.6705050 | 2.1438854 | 2.1953115 | 2.1947616 | 2.2023810 | 2.1985242 |
| $t_{EI}{}^K$ | b | 2.1043494 | 2.1985634 | 2.2003673 | 2.2003683 | 2.2003840 | 2.2003803 |
| $t_{MM}{}^0$ | 1.0740540 | 2.1478812 | 2.1274896 | 2.1913031 | 2.2012906 | 2.2001552 | 2.2003152 |
| $t_{MM}{}^K$ | 1.8421008 | 2.2012430 | 2.1976699 | 2.2003433 | 2.2003831 | 2.2003827 | 2.2003827 |
| $t_{AF}{}^{AF}$ | 2.1421248 | 2.1932379 | 2.20011387 | 2.2003761 | 2.2003816 | 2.2003827 | 2.2003827 |
| $t_{OAF}$ | 2.1415157 | 2.1871933 | 2.1997718 | 2.2003772 | 2.2003801 | 2.2003826 | 2.2003827 |

[a] Numbers in parentheses are multiplicative powers of 10.
[b] Undefined.

correction to $t_l^0$, one finds a larger basis set is required than for variationally corrected calculations. That is, zero-order calculations converge slower than first-order ones.

Note that neither the Euler integral method nor the method of moments eliminates pseudoresonances and pseudoantiresonances. Table III gives values of $t_{EI}^0$ and $t_{MM}^0$ obtained as discussed above. The results were corrected using (24) to give an indication of the error. The comparison of $t_{EI}^0$ and $t_{MM}^0$ to $t_{MN}^0$ is particularly interesting since these three methods are all zero order. Of these three methods $t_{EI}^0$ is much worse than the other two. Table IV compares some of the methods as a function of the number of square-integrable basis functions. At the energy shown (and at other energies we have examined) $t_{EI}^0$ converges poorly whereas $t_{MM}^0$ converges very well. It is interesting that $t_{EI}^K$ is often accurate when $t_{EI}^0$ is very inaccurate.

Table IV shows that in general the variationally corrected results converge much faster than the zero-order results. However, all the methods except the Euler integral method yield tangents that converge to within $7 \times 10^{-5}$ of the exact tangent of the phase shift at $n = 14$ (not shown in table).

It is of interest to compare the Born approximation results with the results obtained by the algebraic variational methods with $n = 0$. The Born approximation uses the trial function $S_l(r)$. Thus $t_B^0 = 0$ and the Born approximation result $t_B^K$ is computed from the one integral $M_{SS}$. At $k = 0.55a_0^{-1}$, this method yields $t_B^K = 0.99547511$. The table shows that all the other methods, even with $n = 0$ are superior to the Born approximation. In comparison of the computational effort required note that the Born approximation requires the evaluation of one free-free integral, the zero-order Kohn method requires two, and the variationally corrected Kohn method requires four. With $n = 0$, the algebraic variational methods are much more sensitive to the choice of $C_l(r)$ than with large $n$. As $n$ is increased the choice of $C_l(r)$ becomes less important.

As the size of the basis set is increased the widths of the regions affected by spurious singularities become smaller and eventually go to zero in the limit $n \rightarrow \infty$. At any given energy the calculations eventually (as $n$ is increased) converge monotonically (a bound principle becomes applicable) to the exact answer. These and other details about bounds and convergence may be found elsewhere.[31,43,58–65]

### I. The Optimized Anomaly-Free Method

One possible difficulty with the anomaly-free method was pointed out in Section II.H. Another difficulty is that the calculated phase shifts are not continuous functions of energy or of changes in the parameters in the basis set. Nesbet and Oberoi[48,49] have proposed a method they call the optimized anomaly-free method which is supposed to cure some of the difficulties of the anomaly-free method. It is formulated in such a way that it gives no pseudoresonances and no pseudoantiresonances. Some results computed by their formulas are included in Tables I through IV as $t_{\text{OAF}}$.

### J. The Minimum Variance Method and the Least-Squares Method

Recently Bardsley et al.[65] have proposed calculations based on minimization of the variance integral

$$U(X_l^0) = \int_0^\infty dr\, |w(r) L_l X_l^0(r)|^2 \tag{157}$$

or the variance sum

$$U_R(X_l^0) = \sum_i |w(r_i) L_l X_l^0(r_i)|^2 \tag{158}$$

where $w(r)$ is an arbitrary weighting function. Earlier Miller[66] had considered (157) with the special choice $w(r) = 1$. $U(X_l^0)$ and $U_R(X_l^0)$ are nonnegative; these schemes have the advantage of containing an internal criterion by which one can choose between two different basis sets used for $X_l^0$, that is, one chooses whichever of the two basis sets leads to a lower value of $U(X_l^0)$ or $U_R(X_l^0)$. The minimum-norm method also has this ability to choose between two different basis sets; that is, one chooses the basis set that yields the lowest eigenvalue of the matrix **Q**. The methods of the Kohn and Rubinow type also have a criterion for choosing between basis sets; that is, one chooses the basis set for which $|\text{DET } M|$ is smaller, or for which the variational correction to the phase shift is smaller. This criterion, however, is not quite as satisfying as the criterion of minimum variance or of minimum eigenvalues of the matrix **Q** since the latter converge monotonically as additional basis functions are added but $|\text{DET } M|$ and the corrections to the phase shift do not.

Ládanyi et al.[53] have considered a generalization of the method of moments based on use of the trial function

$$X_l^0(r) = \alpha_{l0}{}^0 S_l(r) + \alpha_{l1}{}^0 C_l(r) + \sum_{a=1}^n c_a{}^l \eta_a{}^l(r) \tag{159}$$

with

$$(\alpha_{l0}{}^0)^2 + (\alpha_{l1}{}^0)^2 = 1 \tag{160}$$

and minimization of

$$\lambda(X_l^0) = \sum_{i=0}^{k>n} \left| \int_0^\infty dr w_i(r) L_l X_l^0(r) \right|^2 \tag{161}$$

They call this the least-squares method. This method has a number of features in common with the method of minimization of the variance integral.

## K. Methods Involving Artificial Channel Radii

Another way to do scattering calculations by expanding the wave function in basis functions is to divide all space into an internal region and an external region. The wave function is expanded in a set of square-integrable basis functions in the internal region ($r < a$, where $a$ is the so-called "channel radius") and in terms of functions like $S_l$ and $C_l$ in the external region ($r > a$). The boundary is chosen so that the potential is zero or simple for $r > a$. Thus the difficult part is to solve for the wave function in the internal region, and for this part we can use techniques similar to those used in bound-state problems. Eventually the solutions in the internal and the external region are matched. Techniques that incorporate such an artificial "channel radius" are very popular in nuclear physics and are here called artificial channel radius theories.

The oldest of the artificial channel radius theories is the theory of Kapur and Peierls.[67] The most popular of the artificial channel radius theories is the derivative matrix technique of Wigner and Eisenbud.[68] The derivative matrix is often called the $R$ matrix (it should not be confused with the reactance or reaction matrices; to avoid this confusion we call it the $NR$ matrix). For single-channel scattering it is a $1 \times 1$ matrix (i.e., a number).

In the derivative matrix method one uses in the internal region the set of $n$ basis functions $\mu_{lj}(r)$ which have been orthonormalized as follows

$$\int_0^a dr \mu_{li}(r)\mu_{lj}(r) = \delta_{ij} \qquad i, j = 1, 2, \ldots, n \tag{162}$$

and which satisfy the boundary conditions

$$\mu_{lj}(0) = 0 \qquad j = 1, 2, \ldots, n \tag{163}$$

$$\frac{a}{\mu_{lj}(a)} \left.\frac{d\mu_{lj}}{dr}\right|_{r=a} = b \qquad j = 1, 2, \ldots, n \tag{164}$$

where $b$ is a constant.

Then one diagonalizes the matrix $\mathbf{H}^{\mu\mu l}$ where

$$H_{ij}^{\mu\mu l} = \int_0^a dr \mu_{li}(r)\left[-\frac{\hbar^2}{2m}\frac{d^2}{dr^2} + V(r) + \frac{l(l+1)\hbar^2}{2mr^2}\right]\mu_{lj}(r) \tag{165}$$

to obtain the eigenfunctions

$$w_{lk}(r) = \sum_j c_{ljk}\mu_{lj}(r) \tag{166}$$

and eigenvalues $E_{lk}$. Note the diagonalization must be done only once, not at every energy. The solution at given energy is expanded in the internal region as

$$X_l^0(r) = \sum_k A_k w_{lk}(r) \tag{167}$$

where we have suppressed a subscript $l$ on $A_k$. To determine the coefficients we require

$$\int_0^a dr \mu_{lj}(r) L_l X_l^0(r) = 0 \qquad j = 1, 2, \ldots, n \tag{168}$$

which, using integration-by-parts twice and (165) through (167), yields

$$X_l^0(a) = NR\left[ a \left.\frac{dX_l^0}{dr}\right|_{r=a} - bX_l^0(a) \right] \tag{169}$$

where $NR$ is the element of the derivative matrix and is given by

$$NR = \left(\frac{\hbar^2}{2ma}\right) \sum_{k=1}^{n} \frac{[w_{lk}(a)]^2}{E_{lk} - E} \tag{170}$$

The most common practice in derivative matrix theory is to use in the external region a linear combination of an incoming wave $I_l(r)$ and an outgoing wave $O_l(r)$ satisfying complex boundary conditions. However, the theory has also been stated in terms of the more convenient functions satisfying boundary conditions like (43) and (44). Thus we assume that in the external region

$$X_l^0(r) = \mathscr{S}_l(r) + t_l^0 \mathscr{C}_l(r) \tag{171}$$

where $\mathscr{S}_l(r)$ and $\mathscr{C}_l(r)$ have the asymptotic forms

$$\mathscr{S}_l(r) \underset{r\to\infty}{\sim} a_0^{-1/2} kr j_l(kr) \tag{172}$$

$$\mathscr{C}_l(r) \underset{r\to\infty}{\sim} a_0^{-1/2} kr \xi_l(kr) \tag{173}$$

and in addition satisfy

$$L_l \mathscr{S}_l = 0 \tag{174}$$

$$L_l \mathscr{C}_l = 0 \tag{175}$$

in the external region. Comparing (171) to (169) yields

$$\mathscr{S}_l(a) - NR\left[ a \left.\frac{d\mathscr{S}_l}{dr}\right|_{r=a} - b\mathscr{S}_l(a) \right] + t_l^0 \left\{ \mathscr{C}_l(a) - NR\left[ a \left.\frac{d\mathscr{C}_l}{dr}\right|_{r=a} - b\mathscr{C}_l(a) \right] \right\} = 0 \tag{176}$$

which can be solved for $t_l^0$. The derivative matrix method has the advantage of requiring no free-free integrals but the disadvantage that the square-integrable basis will probably have to be very large. This disadvantage results from the fact that the square-integrable basis must be used to expand the scattering wave function in the whole internal region but the channel radius may have to be large.

An alternative method of derivation of the derivative matrix method can be given[49,69] which shows that the variational correction to $t_l^0$ computed from (176) vanishes and that $t_l^0$ is stationary with respect to variation of the coefficients in the trial wave function. These properties are not obvious in the usual type of derivation (such as given above).

However, the derivative matrix method still has an important disadvantage,[49,69,70] namely, that the wave function has a discontinuous derivative at the channel radius. The method converges to the correct answer and the discontinuity vanishes as the number of basis functions is increased but in practical calculations the finite discontinuity in the derivative is undesirable.

A method that has a number of the derivative matrix method's advantages, including division into internal and external regions and variationally correct results, has been proposed by Crawford.[71] Crawford uses an analogue of the Kohn variational theory in which the surface terms in the derivation of the variational expression are calculated at a finite channel radius rather than in the asymptotic limit.[69] In this method in the external region we write

$$X_l^0(r) = \left(\frac{ma_0}{\hbar k}\right)^{1/2} [\alpha_{l0}{}^0 \mathscr{S}_l(r) + \alpha_{l1}{}^0 \mathscr{C}_l(r)] \tag{177}$$

where $\mathscr{S}_l(r)$ and $\mathscr{C}_l(r)$ are defined by (172) to (175). Note that the $X_l^0(r)$ of (177) has different units from all our previous $X_l^0(r)$ and that

$$t_l^0 = \frac{\alpha_{l1}{}^0}{\alpha_{l0}{}^0} \tag{178}$$

Then the expression

$$\left(\frac{\hbar}{2}\right)\alpha_{l0}{}^0 t_l \alpha_{l0}{}^0 = -J(X_l^0) + \left(\frac{\hbar}{2}\right)\alpha_{l0}\alpha_{l1}{}^0 \tag{179}$$

where

$$J(X_l^0) = \int_0^a dr X_l^0(r) L_l X_l^0(r) \tag{180}$$

is stationary under a variation that preserves $\alpha_{l0}$, that is,

$$\frac{\partial t_l}{\partial \alpha_{l1}{}^0} = 0 \tag{181}$$

Note that (178) and (179) are equivalent to (24) except that we have changed the normalization. In the interior region we expand

$$X_l^0(r) = \sum_i c_i \eta_i(r) \tag{182}$$

Now we require continuity of the trial function and its first derivative at $r = a$, that is,

$$\left(\frac{ma_0}{\hbar k}\right)^{1/2} [\alpha_{l0}{}^0 \mathscr{S}_l(a) + \alpha_{l1}{}^0 \mathscr{C}_l(a)] = \sum_i c_i \eta_i(a) \tag{183}$$

$$\left(\frac{ma_0}{\hbar k}\right)^{1/2} \left[ \alpha_{l0}{}^0 \left.\frac{d\mathscr{S}_l}{dr}\right|_{r=a} + \alpha_{l1}{}^0 \left.\frac{d\mathscr{C}_l}{dr}\right|_{r=a} \right] = \sum_i c_i \left.\frac{d\eta_i}{dr}\right|_{r=a} \tag{184}$$

These two equations can be solved for

$$\alpha_{l0}{}^0 = \sum_i \beta_{l0i} c_i \tag{185}$$

and

$$\alpha_{l1}{}^0 = \sum_i \beta_{l1i} c_i \tag{186}$$

where

$$\beta_{l0i} = -\left(\frac{\hbar}{kma_0}\right)^{1/2} \left[ \frac{d\mathscr{C}_l}{dr} \eta_i(r) - \mathscr{C}_l(r) \frac{d\eta_i}{dr} \right]\Bigg|_{r=a} \tag{187}$$

and

$$\beta_{l1i} = \left(\frac{\hbar}{kma_0}\right)^{1/2} \left[ \frac{d\mathscr{S}_l}{dr} \eta_i(r) - \mathscr{S}_l(r) \frac{d\eta_i}{dr} \right]\Bigg|_{r=a} \tag{188}$$

In deriving this result we use the constancy of the Wronskian of the external solutions $\mathscr{S}_l(r)$ and $\mathscr{C}_l(r)$, that is,

$$\mathscr{S}_l(r) \frac{d\mathscr{C}_l}{dr} - \frac{d\mathscr{S}_l}{dr} \mathscr{C}_l(r) = -k \tag{189}$$

We can rewrite (185) and (186) using the row vectors $\boldsymbol{\beta}_{l0}$ and $\boldsymbol{\beta}_{l1}$ and the column vector $\mathbf{c}$ as

$$\alpha_{l0}{}^0 = \boldsymbol{\beta}_{l0} \mathbf{c} \tag{190}$$

$$\alpha_{l1}{}^0 = \boldsymbol{\beta}_{l1} \mathbf{c} \tag{191}$$

Then we can write the stationary expression as

$$\sum_{ij} \beta_{l0i} c_i t_l \beta_{l0j} c_j = -\left(\frac{2}{\hbar}\right) \sum_{ij} c_i c_j L_{ij}{}^{\eta\eta} + \sum_{ij} \beta_{l0i} c_i \beta_{l1j} c_j \tag{192}$$

where

$$L_{ij}{}^{\eta\eta} = \int_0^a dr \eta_i{}^*(r) L_l \eta_j(r) \tag{193}$$

The matrix equivalent of the stationary expression is thus

$$\mathbf{c}^{\dagger}\boldsymbol{\beta}_{l0}{}^{\dagger}t_l\boldsymbol{\beta}_{l0}\mathbf{c} = \mathbf{c}^{\dagger}\mathbf{B}\mathbf{c} \tag{194}$$

where the dagger indicates the transpose matrix and

$$\mathbf{B} = -\left(\frac{2}{\hbar}\right)\mathbf{L}^{\eta\eta} + \boldsymbol{\beta}_{l0}{}^{\dagger}\boldsymbol{\beta}_{l1} \tag{195}$$

The stationary value of $t$ is given by (194) when the coefficients $c_i$ are such that the right-hand side is stationary with respect to all variations of $c_i$ subject to the constraint $\alpha_{l0}{}^0 = \sum_i \beta_{l0i}c_i$. Thus we introduce a Lagrange multiplier $\lambda$ and consider

$$J'(X_l{}^0) = \mathbf{c}^{\dagger}\mathbf{B}\mathbf{c} + \lambda\boldsymbol{\beta}_{l0}\mathbf{c} \tag{196}$$

Requiring

$$\frac{\partial J'(X_l{}^0)}{\partial c_i} = 0 \tag{197}$$

yields

$$\mathbf{c}^{\dagger}\mathbf{B} = -\lambda\boldsymbol{\beta}_{l0} \tag{198}$$

that is,

$$\mathbf{B}^{\dagger}\mathbf{c} = -\lambda\boldsymbol{\beta}_{l0}{}^{\dagger} \tag{199}$$

that is,

$$\mathbf{c} = -\lambda(\mathbf{B}^{\dagger})^{-1}\boldsymbol{\beta}_{l0}{}^{\dagger} \tag{200}$$

Assuming $\mathbf{B}$ is symmetric we then find that the stationary value of $t$ satisfies

$$t_l\boldsymbol{\beta}_{l0}{}^{\dagger}\boldsymbol{\beta}_{l0} = \mathbf{B} \tag{201}$$

That is,

$$t_l = (\boldsymbol{\beta}_{l0}\mathbf{B}^{-1}\boldsymbol{\beta}_{l0}{}^{\dagger})^{-1} \tag{202}$$

Using (190) and (200), that is,

$$\alpha_{l0}{}^0 = (-\lambda)\boldsymbol{\beta}_{l0}(\mathbf{B}^{-1})\boldsymbol{\beta}_{l0}{}^{\dagger} \tag{203a}$$

$$= (-\lambda)t_l{}^{-1} \tag{203b}$$

we find $\lambda = -t_l\alpha_{l0}{}^0$. (Thus we can find $\lambda$ for any choice of the arbitrary $\alpha_{l0}{}^0$.) This method will be called the Kohn-Crawford procedure.

Oberoi and Nesbet[72] have suggested a modification of the Kohn and Rubinow methods in which the functions $S_l(r)$ and $C_l(r)$ are replaced by $\mathscr{S}_l(r)$ and $\mathscr{C}_l(r)$, respectively, for $r \geqslant a$. Oberoi and Nesbet call these functions numerical asymptotic functions. The functions are still to be arbitrary at small $r$ except for the zero value at the origin. This procedure simplifies the free-free and bound-free integrals and has similar advantages to the Kohn-Crawford procedure.

### L. Methods Involving Only Square-Integrable Basis Functions but No Artificial Channel Radii

There are many approaches to the treatment of resonances in which one approximates the scattering wave function in the resonance energy region by a square-integrable function. Recently some methods have been proposed for treating scattering problems in resonant or nonresonant energy regions by doing calculations involving only square-integrable basis functions. Schlessinger and Schwartz[73] proposed solving the scattering problem by solving a nonhomogeneous modification of (3) for negative energies and using rational fractions to extrapolate the partial wave contribution to the scattering amplitude to positive energies.

Reinhardt and co-workers[74] have proposed that a square-integrable basis set be used to calculate the Fredholm determinant for complex $E$ and that rational fractions be used to extrapolate it to the real axis. The phase shift can be calculated from the extrapolated determinant. This method is closely related to the methods of Schlessinger and Schwartz, but appears to have some advantages. Doolen *et al.*[75] have proposed yet another way to use complex variable analyticity to calculate scattering amplitudes, namely the rotation method involving complex coordinates and complex energies.

### M. The Schwinger Variational Method

Another of the standard variational methods of scattering theory which can be applied using basis functions is the Schwinger variational method.[76,77] This method can be used in an algebraic variational calculation with a trial function of the form (40) or with a trial function that does not even satisfy correct scattering boundary conditions, for example, a power series.[78] It has the disadvantage that it requires more complicated integrals than any of the methods discussed so far; that is, it requires integrals involving the Green's function. This is the reason it has received less attention. Kato[27] and Schwartz[79] have given different methods for eliminating the integrals involving the Green's function. Kato's method introduces different complicated integrals and Schwartz's method involves a special choice of trial function. Overall, the Schwinger method still appears to be less well suited to algebraic variational calculations than the methods considered above.

### N. Other Methods

Malik[51] has introduced a variational method that is identical to the Kohn method except that (49) is replaced by

$$t_l^{\,0} = -\frac{2ma_0^{1/2}}{\hbar^2 k}\int_0^\infty dr X_l^{\,0}(r) L_l(kr)^{-1} j_l(kr) \tag{204}$$

John[42] has considered methods in which various new combinations of $n + 1$ of the sets of $n + 3$ equations (48), (50b), (49) or (54), and (204) are used. Calculations[42] indicate these methods are generally in good agreement with the Kohn, Rubinow, and Hulthén methods.

Knudson and Kirtman[80] have presented a variation-perturbation treatment of scattering in which the Hulthén method is used to obtain $X_l^0$. This causes the first-order correction to the phase shift to vanish (as discussed above). Knudson and Kirtman consider the second- and third-order corrections to the phase shift.

Kohn[81] has considered the scattering variational principle in momentum space.

## III. MULTICHANNEL SCATTERING

### A. Introduction

For scattering of composite particles we use multichannel scattering theory. For such problems we must determine the dependence of the wave function both on the internal coordinates of the composite particles and also on the coordinates that describe their relative motion. These two dependencies are coupled. Most of the variational methods for central potential scattering discussed in Sections II.C, II.D, and II.F through II.N can be extended to treat the relative motion part of the multichannel scattering problem, and many have been so extended. Coupling each of these treatments to the former part of the problem (i.e., with the treatment of the internal motion) can be done in a large variety of ways, leading to an even larger number of ways in which the combined problem can be solved. Many of the earliest treatments considered only special choices of trial function or considered only special cases, for example, problems involving only two channels. In this section we consider some of the more general and systematic formalisms that can be applied in a straightforward way to a large number of multichannel scattering processes. (The very earliest work on multichannel algebraic variational scattering calculations in chemical physics was carried out by Huang[8] and Massey and Moiseiwitsch.[37] In these calculations only one channel is open. Massey and Moiseiwitsch[82,83] were the first to treat a case in which more than one channel is open. Reviews of the early work were provided by Huck[84] and Mott and Massey.[85])

### B. Wave Functions and Kohn, Rubinow, Minimum-Norm, and Anomaly-Free Methods

The wave function is expanded in eigenfunctions of total angular momentum, and (as in single-channel scattering) each total angular momentum may be considered separately. In a multichannel scattering calculation with $N$ open two-body channels of a given total angular momentum, the component of the wave function with that total angular momentum has the asymptotic property (we restrict ourselves to the case where there are no open three-body channels)

$$\psi^p \underset{r_i\to\infty}{\sim} \sum_{i=1}^{N} Y_{ip}(r_i) f_i(x_i) \tag{205}$$

where

$$Y_{ip}(r_i) \equiv \alpha_{0ip} A_{i0}(r_i) + \alpha_{1ip} A_{i1}(r_i) \tag{206}$$

$$H_i(x_i) f_i(x_i) = E_i f_i(x_i) \tag{207}$$

and the channel Hamiltonians $H_i$ are defined in terms of the total Hamiltonian $H$ by

$$H_i(x_i) = \lim_{r_i\to\infty} [H(r_i, x_i) - T(r_i)] \tag{208}$$

where $T(r_i)$ is the kinetic energy of relative motion in channel $i$. In channel $i$, $r_i$ is the radial coordinate of relative motion of the two subsystems, $x_i$ denotes the collection of the remaining coordinates in the barycentric Hamiltonian in that channel, and $f_i(x_i)$ are the eigenfunctions of $H_i(x_i)$, that is, the "internal eigenfunctions." The "free functions" $A_{i0}$ and $A_{i1}$ have the asymptotic properties

$$A_{i0}(r_i) \underset{r_i\to\infty}{\sim} r_i^{-1} a_0^{-1/2} \sin\theta_i \tag{209}$$

$$A_{i1}(r_i) \underset{r_i\to\infty}{\sim} r_i^{-1} a_0^{-1/2} \cos\theta_i \tag{210}$$

where

$$\theta_i = k_i r_i - \tfrac{1}{2} l_i \pi - \nu_i \ln 2k_i r_i + \sigma_{l_i} \tag{211}$$

$l_i$ is the orbital angular momentum of relative motion in channel $i$, $k_i$ is the wave number wave in channel $i$, that is,

$$k_i = \hbar^{-1} \mu_i v_i \tag{212}$$

$\mu_i$ is the reduced mass, and $v_i$ is the velocity for relative motion in channel $i$, that is,

$$v_i = \left[\frac{2(E - E_i)}{\mu_i}\right]^{1/2} \tag{213}$$

$$\nu_i = \frac{Z_{A_i} Z_{B_i}}{k_i} \tag{214}$$

$Z_{A_i}$ and $Z_{B_i}$ are the charges on the subsystems in channel $i$, and

$$\sigma_{l_i} = \arg \Gamma(l_i + 1 + i\nu_i) \tag{215}$$

[Sometimes it is convenient to add a factor of $v_i^{-1/2}$ to the right-hand sides of (209) and (210). We do not do that in this article.] The problem is to approximate solutions to the Schrödinger equation with boundary conditions (205) and to examine the $Y_{im}$ to find $\{\alpha_{\beta im}\}$. From the latter quantities we can find the approximate reactance matrix and hence the approximate cross-sections. By using numerical techniques we attempt to make the differences from the exact reactance matrix elements and cross-sections as small as desired.

One approach to solving this problem is the eigenfunction-expansion method (also called the coupled-channels method or the close-coupling method). In this approach the wave function is expanded as

$$\psi^{p^0} = \sum_{i=1}^{P} X_{ip}{}^0(r_i) f_i(x_i) \tag{216}$$

where

$$X_{ip}{}^0(r_i) \underset{r_i \to \infty}{\sim} Y_{ip}{}^0(r_i) \tag{217}$$

or $X_{ip}{}^0(r_i)$ may be square-integrable. The superscript 0 indicates a trial value or a quantity appearing in a trial function. Coupled differential equations[86–107]

$$\sum_{j=1}^{P} [H_{ij}{}^{op} - E\,\delta_{ij}] X_{jp}{}^0(r_j) = 0 \qquad i = 1, 2, \ldots, P \tag{218}$$

$$H_{ij}{}^{op} \equiv \int dx_i\, f_i(x_i) H(r_i, r_i) f_j(x_j) \tag{218a}$$

for the $X_{jp}{}^0$ are derived by requiring that $(H - E)\psi^{p^0}$ have no component in the space spanned by the $\{f_i(x_i)\}_1{}^P$. The same differential equations may be derived by considering the *Kohn variational functional*[6,14,26]

$$I_{pq} \equiv \langle \psi^{p^0} |\, H - E \,| \psi^{q^0} \rangle \tag{219}$$

and using the *Kohn variational principle* for the reactance matrix elements to require that there be no first-order corrections to the approximate reactance matrix elements under point-by-point variation of the $X_{ip}{}^0(r_i)$. In the case where rearrangement channels are included in the expansion (216), that is, when all $r_i$ are not the same, the differential equations become integrodifferential equations. If the set of eigenstates of $H_i$ included in the trial wave function expansion includes at least all the open channels, then the minimum principles[59,108,109] first developed by Spruch and co-workers may be applied. In the most useful version of these principles, one proves that the sum of the eigenphase shifts increases monotonically as the basis set is increased. Notice the difference of this minimum principle from the bound principles of single-channel scattering. In single-channel scattering we are concerned with bounding errors caused by approximations to the continuum wave function $X_i(r)$. To use the

Spruch-type minimum principle, however, we must solve for the continuum functions $X_{jp}(r_j)$ by essentially exact numerical techniques and one is concerned with errors caused by approximating the target part of the wave function. In algebraic variational methods we desire to avoid the numerical integration of (coupled integro) differential equations and thus the Spruch-type minimum principle is not directly applicable. We return to this point in Section III.E below.

Note that for scattering processes involving identical particles all the $\alpha_{\beta ip}$ are not independent. If channel $j$ is related to channel $i$ by a permutation of identical particles then $\alpha_{\beta jp} = \pm\alpha_{\beta ip}$. Thus for example, in electron-hydrogen atom scattering where we may assume the nucleus is infinitely heavy so $r_2 = x_1$ and $r_1 = x_2$, we use $(1 \pm P_{12}) \sum_{i=1}^{N} X_{ip}{}^0(r_1)f_i(r_2)$ where the permutation operator $P_{12}$ interchanges $r_1$ and $r_2$ rather than using $\sum_{i=1}^{N} [X_{ip}{}^0(r_1)f_i(r_2) + X_{N+i,p}^0(r_2)f_i(r_1)]$. In this way we include all the arrangement channels but do not increase the number of independent coefficients $\alpha_{\beta ip}$. In a similar way (for any scattering problem) we can multiply any trial wave function considered in this section by a projection operator or permutation operator to enforce correct permutational symmetry.[111] We do not include these operators explicitly in the equations but it can be done when necessary.

In general, the number $P$ of open channels included in the trial wave function expansion must be equal to or larger than the number $N$ of open channels for the expansion (216) to yield an accurate approximation. In some cases, especially where rearrangements are possible, letting the expansion (216) include all square-integrable internal eigenfunctions of all two-body channel Hamiltonians does not yield an accurate approximation. Rather than include continuum internal eigenfunctions in (216) which leads to difficulties in satisfying the scattering boundary conditions,[110] two alternative approaches have been used. One approach[112–115] is to add terms of the form $Z_{jp}(r_j)g_j(x_j)$ to the expansion where $g_j$ are square-integrable but are not eigenfunctions of $H_j$ and are not necessarily linear combinations of square-integrable eigenfunctions of $H_j$. These states are called pseudostates. The other approach, the correlation method[116–121] of Gailitis, Burke and Taylor, and Miller, is to use an expansion of the form

$$\psi^{p^0} = \sum_{i=1}^{P} X_{ip}{}^0(r_i)f_i(x_i) + \sum_{m=1}^{M} c_{mp}W_m(r_i, x_i) \tag{220}$$

where the $W_m(r_i, x_i)$ are square-integrable functions added to make the basis set complete. The $W_m(r_i, x_i)$ are called correlation functions. Miller[120] and Hahn[122] showed how to use a trial function like (220) to treat rigorously rearrangements of particles with arbitrary masses without using projection operators. (Their procedures are incorporated into Sections III.C and III.D below.) Note that the two sums in (220) are not orthogonal. Requiring that $(H - E)\psi^{p^0}$ have no component in the spaces spanned by the $\{f_i\}$ and the $\{W_m\}$, we can derive a set of coupled algebraic and (integro) differential equations for the $c_{mp}$ and $X_{ip}{}^0(r_i)$. Using projection operator techniques[123] we can derive an exactly equivalent set of equations in which the effect of the correlation functions is merely to add extra nonlocal potential matrix elements to the $H_{ij}$ in (218) for the $X_{ip}{}^0(r_i)$. The same (integro) differential equations, containing the extra potential terms, for the $X_{ip}{}^0(r_i)$ may be derived by considering the *Kohn variational functional* (219) and using the *Kohn variational principle* for the reactance matrix elements to require that there be no first-order corrections[26] to the approximate reactance matrix elements under point-by-point variation of the $X_{ip}{}^0(r_i)$ and variation of the coefficients $c_{mp}$. When at least all the open channels are included in the sum over $i$ in (220), it can be called the generalized variational bound method because of the minimum principle which is then satisfied.[116–120,122]

The Hulthén, Kohn, and Harris algebraic variational methods of potential scattering theory were extended to elastic scattering of composite particles using a trial function like[10,11,37,38,39,124–127]

$$\psi^{p^0} = Y_{1p}{}^0(r_i)f_1(x_i) + \sum_{m=1}^{M} c_{mp}W_m(r_i, x_i) \tag{221}$$

We next consider algebraic variational methods that may be systematically applied to elastic and inelastic scattering.

Harris and Michels[15,16] have considered a trial function of a form closely related to (220). They considered the trial function

$$\psi^{p^0} = \sum_{i=1}^{P} Y_{ip}{}^0(r_i)f_i(x_i) + \sum_{m=1}^{M} c_{mp}W_m(r_i, x_i) \tag{222}$$

and they specifically chose the first sum in (222) to be orthogonalized to the second sum. Whereas in (220) the $X_{ip}(r_i)$ are subjected to point-by-point variation, in (222) the $Y_{ip}(r_i)$ may be varied only in the sense of using various coefficients $\alpha_{0ip}$, $\alpha_{1ip}$ [see (206)]. Harris and Michels solve for trial-function values of $\alpha_{0ip}$, $\alpha_{1ip}$, and $c_{mp}$ by requiring that $(H - E)\psi^{p^0}$ have no component in the space spanned by the $\{W_m(r_i, x_i)\}_1{}^M$ and by minimizing the components of $(H - E)\psi^{p^0}$ in the spaces spanned by the $\{A_{i0}(r_i)f_i(x_i)\}_1{}^P$ and the $\{A_{i1}(r_i)f_i(x_i)\}_1{}^P$. Thus this part of their procedure is a least-squares method in the general sense of the term. Then they substitute this trial function into the Kohn or Rubinow variational principle for the reactance matrix elements to obtain improved estimates[128] of the $\alpha_{0ip}$ and $\alpha_{1ip}$. In Section III.C we illustrate multichannel algebraic variational techniques by using the Kohn and Rubinow variational methods[6,14] to solve for the coefficients in the trial function (222). Then we substitute the trial function into the Kohn and Rubinow variational principle[6,14] for the reactance matrix elements and the reciprocals of the reactance matrix elements, respectively, to obtain improved estimates of the $\alpha_{0ip}$ and $\alpha_{1ip}$. A further difference in Section III.C as compared to Harris and Michels' work is that we do not require orthogonality in the two sums (see also Appendix 1). The procedure of Section III.C is applicable to both nonrearrangements and rearrangements.

Recently Nesbet[14] considered conversion of the close-coupling method including only open channels [see (216)–(218)] from a problem in coupled differential or integrodifferential equations to an algebraic problem by rewriting the trial function (216) as

$$\psi^{p^0} = \sum_{i=1}^{P} \left[ Y_{ip}{}^0(r_i) + \sum_{a=1}^{ni} c_{ap}{}^i \eta_a{}^i(r_i) \right] f_i(x_i) \tag{223}$$

Then the point-by-point variation of the functions $X_{ip}{}^0(r_i)$ is replaced by the variation of the coefficients $c_{ap}{}^i$, $\alpha_{0ip}{}^0$, and $\alpha_{1ip}{}^0$. Nesbet used the Kohn, Rubinow, and anomaly-free variational methods to solve for the coefficients in the trial function (223). He then used the Kohn, Rubinow, and anomaly-free variational principles to determine improved values of $\alpha_{0ip}$ and $\alpha_{1ip}$. Note that the anomaly-free method is just a choice between the Kohn and Rubinow methods based upon a criterion given by Nesbet. This type of treatment was extended easily[17] to include closed channels, that is, to use the trial function

$$\psi^{p^0} = \sum_{i=1}^{P} Y_{ip}{}^0(r_i) f_i(x_i) + \sum_{i=1}^{Q} \sum_{a=1}^{ni} c_{ap}{}^i \eta_a{}^i(r_i) f_i(x_i) \tag{224}$$

where $Q > P$. In Section III.D we consider the conversion of the correlation method for nonrearrangement or rearrangement scattering [see (220)] from a problem of coupled differential or integrodifferential equations to an algebraic problem using the Kohn, Rubinow, and anomaly-free variational methods. The extension of these variational methods to a correlation-type trial function has also been considered by Nesbet and Lyons[129] and Chung and Chen.[130,131]

The above discussion and references should make it clear that the methods presented in Sections III.C and III.D depend especially on the earlier developments of Gailitis, Burke and Taylor, Miller, Harris and Michels, and Nesbet.

## C. Application of the Kohn Variational Method to the Harris and Michels-type Trial Function

We first consider the case where

$$\alpha_{0ip} = \alpha_{0ip}^{0} = \delta_{ip} \qquad i, p = 1, 2, \ldots, P \tag{225}$$

where the first quantity is the exact value of $\alpha_{0ip}$ and the second is the trial value of $\alpha_{0ip}$. Then by the definition of the reactance matrix, it is given by[132]

$$\mathbf{R} = \mathbf{V}^{1/2}\boldsymbol{\alpha}_1\mathbf{V}^{-1/2} \tag{226a}$$

or

$$R_{ij} = \left(\frac{v_i}{v_j}\right)^{1/2} \alpha_{1ij} \qquad i, j = 1, 2, \ldots, P \tag{226b}$$

where $V_{ij} \equiv v_i\, \delta_{ij}$. Note that $i$ is the final channel and $j$ is the initial channel. Since the reactance matrix is real, we restrict $\alpha_{1ij}$ to real values. We now consider the variational determination of the coefficients $\alpha_{1ip}^{0}$ and $c_{mp}$ in the trial function given by (222), (206), and (225). Under variation of all the $\alpha_{1ip}$, $\alpha_{1iq}$, $c_{mp}$, and $c_{mq}$ ($i = 1, 2, \ldots, P$; $m = 1, 2, \ldots, M$), the *Kohn variational principle* for the reactance matrix elements in the form in which it is commonly used (see, for example, Refs. 91, 97, 117, 119, and 15) is

$$\delta\left[I_{pq} - \left(\frac{\hbar v_p}{2a_0}\right)\alpha_{1pq}^{0}\right] = 0 \qquad p, q = 1, 2, \ldots, P \tag{227}$$

That is, the stationary value of $\alpha_{1pq}$ is

$$\alpha_{1pq}^{1} = \alpha_{1pq}^{0} - \left(\frac{2a_0}{\hbar v_p}\right)(I_{pq}) \tag{228}$$

where the superscripts 1 and 0 denote variationally corrected and trial values, respectively. In the methods using point-by-point variation (functional variation) of the $X_{ip}{}^{0}(r_i)$ one sets $I_{pq}$ equal to 0 by requiring the $X_{ip}{}^{0}(r_i)$ to satisfy (integro) differential equations.[91,97,117,119,120] For example, if the trial function is (216), the (integro) differential equations are (218).[91,97] In that case $\alpha_{1pq}^{1}$ equals $\alpha_{1pq}^{0}$; that is, the variationally correct reactance matrix elements are given directly by the asymptotic form of the trial function. In the algebraic variational methods, such as the one considered here, we do not make the correction to the trial reactance matrix elements vanish completely. Following the Kohn variation method we note that the stationary condition (227) is satisfied with respect to the restricted functional form of the trial function when

$$\frac{\partial I_{pq}}{\partial c_{mr}} = 0 \qquad \begin{aligned} r &= p, q \\ p, q &= 1, 2, \ldots, P \\ m &= 1, 2, \ldots, M \end{aligned} \tag{229}$$

and when

$$\frac{\partial I_{pq}}{\partial \alpha_{1ij}^0} = \frac{\hbar v_i}{2a_0}(\delta_{ip}\delta_{jq}) \qquad \begin{array}{c} i, p, q = 1, 2, \ldots, P \\ j = p, q \end{array} \tag{230}$$

Using these equations to determine values for the parameters $\alpha_{1pq}^0$, we can then use (228) to obtain an improved approximation. The use of (229) and (230) to obtain values of a particular coefficient $\alpha_{1pq}^0$ proceeds as follows:

$$I_{pq} = I_{pq}{}^{FF} + I_{pq}{}^{CF} + I_{qp}^{CF^*} + I_{pq}{}^{CC} \tag{231}$$

where

$$I_{pq}{}^{FF} = \sum_{ij} \langle Y_{ip}{}^0 f_i | H - E | Y_{jq}{}^0 f_j \rangle \tag{232}$$

$$I_{pq}{}^{CF} = \sum_{mi} c_{mp}{}^* \langle W_m | H - E | Y_{iq}{}^0 f_i \rangle \tag{233}$$

$$I_{pq}{}^{CC} = \sum_{mn} c_{mp}{}^* c_{nq} \langle W_m | H - E | W_n \rangle \tag{234}$$

Then requiring $\partial I_{pq}/\partial c_{mp}{}^* = 0$ yields

$$\sum_n \langle W_m | E - H | W_n \rangle c_{nq} = \sum_i \langle W_m | H - E | Y_{iq}{}^0 f_i \rangle \tag{235}$$

$$m = 1, 2, \ldots, M$$
$$q = 1, 2, \ldots, P$$

Or in a more compact notation,

$$I_{pq}{}^{CF} = \sum_m c_{mp}{}^* J_{mq} \tag{236}$$

$$I_{pq}{}^{CC} = -\sum_{mn} c_{mp}{}^* c_{nq} N_{mn} \tag{237}$$

where

$$J_{mq} \equiv \sum_i \langle W_m | H - E | Y_{iq}{}^0 f_i \rangle \tag{238}$$

and

$$N_{mn} \equiv \langle W_m | E - H | W_n \rangle \tag{239}$$

Then requiring $\partial I_{pq}/\partial c_{mp}{}^* = 0$ yields

$$\sum_n N_{mn} c_{nq} = J_{mq} \qquad \begin{array}{c} m = 1, 2, \ldots, M \\ q = 1, 2, \ldots, P \end{array} \tag{240}$$

or

$$\mathbf{c} = \mathbf{N}^{-1}\mathbf{J} \tag{241}$$

Notice that, although we obtained (240) by the Kohn variational principle, the result is the same as if we had arbitrarily required $(H - E)\psi^{q^0}$ to have no component in the space spanned by the $\{W_m\}$. Substituting (241) into

(231) yields

$$I_{pq} = I_{pq}{}^{FF} + \sum_m (\mathbf{N}^{-1}\mathbf{J})_{mp}{}^* J_{mq} + \sum_m (\mathbf{N}^{-1}\mathbf{J})_{mq} J_{mp}{}^* - \sum_{mn} (\mathbf{N}^{-1}\mathbf{J})_{mp}{}^* (\mathbf{N}^{-1}\mathbf{J})_{nq} N_{mn} \quad (242)$$

$$= I_{pq}{}^{FF} + 2\sum_{mn} (\mathbf{N}^{-1})_{mn}{}^* J_{np}{}^* J_{mq} - \sum_{mnor} (\mathbf{N}^{-1})_{mo}{}^* J_{op}{}^* (\mathbf{N}^{-1})_{nr} J_{rq} N_{mn} \quad (243)$$

Now the last term of (243) is equal to

$$-\sum_{mor} (\mathbf{N}^{-1})_{mo}{}^* J_{op}{}^* \delta_{mr} J_{rq} \quad (244a)$$

$$= -\sum_{mo} (\mathbf{N}^{-1})_{om} J_{op}{}^* J_{mq} \quad (244b)$$

Therefore,

$$I_{pq} = I_{pq}{}^{FF} + \sum_{mn} J_{np}{}^* (\mathbf{N}^{-1})_{nm} J_{mq} \quad (245)$$

$$= \sum_{ij} \langle Y_{ip}{}^0 f_i | \, M \, | Y_{jq}{}^0 f_j \rangle \quad (246)$$

where

$$M = H - E - \sum_{mn} (H - E) \, |W_m\rangle \langle W_m| \, (H - E)^{-1} \, |W_n\rangle \langle W_n| \, (H - E) \quad (247)$$

Alternatively from (206), (225), and (246) we obtain

$$I_{pq} = M_{pq}{}^{00} + \sum_j M_{pj}{}^{01} \alpha_{1jq} + \sum_j \alpha_{1jp}^0 M_{jq}{}^{10} + \sum_{ij} \alpha_{1ip}^0 M_{ij}{}^{11} \alpha_{1jq}^0 \quad (248)$$

where

$$M_{ij}{}^{\alpha\beta} = \langle A_{i\alpha}{}^0 f_i | \, M \, | A_{j\beta}{}^0 f_j \rangle \quad (249)$$

with $\alpha, \beta = 0, 1$. Note that

$$\frac{\partial I_{pq}}{\partial \alpha_{1ij}} = \delta_{jp} \left( M_{iq}{}^{10} + \sum_k M_{ik}{}^{11} \alpha_{1kq}^0 \right) + \delta_{jq} \left( M_{pi}{}^{01} + \sum_k M_{ki}{}^{11} \alpha_{1kp}^0 \right) \quad (250)$$

Now (230) requires $\partial I_{pq} / \partial \alpha_{1ip} = 0$ for $p \neq q$ which yields

$$\sum_k M_{ik}{}^{11} \alpha_{1kq}^0 = -M_{iq}{}^{10} \qquad q \neq p \quad (251)$$

Further we include the case $p = q$ by using, again from (230):

$$\frac{\partial I_{pq}}{\partial \alpha_{1iq}^0} = \frac{\hbar v_i}{2a_0} \delta_{ip} \quad (252)$$

This yields

$$\delta_{pq} \left( M_{iq}{}^{10} + \sum_k M_{ik}{}^{11} \alpha_{1kq}^0 \right) + M_{pi}{}^{01} + \sum_k M_{ki}{}^{11} \alpha_{1kp}^0 = \left( \frac{\hbar v_i}{2a_0} \right) \delta_{ip} \quad (253)$$

But[14]

$$M_{pi}{}^{01} - M_{ip}{}^{10} = \left(\frac{\hbar v_i}{2a_0}\right)\delta_{ip} \tag{254}$$

and

$$M_{ki}{}^{11} = M_{ik}{}^{11} \tag{255}$$

Thus (231) becomes

$$(1 + \delta_{pq})\left(M_{ip}{}^{10} + \sum_k M_{ik}{}^{11}\alpha^0_{1kp}\right) = 0 \tag{256}$$

Equation (256) is not different from (251) so we can write for all $i$ and $j$:

$$\sum_k M_{ik}{}^{11}\alpha^0_{1kj} = -M_{ij}{}^{10} \tag{257}$$

or

$$\boldsymbol{\alpha}_1{}^0 = -(\mathbf{M}^{11})^{-1}\mathbf{M}^{10} \tag{258}$$

This equation can be solved for the trial coefficients $\alpha^0_{1iq}$. (These can be used with (206) and (241) to calculate the trial $\{c_{mp}\}$ if it is desired to use the trial wave functions. If one wants only to calculate the reactance and scattering matrices, then one need not explicitly obtain the $\{c_{mp}\}$.) Equation (258) may now be substituted into (248) to obtain

$$\mathbf{I} = \mathbf{M}^{00} + \mathbf{M}^{01}\boldsymbol{\alpha}_1{}^0 \tag{259}$$

Now (258) and (259) can be used on the right-hand side of (228) to obtain an improved estimate of $\boldsymbol{\alpha}_1$. Then (226) can be used to obtain the reactance matrix. Partial cross-sections may be obtained from the reactance matrix using well known formulas.

In the event det $\mathbf{M}^{11}$ is very small, (257) is numerically ill-conditioned. Nesbet suggested when a similar problem occurs in the algebraic solution of the close-coupling equations[14] that if det $\mathbf{M}^{11}$ is smaller than det $\mathbf{M}^{00}$ we should solve for the reciprocals of the reactance matrix elements instead of for the reactance matrix elements. Thus

$$\alpha^0_{1ip} = \delta_{ip} \tag{260}$$

Then the reactance matrix elements are given by[14]

$$R_{ip}{}^{-1} = \left(\frac{v_i}{v_p}\right)^{1/2}\alpha^0_{0ip} \tag{261}$$

and the variational principle is[14]

$$\delta\left[I_{pq} + \left(\frac{\hbar}{2a_0}\right)v_p\alpha^0_{0pq}\right] = 0 \tag{262}$$

which has been called the inverse Kohn or the Rubinow variational principle. Then

$$\boldsymbol{\alpha}_0{}^1 = \boldsymbol{\alpha}_0{}^0 + \left(\frac{2a_0}{\hbar}\right)\mathbf{V}^{-1}\mathbf{I} \tag{263}$$

The Rubinow variational method leads to

$$\boldsymbol{\alpha}_0{}^0 = -(\mathbf{M}^{00})^{-1}\mathbf{M}^{01} \tag{264}$$

where the details of the derivation are similar to those before. The choice as to whether to use the Kohn or Rubinow variational principle is sometimes arbitrary. In the anomaly-free method[14] the former is used if det $\mathbf{M}^{00}$ is smaller than det $\mathbf{M}^{11}$ and vice versa.

Note that if the minimum-norm method is used to determine $\boldsymbol{\alpha}_0$ and $\boldsymbol{\alpha}_1$, neither (225) nor (260) holds. In that case we use

$$\mathbf{R} = \mathbf{V}^{1/2}\boldsymbol{\alpha}_1\boldsymbol{\alpha}_0{}^{-1}\mathbf{V}^{-1/2} \tag{265}$$

instead of (226) or (261).

The method presented in this section is not an alternate derivation of the Harris-Michels results. The reactance matrices and wave functions obtained from these equations are different from those obtained from the Harris-Michels equations. The main advantage of the present method is that the equations are derived using the usual scattering variational principle (the Kohn variational principle). Thus the procedure represents the most general application of the Kohn variational method. It should be useful because the results can be compared more directly to other results obtained by the Kohn variational method but using a more restricted trial function. A further value of the equations here is in tying together three different approaches: the applications of the Kohn variational method to the algebraic close-coupling equations by Nesbet and to correlation functions by Miller and others, and the use of the Harris and Michels trial function (which, as discussed in Section III.D, is the methodologically simplest version of the algebraic correlation method trial function).

### D. Algebraic Correlation Method

We now consider application of the Kohn variational principle to the trial function

$$\psi^{p^0} = \sum_{i=1}^{P}\left[Y_{ip}{}^0(r_i) + \sum_{a=1}^{ni} c_{ap}{}^i\eta_a{}^i(r_i)\right]f_i(x_i) + \sum_{m=1}^{n_Q} c_{mp}W_m(r_i, x_i) \tag{266}$$

where the $\alpha^0_{0ip}$, $\alpha^0_{1ip}$, $c_{ap}{}^i$, and $c_{mp}$ are coefficients to be determined. This may be considered to be an algebraic version of the correlation method discussed previously by Gailitis, Burke and Taylor, and Miller (see

Section III.B). The Harris and Michels-type trial function may be considered to be a special case of (266) in which only the free terms and the correlation terms are retained. Conversely, (266) may be considered to be a special case of the Harris and Michels-type trial function in which some of the correlation functions $W_m$ are of the separable form $\eta_a{}^i f_i$. The treatments here of the trial functions (222) and (266) are numerically equivalent when the $W_m$'s in (222) are chosen in such a way as to make (222) and (266) identical. The new terms (of the form $c_{ap}{}^i \eta_a{}^i$) in (266), as opposed to (222), will be called "bound terms." The new problem posed by explicit use of (266) is the determination of their coefficients. This may be done using the procedure (the Kohn variational principle) used by Nesbet[14] for the algebraic close-coupling problem. This leads to the same type of equations as in the previous section but with the following important changes:

$$J_{mq} \rightarrow \sum_i \left\langle W_m \middle| H - E \middle| \left( Y_{iq}{}^0 + \sum_a c_{aq}{}^i \eta_a{}^i \right) f_i \right\rangle \tag{267}$$

$$M_{ij}{}^{\alpha\beta} \rightarrow \langle A_{i\alpha}{}^0 f_i | \, U \, | A_{j\beta}{}^0 f_j \rangle \tag{268}$$

where

$$U = M - \sum_{qr} \sum_{ab} M \, |\eta_a{}^q f_q\rangle \langle \eta_a{}^q f_q | \, M^{-1} \, |\eta_b{}^r f_r\rangle \langle \eta_b{}^r f_r | \, M$$

Finally the coefficients of the bound terms in the trial function are given by

$$c_{aq}{}^i = \sum_r (\alpha^0_{0rq} c_{a0}{}^{ir} + \alpha^0_{1rq} c_{a1}{}^{ir}) \tag{269}$$

where the $c_{a\alpha}{}^{ir}$ satisfy

$$\sum_s \sum_b \langle \eta_a{}^p f_p | \, M \, |\eta_b{}^s f_s \rangle c_{b\alpha}{}^{sq} = -\langle \eta_a{}^p f_p | \, M \, | A_{q\alpha}{}^0 f_q \rangle \qquad \alpha = 0, 1 \tag{270}$$

where the trial values $\alpha^0_{\alpha n i}$ should be used in (270).

The generalization of this treatment to include closed channels in the close-coupling part of the trial wave function [i.e., to replace the first part of (266) by (224)] can be accomplished in a straightforward way.

The reasons why one might wish to single out certain of the square-integrable basis functions for inclusion in the bound-term sum instead of the correlation-term sum are various. One of them is for explicit close-coupling type of interpretations of the scattering process in terms of contributions from various target eigenstates. Of course such interpretations can also be made using the Harris and Michels-type trial function, but in that case the interpretation is less direct because the $W_m$ must first be projected on the $f_i$. A second reason is for ease in checking the computations (both for coding errors and for completeness of a basis set) using the algebraic variational method against numerical integration of the correlation method integrodifferential equations. It appears that such comparison (see, e.g., Ref. 17) of the algebraic methods with numerical integration methods is very useful, at least until more experience is gained with the algebraic methods. In the form of (266), the wave functions can be compared most directly. Further, in this form one can compare other elements of related treatments (e.g., the potential matrix elements of the algebraic correlation method with their analogues in the algebraic close-coupling method) to discover how additional flexibility in the wave function affects the results in detail. A third use for the equations in this section is that they provide a link to make explicit the connection between the correlation method of Miller and others and the method of Harris and Michels. The relation between this section and the Harris and Michels method is that the trial functions (but not the methods of determining the coefficients) are formally equivalent. The relation between this section and the correlation method is discussed in Section III.B.

### E. Discussion of Above Methods

The schemes of Sections III.C and III.D combine the use of correlation terms in the trial wave function expansion with an algebraic variational approach to the calculation. The advantages of correlation terms have been discussed in the references to Section III.B and the advantages of algebraic variational methods have been discussed in Section I, but it is perhaps worthwhile to discuss these points further.

Correlation terms may be used to put special physical features of the compound system into the trial function. This often requires expressing these terms in a coordinate system different from the one in which the scattering boundary conditions achieve their most natural form. For example, for electron-atom scattering it may be desired to express the $W_m$ in terms of interelectronic coordinates ($r_{ij}$), or for chemical reactions it may be desired to express the $W_m$ in terms of transition-state normal coordinates. Another example, which has applications for all kinds of collisions, is to use the correlation terms to include so-called perturbed stationary states (also called polarized orbitals in electron scattering or molecular states in high-energy atom-atom scattering). Another example of their use is to include specific compound resonance states. Or we may think of the purpose of correlation terms as a way to effectively include virtual target continuum contributions even when the target does not really break up. In the Harris and Michels-type function the correlation terms not only carry out these special purposes but also perform the role of the target-eigenstate terms of the close-coupling approximation. In general we can think of these terms as providing a convenient method of making the expansion of the trial wave function effectively complete. The advantage of the algebraic approach is that it obviates the need to solve very large sets of coupled (integro) differential equations for the wave function. Even with the latest techniques this can be very time-consuming. Alternatively we can think of the algebraic method as being a new and the very latest method of solving the coupled (integro) differential equations.

It should be clear that if the form (222) of the wave function spans the same space as the form (223) then the former will lead to the same algebraic problem. The advantages of (222) over (223) are (*a*) the important possibility of using a more general trial function as discussed in the preceding paragraph and (*b*) the minor advantage of reducing the number of subscripts on the coefficients.

We did not require the correlation terms to be orthogonal to the free functions. Although Harris and Michels explicitly made their correlation terms orthogonal to their free functions,[15] this is not really necessary even for their method. This is proved in Appendix 1.

The variational principles used here allow us to variationally improve the reactance matrix elements (or their reciprocals) but they do not allow us to improve the rest of the trial wave function. Unlike the case where point-by-point variations are allowed, the algebraic method does not allow us to obtain any trial wave function whose asymptotic form corresponds to the variationally correct $R_{ip}$. The advantage of variationally improved results is that errors in the $R_{ip}$ are second order in the errors in the wave function. Thus the calculated scattering cross-sections are not as sensitive to the deficiencies of the basis set as when nonvariational techniques are used and are not as sensitive to these deficiencies as the trial wave function.

We should point out that for the combination of reasons discussed in the last three paragraphs the extension of the Harris, Michels, and Nesbet methods presented here is probably at this time one of the most economical and straightforward general ways to obtain the wave function for most scattering problems, including inelastic electron scattering and chemical reactions. If one desires to obtain directly the scattering probabilities (not the wave function), one of the methods dealing directly with the transition matrix[107,133] may be preferable. For example, Baer and Kouri[133] have developed an algebraic technique for solving for "channel operators" and amplitude densities (from which the cross-sections and wave functions may then be calculated). It is being extended to include the use of correlation functions by using a variational principle.[134] Reinhardt has also presented a new method that requires neither expansion of the wave function in a basis nor numerical integration of coupled (integro) differential equations.[135] In Section III.F we consider other algebraic variation methods which involve expansion of the wave function in a basis set.

Next we consider a few illustrative calculations on elastic electron-hydrogen atom scattering using algebraic close-coupling methods. We consider the singlet electron-spin state. Then we need not explicitly include

TABLE V

Basis Set for Electron-Hydrogen Atom Scattering Calculations of Figs. 5 to 11[a]

| $a$[b] | $\zeta_a$ |
|---|---|
| 1 | 0.005 |
| 2 | 0.01 |
| 3 | 0.02 |
| 4 | 0.05 |
| 5 | 0.15 |
| 6 | 0.4 |
| 7 | 0.7 |
| 8 | 1.1 |
| 9 | 1.5 |
| 10 | 3.0 |

[a] The calculations were performed by the algebraic $1s$-$2s$-$2p$ close-coupling method (including exchange) as described by Seiler et al.[17] as modified in Ref. 146. Two free functions $[(1-e^{-r})^3 j_1(kr)$ and $(1-e^{-r})^3 \xi_1(kr)]$ were used, and the short-range parts of the four channel functions were each expanded in terms of 10 Slater-type functions of the form $r^l e^{-\xi_a r}$, with $a = 1, 2, \ldots, 10$. Note that distances are in bohrs.

[b] See (224) where $P = 4$, $n_i = 10$, and $Q = 4$.

spin in the wave function, but the wave function must be symmetric under interchange of the two electrons' spatial coordinates. These calculations use as trial function the symmetrizer times the function of (222) with $P = 1$ and $M = 40$. We consider the case where the total angular momentum is $L = 1$. The basis set is explained in Table V. The bound-bound Hamiltonian matrix

$$H_{pq}^{\eta\eta} = \langle W_p | H | W_q \rangle \tag{271}$$

has 40 eigenvalues $E_\mu$. The lowest twelve are listed in Table VI. The analysis of Section II.H, which is based on Nesbet's work,[13] may be extended to the multichannel scattering problem, and we would expect the Kohn and Rubinow methods to show spurious structure near but not exactly at the energies where $E$ equals one of the $E_\mu$.[136,62]

Fig. 6 illustrates the poles in the $M_{11}^{\alpha\beta}$ which occur near the fifth, sixth, and seventh eigenvalues of $\mathbf{H}^{\eta\eta}$. Fig. 7$a$ shows $t_R^0$ and $t_R^R$ for the energy region near the fifth and sixth eigenvalues of $\mathbf{H}^{\eta\eta}$. $t_R^0$ shows two pseudo-antiresonances and $t_R^R$ shows two pseudoresonances. Except in these two regions $t_R^0$ is too close to $t_R^R$ to be distinguished on the plot. Further, at

TABLE VI

Lowest Twelve Eigenvalues $E_\mu$ of $\mathbf{H}^{\prime\prime\prime}$ for $L = 1$, $S = 0$ for the Basis Set of Table V

| $\mu$ | $E_\mu$ (Ry) | $k_\mu^2$ ($a_0^{-2}$) |
|---|---|---|
| 1 | −0.9999842 | 0.0000158 |
| 2 | −0.9999081 | 0.0000919 |
| 3 | −0.9995379 | 0.0004621 |
| 4 | −0.9973415 | 0.0026585 |
| 5 | −0.9838141 | 0.0161859 |
| 6 | −0.9321195 | 0.0678805 |
| 7 | −0.7936308 | 0.2063692 |
| 8 | −0.4424070 | 0.5575930 |
| 9 | −0.2517707 | 0.7482293 |
| 10 | −0.2500602 | 0.7499398 |
| 11 | −0.2499964 | 0.7500036 |
| 12 | −0.2499806 | 0.7500194 |

the fifth eigenvalue of $\mathbf{H}^{\prime\prime\prime}$ the pseudoantiresonance of $t_R{}^0$ is so narrow it is hidden behind $t_R{}^R$ on the plot.

Although there has been much discussion of how to tell spurious singularities from real resonances, one very important practical method has not been mentioned in the literature. At a real resonance, both the uncorrected and the corrected phase shift increase by $\pi$. If, as in the present example, either the corrected or the uncorrected phase shift decreases by $\pi$, then the resonance is not real.

Fig. 7*b* shows $t_K{}^0$ and $t_K{}^K$ in the same energy region. Figs. 7*a* and 7*b* have been drawn so they partially overlap; thus it is clearly seen that at the positions of the two pseudoresonances of Fig. 7*a* (where Fig. 5 shows $M_{11}{}^{00}$ and $M_{11}{}^{01}$ have correlated zeroes), $t_K{}^0$ and $t_K{}^K$ are smooth. Instead $t_K{}^0$ and $t_K{}^K$ show spurious structure where $M_{11}{}^{10}$ and $M_{11}{}^{11}$ show correlated zeroes. Actually $M_{11}{}^{10}$ shows four zeroes in this energy region. The third of these (at $k^2 = 0.088a_0{}^{-2}$) does not cause any extra spurious structure but is associated with the phase shift passing smoothly through zero. Similarly the third zero of $M_{11}{}^{00}$ (at $k^2 = 0.112a_0{}^{-2}$) does not cause a third region of spurious structure in $t_R{}^0$ or $t_R{}^R$ but is associated with the phase shift passing smoothly through zero. This again emphasizes the need to consider correlated zeroes of the $M_{11}{}^{\alpha\beta}$. It is also interesting to notice that the reason the spurious structure associated with $t_K{}^0$ and $t_K{}^K$ occurs at an energy so much different from the energy of the seventh eigenvalue of $\mathbf{H}^{\prime\prime\prime}$ is that the "background" $M_{11}{}^{10}$ and $M_{11}{}^{11}$ are near zero. Nevertheless, Fig. 5 shows the spurious structure is unequivocably attributable to the influence of the seventh poles of $M_{11}{}^{10}$ and $M_{11}{}^{11}$.

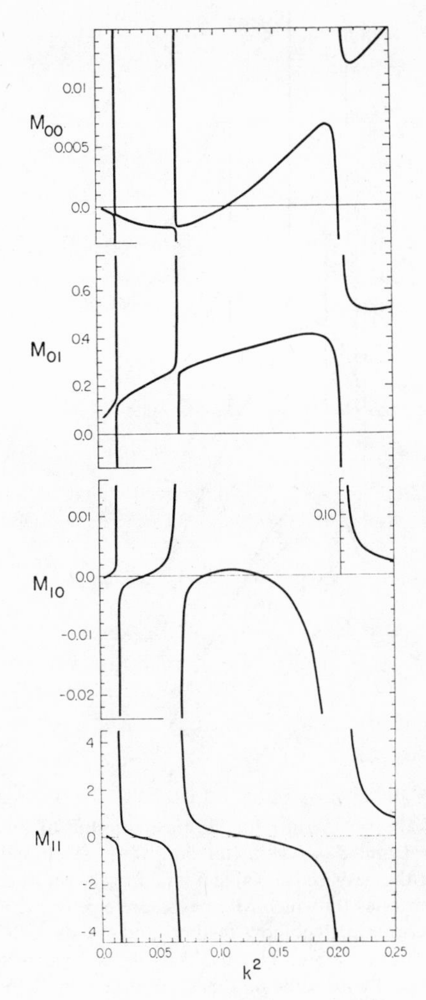


**Fig. 6.** The quantities $M_{11}{}^{00}$, $M_{11}{}^{01}$, $M_{11}{}^{10}$, and $M_{11}{}^{10}$ (in Rydbergs) as functions of $k^2$ (in atomic units) for elastic scattering of an electron off the ground state hydrogen atom with $L = 1$ and $S = 0$. The calculations were made with the basis set of Table V and are shown in the vicinity of the fifth, sixth, and seventh energies where $\mathbf{M}^{\eta\eta}$ has a zero eigenvalue. In the figures the $M_{11}{}^{\alpha\beta}$ are labeled $M_{\alpha\beta}$ as in Ref. 13.

Fig. 7*c* shows the Hulthén method results in this energy region. As expected the Hulthén method predicts complex phase shifts in two energy intervals. Near the sixth eigenvalue of $\mathbf{H}^{\prime\prime\prime}$, the figure also shows a sudden change in slope of the phase shift predicted by this method just to either side of the interval where complex values are obtained (nothing is plotted for this interval). This region is shown with an expanded abscissa in Fig. 8.

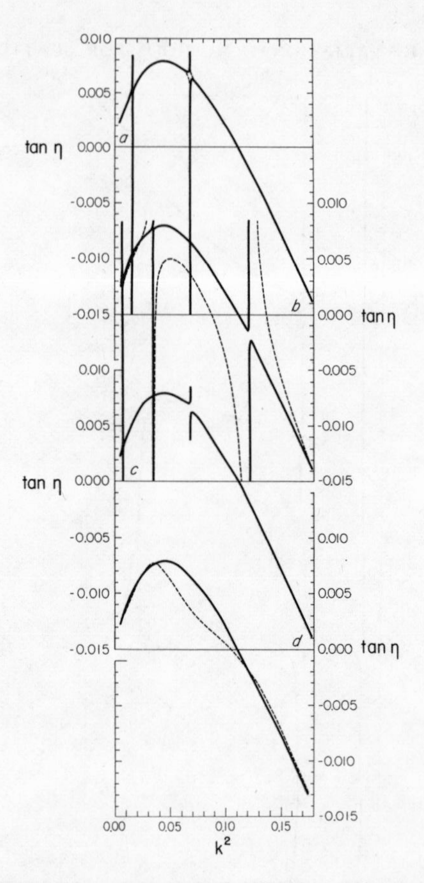


**Fig. 7.** The tangent of the phase shift for elastic scattering of an electron off the hydrogen-atom with $L = 1$ and $S = 0$ as a function of $k^2$ (in atomic units). The calculations were made with the basis set of Table V and are shown in the vicinity of the fifth, sixth, and seventh energies for which $\mathbf{M}^{\eta\eta}$ has a zero eigenvalue. (*a*) The solid line is the result $t_R{}^R$ of the corrected Rubinow method and the dashed line is the result $t_R{}^0$ of the uncorrected Rubinow method. The results using the uncorrected and corrected Rubinow method are practically the same on this scale, except in the regions of the two pseudoresonances where the two results tend to infinity in opposite directions; that is, the corrected Rubinow result shows pseudoresonances and the uncorrected Rubinow result shows pseudoantiresonances. (*b*) The solid line is the corrected Kohn result $t_K{}^K$ and the dashed line is the uncorrected Kohn result $t_K{}^0$. The corrected Kohn result shows three pseudoresonances and the uncorrected Kohn result shows two pseudoresonances followed by a pseudoantiresonance. (*c*) The solid line is the result $t_H$ of the Hulthén method. The only spurious structure that appears here is in the region of the sixth zero eigenvalue. The spurious structure in the region of the fifth zero eigenvalue is too small to be seen on this scale, and the structure due to the seventh zero eigenvalue occurs at higher $k^2$. (*d*) The solid line represents the results $t_{MN}{}^K$ and $t_{MN}{}^{MN}$ of the minimum-norm method corrected with either the Kohn or the minimum-norm variational expression (the two results are indistinguishable). The dashed line represents the result $t_{MN}{}^0$ of the uncorrected minimum-norm method.

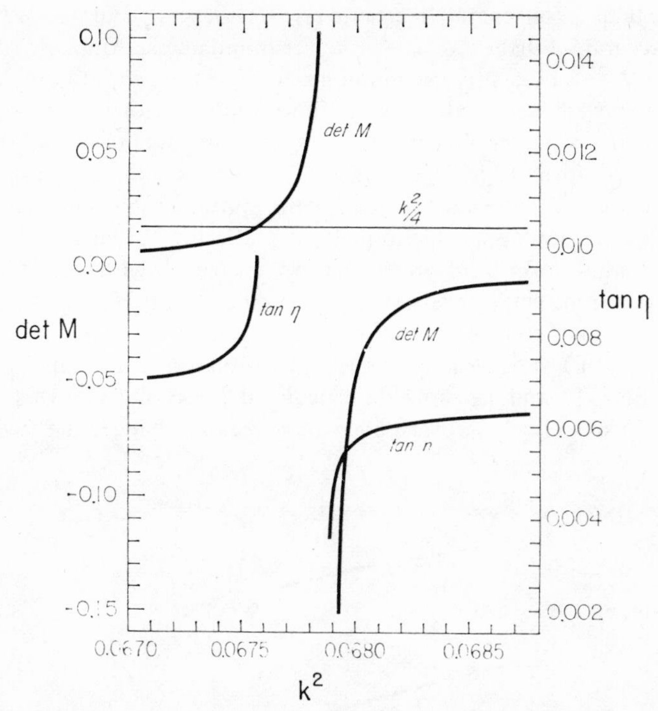


**Fig. 8.** The three solid curves represent the left- and right-hand sides of inequality (77) in $Ry^2$, and the tangent of phase shift obtained by the Hulthén method as functions of $k^2$ (in atomic units) for the electron-hydrogen scattering problem of Fig. 7 in the vicinity of the sixth energy where $\mathbf{M}^{nn}$ has a zero eigenvalue. The Hulthén result for the tangent of the phase shift becomes complex at the energy where the curves whose numerical values in Rydberg atomic units are DET $M$ and $k^2/4$ cross and remain complex until sixth energy where $\mathbf{M}^{nn}$ has a zero eigenvalue.

We see the pole in DET $M_{11}$ and the interval where complex values are obtained. This figure illustrates the discussion of (77). Hopefully, the figure will make the previous discussions[13,16] more clear.

Fig. 7*d* shows the minimum norm results in the vicinity of the fifth and sixth eigenvalues of $\mathbf{H}^{nn}$. The zero-order minimum norm result shows no spurious structure but is very inaccurate. In a certain sense, this can be blamed on (110). For example, at the sixth eigenvalue of $\mathbf{H}^{nn}$, $t_R{}^0$, $t_K{}^0$, and $t_{MN}{}^0$ are all equal. But the slopes of phase shifts are very different. The requirement of equality of the phase shifts at the eigenvalue of $\mathbf{H}^{nn}$ does not give much clue to the value of $t_R{}^0$ elsewhere because it has a pseudo-antiresonance near there and so it assumes all values from $-\infty$ to $+\infty$ in this region. But $t_{MN}{}^0$ is smooth and so the fact that it is bad at the eigenvalue of $\mathbf{H}^{nn}$ means it is also bad far from the eigenvalue. Thus $t_{MN}{}^0$ is less accurate than $t_R{}^0$ in general. In general $t_{MN}{}^0$ is not very accurate but $t_{MN}{}^{MN}$ is very accurate. Further we have not ever found any spurious structure in $t_{MN}{}^0$ or $t_{MN}{}^{MN}$. Evidently the minimum-norm method with the variational correction[50] we have called the minimum-norm variational expression has overcome the problem of correlated zeroes in the numerator and denominator of the formula for the tangent of the phase shift. Unfortunately $t_{MN}{}^{MN}$ is the only method (of those being considered in this section) for which an extension applicable to inelastic scattering has never been given. The minimum-norm zero-order method corrected with the Kohn or Rubinow variational expression as recommended in Refs. 15 and 16 sometimes shows spurious structure (see Section II.H).

Figs. 9 and 10 show $M_{11}{}^{\alpha\beta}$ in the energy region near the ninth eigenvalue of $\mathbf{H}^{\eta\eta}$. Figs. 11 and 12 show the calculated phase shifts in this energy region. This eigenvalue corresponds to a real resonance, the lowest $^1$P

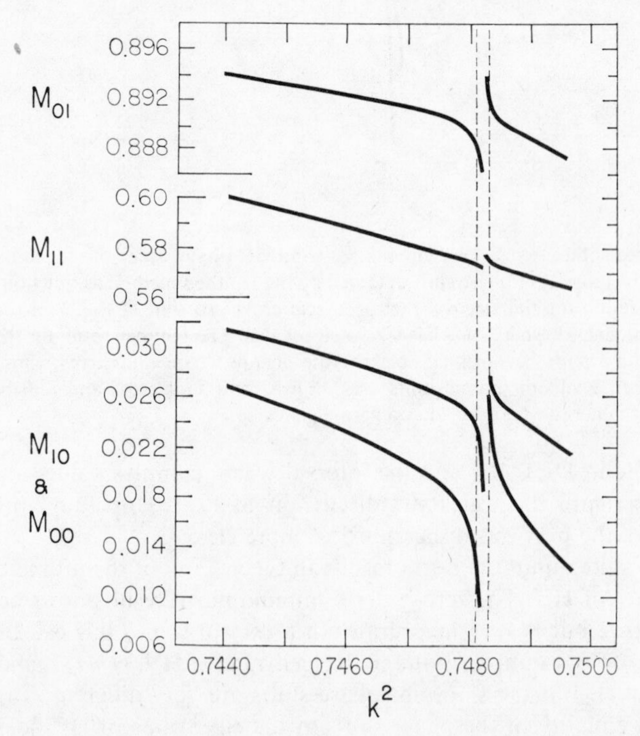


**Fig. 9.** The quantities $M_{11}{}^{00}$, $M_{11}{}^{01}$, $M_{11}{}^{10}$, and $M_{11}{}^{11}$ (in Ry) as functions of $k^2$ (in atomic units) for elastic scattering of an electron off the ground state hydrogen atom with $L = 1$ and $S = 0$. The calculations were made with the basis set of Table V and are shown in the vicinity of the ninth energy at which $\mathbf{M}^{\eta\eta}$ has a zero, that is, in the vicinity of the first $^1$P resonance.

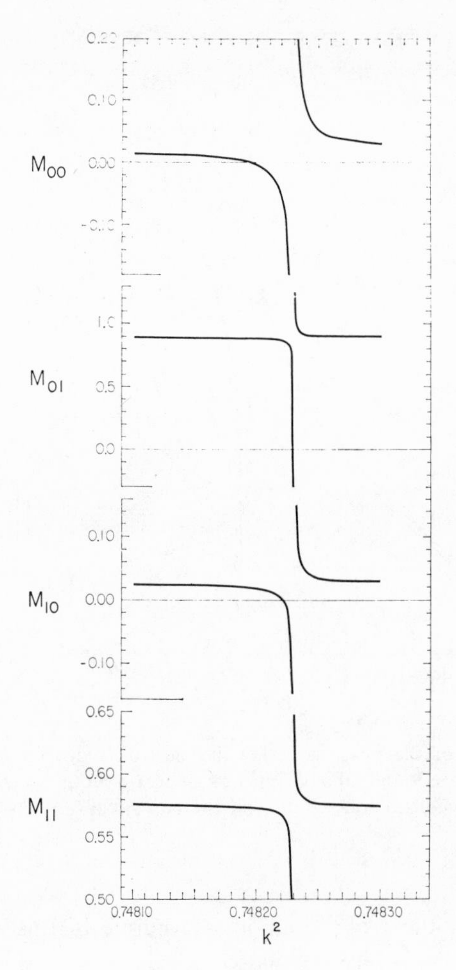


**Fig. 10.** The quantities $M_{11}^{00}$, $M_{11}^{01}$, $M_{11}^{10}$, and $M_{11}^{11}$ (in Ry) as functions of $k^2$ (in atomic units) for the energy region bounded by the dashed lines in Fig. 9.

resonance.[17,93,118] (Note: this should not be confused with the second-lowest $^1S$ resonance,[17,93,118,129,137] which is at just about the same energy.) In this case $M_{11}^{11}$ and $M_{11}^{01}$ each shows its zero very close to the eigenvalue of $\mathbf{H}^{nn}$. The zeroes of $M_{11}^{00}$ and $M_{11}^{10}$ are farther from the eigenvalue of $\mathbf{H}^{nn}$. These features are consequences of the background phase shift being near zero and the resonance being very narrow. Notice that all seven calculations shown in Fig. 12 show the resonance correctly; none of the methods shows antiresonance behavior when the resonance is real. In

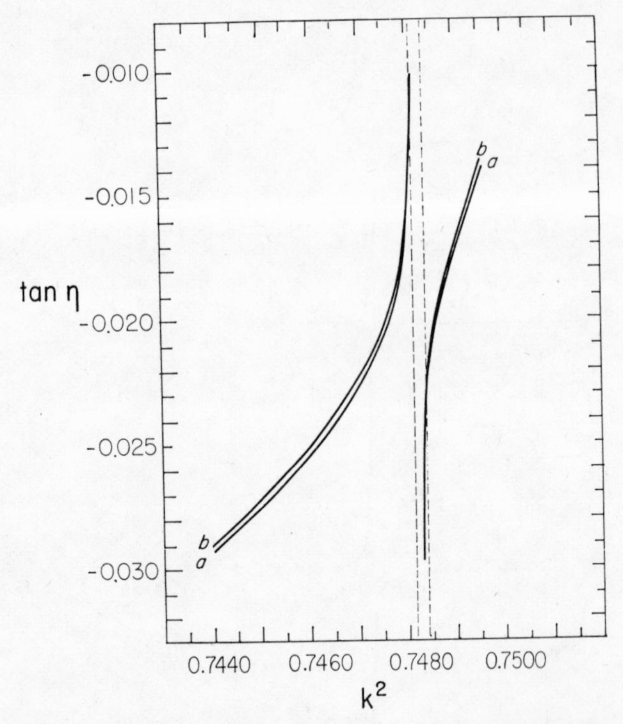


**Fig. 11.** The tangent of the phase shift for the electron-hydrogen atom problem and energy region of Figs. 9 and 10 as a function of $k^2$ (in atomic units). The curve marked $a$ is the corrected Rubinow result, and the curve marked $b$ is the corrected Kohn result.

addition, notice that the Kohn method is accurate in this region even though $M_{11}{}^{11}$ has a zero in this region.[138]

The same type of analysis of spurious singularities and real resonances may be given for cases where more than one channel is open. Although the general principles are the same the details are more complicated.

Other valuable analyses of the spurious singularities in the Kohn and Rubinow methods for the case where only one channel is open have been given by Brownstein and McKinley,[43] Kolker,[61] Shimamura,[62] and Payne.[139] In particular Shimamura's analysis leads to $(M + 1) \times (M + 1)$ matrices $\mathbf{N}^S$ and $\mathbf{N}^C$ defined similarly to $\mathbf{N}$ [see (239)] but with $W_{M+1}(r_1, x_1)$ equal to $A_{10}(r_1)$ or $A_{11}(r_1)$, respectively. The spurious singularities of the Kohn method are then identified with energies where det $\mathbf{N}^S$ is zero, and the spurious singularities of the Rubinow method are identified with energies where det $\mathbf{N}^C$ is zero. These criteria are essentially the same as

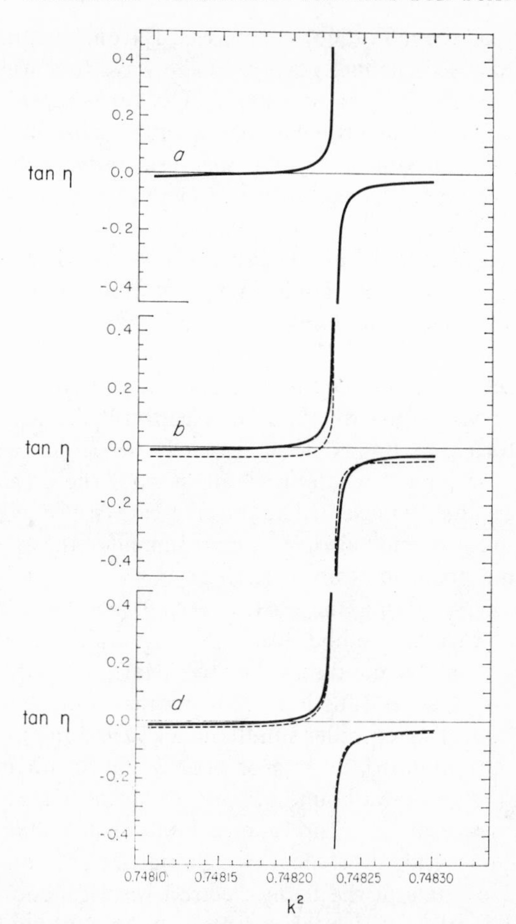


**Fig. 12.** The tangent of the phase shift as a function of $k^2$ (in atomic units) for the energy region bounded by the dashed lines in Fig. 11. (*a*) The results of the uncorrected Rubinow, the corrected Rubinow, and the Hulthén methods are all represented by the one curve. (*b*) The solid curve is the result of the corrected Kohn method and the dashed curve represents the result of the uncorrected Kohn method. (*c*) The solid curve represents the results for both the minimum-norm method corrected with either the Kohn correction or the minimum-norm correction. The dashed curve represents the result of the uncorrected minimum-norm method.

Nesbet's but may sometimes be more convenient. These $(M + 1) \times (M + 1)$ matrices also occur in Kolker's analysis[61] of how many basis functions must be added before convergence becomes monotonic in the Kohn and Harris methods for potential scattering.

It is sometimes useful[19,130,131,140] to separate the problems associated with nonmonotonic convergence as more basis functions are added in the open channels [i.e., as the $n_i$ in (266) are increased] from the problems of convergence as the closed-channel portion of the wave function is made complete [i.e., as $n_Q$ in (266) is increased]. The former problems are analogous to the convergence problems of potential scattering, and the latter errors are those for which the minimum principles of Spruch and others are applicable when the open-channel part of the wave function is treated exactly.

There have been a number of applications of the algebraic variational methods of the type discussed so far in this section to model problems[14,15,48,84,141] and to electron-atom[8,10,11,17–19,37,38,40,42,62,71,83,124–126,129–131,142–146] and positron-atom[11,17,18,37,125,127,144,147,148] scattering but only two applications to other processes; that is, Massey and Ridley[37a] applied the Kohn and Hulthén variational methods to electron-hydrogen molecule scattering and Mortensen and Gucwa applied the Kohn variational method to collinear chemical reactions.[149] In most of these calculations only the corrected results are presented but in a few cases[16,18,126,127,144,139,146] zero-order results are presented also. We next summarize the applications to electron-atom and positron-atom scattering.

For electron-atom scattering the close-coupling method with $P = 1$ is the continuum Hartree method, and the algebraic close-coupling method with $P = 1$ is the continuum matrix Hartree method. Antisymmetrizing the trial wave function leads to the continuum matrix Hartree-Fock method. If we consider simultaneous variation of the bound orbitals and the continuum orbital in a one-configuration wave function we find the variationally correct bound orbitals are the ones that would be obtained in the absence of the continuum orbital; that is, the problem separates into two independent problems which must be solved in order—one for the bound orbitals in the many-electron wave function $f_i(x_i)$ in the absence of the continuum orbital and one for the continuum orbital $X_{ip}{}^0(r_i)$ in the presence of the bound orbitals. One way to obtain this result is to put the system in a large box and normalize all the orbitals by integrating over all space. Then the continuum orbital has vanishingly small amplitude in the region where the bound orbitals are nonzero. One way to introduce correlation of the bound and free orbitals is to introduce extra open-channel configurations or extra square-integrable configurations. But the most general function possible is always a special case of (222). Equation (222) includes as special cases the continuum analogues of both the configuration interaction method and the multiconfiguration self-consistent field methods. In summary, in the first part of the calculation the forms of the terms $f_i(x_i)$ in the wave function at large $r_i$ are determined independently in a bound-state calculation. In the second part of the calculation, the coefficients of terms which do not vanish at large

$r_i$ and the form of the wave function at small $r_i$ are determined. In algebraic variational calculations both these parts of the whole scattering problem consist of determining coefficients of preselected functions.

In classifying the calculations we consider all trial functions as special cases of (222) and we distinguish the methods by the procedure used to find the matrices $\boldsymbol{\alpha}_0{}^0$ and $\boldsymbol{\alpha}_1{}^0$. Calculations in the static approximation are reviewed in Section II. Elastic scattering calculations for energies below the inelastic threshold using other approximations are summarized in Table VII. For elastic scattering a few calculations have been carried out by methods discussed in Section III.F; these are not included in the table. All the inelastic electron-atom and positron-atom scattering calculations using algebraic methods that have ever been reported are summarized in Table VIII. In addition Matese and Oberoi used their calculated wave functions to study photodetachment.[143]

There have also been a number of articles[150] concerned solely with integrals that occur in electron-atom scattering using algebraic variational methods. The methods used for these integrals may also be used for partial-wave Born and Born-Oppenheimer approximation[40,151,152,153] calculations.

### F. Other Methods

One disadvantage of the close-coupling method, the algebraic close-coupling method, the other methods discussed above, and many other methods for solving scattering problems is that the same variational function is used at small $r_i$ (the interaction region, where a large basis set is usually necessary) as is used at large $r_i$ (in the near-asymptotic regions of the various channels, where a small close-coupling basis with no or few closed channels should be sufficient). A few procedures have been devised for decreasing the size of the close-coupling basis set during a calculation as one integrates out toward large $r_i$.[102,103] In addition, other methods similar in spirit to Wigner's derivative matrix ($R$-matrix) method have been used to divide the problem into an interaction region and an external region. Such techniques will probably become very important as more practical calculations are attempted. These techniques can be refined further if necessary. For example, we can apply the Kohn-Crawford variational method[70] to the close-coupling equations but use a successively smaller expansion basis in each of several different regions. If each successively smaller basis set were a subset of the previous one, that is, if the basis sets were nested, such a calculation would not present great difficulties in starting a calculation in region 2 (from $r_i = a$ to $r_i = a'$) using the result of the region 1's calculation at $r_i = a$ as boundary conditions to be satisfied by the trial function in region 2. As the number of

TABLE VII

Algebraic Variational Electron-Atom and Positron-Atom Scattering Calculations (Excluding Static Approximation) When There is Only One Open Channel Using Methods Discussed in Sections III.B to III.E

| Reference(s) | Projectile(s) | Target(s) | $M$ | Method(s)[a] | Where done |
|---|---|---|---|---|---|
| 8 | $e^-$ | H | 2–4 | EI | Williams Bay |
| 37 | $e^-$ | H | 1 | K, H | London |
| 37b | $e^+$ | H | 1–2 | K, H | London |
| 37c | $e^+$ | H | 1–2 | K, H | London |
| 37d | $e^+$ | H | 2–3 | K | New York |
| 38 | $e^-$ | H | 0–2[b] | H | Oslo |
| 124 | $e^-$ | He | 1 | H | Belfast |
| 40 | $e^-$ | H | 1 | K, H | Glasgow, Belfast, and London |
| 10, 11 | $e^-, e^+$ | H | 3–50 | K | Berkeley |
| 125 | $e^-, e^+$ | H | 1–70 | K | Berkeley & Monterey |
| 42 | $e^-$ | H | 1–2 | K, R, H, M, John | Cardiff |
| 126 | $e^-$ | H | $\geqslant 7$ | Harris | East Hartford |
| 126, 16 | $e^-$ | He | $\leqslant 138$ | Harris | East Hartford |
| 16 | $e^-$ | H | 3–34 | H, K, R, MN-K or MN-R | Salt Lake City and East Hartford |
| 139 | $e^-$ | H | 7–34 | K[d] | Iowa City |
| 142 | $e^-$ | H | 45 | AF | Baton Rouge |
| 17 | $e^-, e^+$ | H | 45–48 | AF | Baton Rouge |
| 127 | $e^+$ | H, He | 4–85 | Harris | Greenbelt |
| 143 | $e^-$ | H | 35–50 | AF | Baton Rouge |
| 129 | $e^-$ | H | 17 | K, R, or AF | San Jose |
| 130, 131 | $e^-$ | H | Not given | K | Raleigh and LaJolla |
| 18 | $e^-, e^+$ | H | 50–56 | K, Harris | Tokyo |
| 144 | $e^-, e^+$ | $He^+$, $Li^{2+}$, $Be^{3+}$, $B^{4+}$ | 50–56 | Harris | Tokyo |
| 19 | $e^-$ | H | 2–6 | K, R, AF, MN-K[d] | Minneapolis |
| 147 | $e^+$ | H | Not given[c] | K, R | Baton Rouge |
| 145 | $e^-$ | He | 20–72 | AF | San Jose |
| 71 | $e^-$ | H, $He^+$ | 8–24 | AF | San Jose |
| 148 | $e^+$ | H | 4–23 | K | London |
| 62 | $e^-$ | H | 54–70 | K | Tokyo |

[a] EI, Euler integral; K, Kohn; H, Hulthén, AF, anomaly-free, R, Rubinow, MN-K, minimum norm corrected with the Kohn variational expression; MN-R, minimum-norm corrected with the Rubinow variational expression.

[b] Staver also added a polarization potential to some of his calculations.

[c] In two cases $M$ is given (30 and 100). Note: $M$ is defined in equation (222).

[d] See also Section III.F.

TABLE VIII

Algebraic Variational Electron-Atom and Positron-Atom Scattering Calculations for Cases With More Than One Open Channel Using Methods Discussed in Sections III.B to III.E

| Reference(s) | Projectile(s) | Target(s) | $N_O$ [a] | $M$ | Method(s)[b] | Where done |
|---|---|---|---|---|---|---|
| 83 | $e^-$ | H | 2 | 1 | c | London and Belfast |
| 17 | $e^-$, $e^+$ | H | 2–3 | 30–45 | AF | Baton Rouge |
| 146 | $e^-$ | H | 4 | 28–60 | K, R, AF, MN-K | Minneapolis |
| 129 | $e^-$ | H | 3 | 16 | K, R, or AF | San Jose |
| 147 | $e^+$ | H | 2 | 30–100 | AF | Baton Rouge |
| 148 | $e^+$ | H | 2 | 35–84 | H, K, R, AF | Jerusalem |

[a] $N_O$ is number of open channels included in trial function.

[b] AF, anomaly-free; K, Kohn; R, Rubinow; MN-K, minimum-norm corrected with the Kohn variational expression; H, Hulthén.

[c] Method of Ref. 82.

regions becomes larger it is hard to distinguish such an algebraic variational calculation from a numerical integration. For example, the technique just described resembles in some respects some of the numerical integration techniques now popular for scattering problems which involve obtaining the solution approximately in terms of analytic functions in successive intervals and matching the solutions in different intervals at the boundaries.[154,155]

An even more striking illustration of the difficulty of distinguishing basis set expansion techniques from other methods is the fact that the finite difference methods using central differences can be reformulated as variational calculations using spline-like basis functions.[156] Thus the finite difference methods using central differences[157–159] may be classified as algebraic variational methods.

Having pointed out the difficulty of clearly sorting types, we now discuss a few examples of approaches using basis sets for wave functions which have recently been of interest in chemical physics scattering problems.

The derivative matrix technique and other techniques involving artificial channel boundaries have been extended to the treatment of multichannel scattering problems in many ways and have been much analyzed in nuclear physics.[68–70,160] Although much of the analysis has been formal[161] there have been some computational applications. Recently some specific procedures for incorporating derivative matrix techniques into chemical physics scattering calculations have been suggested. There have been applications to electron-atom scattering, to a model problem, and to collinear chemical reactions.[162–166] Except for considerations of resonance energies and widths,[167] the other artificial channel radii techniques have received less attention in chemical physics so far. However, Crawford has extended the Kohn-Crawford method to the treatment of collinear chemical reactions,[71] and Oberoi and Nesbet have extended the Kohn method with numerical asymptotic functions to electron-atom scattering in the exchange approximation.[72] These methods have the advantages that calculations in the difficult to handle internal region are performed using techniques similar to those used in bound-state calculations, that the trial wave function is continuous and has a continuous first derivative, and that the scattering results are variationally corrected through first order and stationary.

Schlessinger[168] and McDonald and Nuttall and their co-workers[169–171] have extended the method of Schlessinger to multichannel problems. The latter workers use a generalization of the original method in that the wave functions are calculated for complex energies instead of negative energies. These calculations can be carried out algebraically. The calculation of free-free integrals is avoided but a troublesome extrapolation procedure must be introduced. McDonald and Nuttall[75] have extended the rotation method (i.e., the method of complex coordinates) to treat neutron-deuteron scattering. They feel that this method compares favorably to the complex-energy method[169–171] and the Kohn variational method.

Weare and Thiele[172] have performed algebraic variational calculations on collinear atom-diatomic molecule scattering using a method which resembles the Schwinger variational method.

Payne[139,173] developed an algebraic variational method based on compact operators which reduces the scattering problem to a discrete eigenvalue problem. The method is applicable to single-channel and multichannel problems and Payne applied it to elastic electron-hydrogen atom scattering.

Garrett[174] presented two methods for multichannel elastic scattering problems based on the trial function (220) with $P = 1$. In both methods, a trial potential $V_t(r_i)$ is chosen and the equation

$$\left(-\frac{\hbar^2}{2m}\frac{d^2}{dr_i^{\,2}} + \frac{l(l+1)\hbar^2}{2mr_i^{\,2}} + V_t(r_i) - E\right)X_{11}{}^0(r_i) = 0$$

is solved numerically to obtain $X_{11}{}^0(r_i)$ and a trial phase shift $\eta_l{}^t$. In method 1, (229) is used to determine the coefficients $c_{mp}$. Then one checks whether (230) is satisfied. If not, one varies $V_t$, obtains a new $X_{11}{}^0(r_i)$, and checks (230) again. One continues this trial-and-error procedure until (230) is satisfied. Garrett suggested that one also optimize nonlinear parameters in $W_m(r_i, x_i)$ by choosing the values that minimize $|I_{pq}|$. If one chooses the nonlinear parameters so that $I_{pq} = 0$, and if one channel is open, the bounding principle[109] may be applied. In method 2, one does not use (230) but varies $V_t$ until $I_{pq} = 0$. Garrett applied his method to electron and positron scattering by hydrogen atoms.

Knudson and Kirtman extended their variation-perturbation theory formalism, which uses the Hulthén method for the first-order wave function, to multichannel scattering.[175]

Schwartz[79] and Rabitz and Conn[176] have considered variational expressions which use trial $T$ matrices

$$\mathbf{T} = -2i(\mathbf{1} - i\mathbf{R})^{-1}\mathbf{R}$$

instead of trial wave functions. Schwartz's method uses the Schwinger variational expression (see Section II.M) and the method of Rabitz and Conn uses a variational expression due to Newton.[177]

## IV. ADDENDUM

Shortly after this article was submitted for publication, some additional related work became available; it is summarized briefly here, classified by the section to which it is relevant.

Section II.F: Wladawsky[178] proposed a new method—the variational least-squares method—which is very similar to the minimum-norm method.

Section II.H: Kolker[179] provided another discussion of the theorem[43,61] (for central potential scattering) that after enough terms are added to the Kohn trial function the Kohn method provides a lower bound on the tangent of the phase shift.

Section II.J: Read and Soto-Montiel[180] applied the minimum variance method[65] to single-channel scattering problems.

Section II.L: Hazi[181] presented a nonvariational generalization of the original Harris method to the problem of calculating the reactance matrix from square-integrable approximations to the scattering wave function for inelastic scattering. His method involves using an assumed functional form for the $R$-matrix elements and an iterative procedure. Heller and Yamani[182] presented a variational method involving square-integrable trial functions which appears to be an improvement over the derivative matrix method.

Section III.E: Further variational calculations for positron-atom[183] and electron-atom[184] scattering and partial-wave-Born approximation calculations for positron-atom scattering[185] have been reported.

Section III.F: A discussion of some advantages of the derivative matrix method has been given.[186] Koppel and Lin[187] applied the Schwinger variational method to nonalgebraic trial functions obtained by non-variational methods for collinear atom-diatom collisions. Hazi[181] applied his new method to a model multichannel scattering problem.

## APPENDIX 1. ORTHOGONALITY IN THE HARRIS-MICHELS METHOD

Using the notation of Harris and Michels,[16] we label the free functions $\phi_i$ and the correlation terms $\eta_n$.

We define orthogonalized free functions $\tilde{\phi}_i$ as

$$\tilde{\phi}_i = \phi_i - \sum_m S_{mi}^{\eta\phi}\eta_m \tag{A1}$$

where

$$S_{im}^{\phi\eta} = \langle\phi_i \mid \eta_m\rangle \tag{A2}$$

Then

$$M_{ij}^{\tilde{\phi}\tilde{\phi}} = \langle\tilde{\phi}_i \mid H - E \mid \tilde{\phi}_j\rangle \tag{A3}$$

$$M_{in}^{\tilde{\phi}\eta} = \langle\tilde{\phi}_i \mid H - E \mid \eta_n\rangle \tag{A4}$$

etc. Then

$$\mathbf{M}^{\tilde{\phi}\tilde{\phi}} = \mathbf{M}^{\phi\phi} - \mathbf{C} \tag{A5}$$

where

$$\mathbf{C} = \mathbf{S}^{\phi\eta}\mathbf{M}^{\eta\phi} + \mathbf{M}^{\phi\eta}\mathbf{S}^{\eta\phi} - \mathbf{S}^{\phi\eta}\mathbf{M}^{\eta\eta}\mathbf{S}^{\eta\phi} \tag{A6}$$

and

$$\mathbf{M}^{\tilde{\phi}\eta}(\mathbf{M}^{\eta\eta})^{-1}\mathbf{M}^{\eta\tilde{\phi}} = \mathbf{M}^{\phi\eta}(\mathbf{M}^{\eta\eta})^{-1}\mathbf{M}^{\eta\phi} - \mathbf{C} \tag{A7}$$

Thus

$$\mathbf{M}^{\phi\phi} - \mathbf{M}^{\phi\eta}(\mathbf{M}^{\eta\eta})^{-1}\mathbf{M}^{\eta\phi} = \mathbf{M}^{\tilde{\phi}\tilde{\phi}} - \mathbf{M}^{\tilde{\phi}\eta}(\mathbf{M}^{\eta\eta})^{-1}\mathbf{M}^{\eta\tilde{\phi}} \tag{A8}$$

But the Harris-Michels results are calculated from the right-hand side of (A8). Thus the results would be unchanged if we used the nonorthogonalized functions $\phi_i$ instead of the orthogonalized function $\tilde{\phi}_i$.

### Acknowledgments

The authors are grateful to H. J. Kolker, R. K. Nesbet, J. C. Y. Chen, B. Kirtman, R. W. LaBahn, E. Gerjuoy, D. J. Kouri, H. Rabitz, A. F. Wagner, and T. J. George for preprints and to W. H. Miller, D. J. Kouri, F. E. Harris, H. S. Taylor, M. H. Schultz, and E. R. Rodgers for informative discussions. In particular we are very grateful to Professor Harris for emphasizing to us the important distinction between stationary and nonstationary methods.[188]

### References

1. See, for example, J. W. Cooley, *Math. Comp.*, **15,** 363 (1961); D. G. Truhlar, *J. Comput. Phys.*, **10,** 123 (1972).
2. See, for example, S. I. Chan and D. Stelman, *J. Mol. Spectry.*, **10,** 278 (1963); D. F. Zetik and F. A. Matsen, *J. Mol. Spectry.*, **24,** 112 (1967).
3. See, for example, D. R. Hartree, *The Calculation of Atomic Structures*, Wiley, New York, 1967; C. Froese, *Can. J. Phys.*, **41,** 1895 (1963); F. Herman and S. Skillman, *Atomic Structure Calculations*, Prentice-Hall, Englewood Cliffs, N.J., 1963.
4. See, for example, R. K. Nesbet, *Rev. Mod. Phys.*, **35,** 552 (1962); C. C. J. Roothaan and P. S. Bagus, *Methods in Computational Physics*, Vol. II, Academic, New

York, 1963; S. Huzinaga, D. McWilliams, and B. Domsky, *J. Chem. Phys.*, **54,** 2283 (1971).

5. L. Hulthén, *Kgl. Fysiograf. Sallskap. Lund Forh.*, **14,** 257 (1944). The Hulthén method is introduced and applied to scattering of a particle by the Yukawa potential.
6. W. Kohn, *Phys. Rev.*, **74,** 1763 (1948). The development of the Kohn variational method and its application to neutron-proton and neutron-deuteron scattering.
7. L. Hulthén, *Arkiv Mat. Astron. Fys.*, **35A** (25) (1948).
8. S. S. Huang, *Phys. Rev.*, **75,** 980 (1949); **76,** 477, 866, 1878 (1949). Calculations on elastic collisions of electrons with hydrogen atoms. The results of the calculations were not in good agreement with the numerical calculations [P. M. Morse and W. P. Allis, *Phys. Rev.*, **44,** 269 (1933)] because a very restricted functional form was assumed for the wave function and no variational correction to the tangent of the phase shift was used.
9. N. E. Tamm, *Zh. Eksperim. i Teor. Fiz.*, **18,** 337 (1948); **19,** 74 (1949). Development of variational methods for single-channel scattering problems. See also B. L. Moiseiwitsch and G. M. Stacey, *Proc. Phys. Soc.*, **86,** 737 (1965) and Ref. 47.
10. C. Schwartz, *Ann. Phys. N.Y.*, **16,** 36 (1961).
11. C. Schwartz, *Phys. Rev.*, **124,** 1468 (1961). Application of the Kohn method to elastic electron-hydrogen atom collisions using correlation functions.
12. F. E. Harris, *Phys. Rev. Letters*, **19,** 173 (1967). The Harris method is introduced and applied to the $s$-wave scattering from the Yukawa potential.
13. R. K. Nesbet, *Phys. Rev.*, **175,** 134 (1968). A critical examination and comparison of the Kohn, Rubinow, Hulthén, and Harris methods; the development of the anomaly-free method.
14. R. K. Nesbet, *Phys. Rev.*, **179,** 60 (1969). The extension of the Kohn, Rubinow, and anomaly-free methods to multichannel collision processes. Calculations are performed for a two-channel model problem.
15. F. E. Harris and H. H. Michels, *Phys. Rev. Lett.*, **22,** 1036 (1969). The introduction of the minimum-norm method and its application to a two-channel model problem.
16. F. E. Harris and H. H. Michels, *Methods Comp. Phys.*, **10,** 143 (1971). Review and comparison of the minimum-norm method to other variational methods. Equations 75 to 77 of this reference are in error.
17. G. J. Seiler, R. S. Oberoi, and J. Callaway, *Phys. Rev. A*, **3,** 2006 (1971).
18. I. Shimamura, *J. Phys. Soc. Japan*, **30,** 1702 (1971).
19. D. G. Truhlar and R. L. Smith, *Phys. Rev. A*, **6,** 233 (1972).
20. See, for example, N. F. Mott and H. S. W. Massey, *The Theory of Atomic Collisions*, 3rd ed., Clarendon, Oxford, 1965, pp. 522–523; B. H. Bransden, *Atomic Collision Theory*, Benjamin, New York, 1970, p. 17; K. Smith, *The Calculation of Atomic Collision Processes*, Wiley-Interscience, New York, 1971, pp. 50–56.
21. Y. Hahn, T. F. O'Malley, and L. Spruch, *Phys. Rev.*, **128,** 932 (1962).
22. See, for example, N. F. Mott and H. S. W. Massey, *op. cit.*, pp. 523–524; B. H. Bransden, *op. cit.*, pp. 184–193.
23. See, for example, N. F. Mott and H. S. W. Massey, *op. cit.*, pp. 19–25; B. H. Bransden, *op. cit.*, pp. 6–12; K. Smith, *op. cit.*, pp. 10–17.
24. See P. M. Morse and H. Feshbach, *Methods of Theoretical Physics*, Vol. 1, McGraw-Hill, New York, 1953, p. 622.
25. See, for example, N. F. Mott and H. S. W. Massey, *op. cit.*, pp. 113–114; B. H. Bransden, *op. cit.*, pp. 54–57.

26. A *variational principle* is an equation (derived, e.g., by the calculus of variations) that tells us that a certain quantity is stationary with respect to the allowed variations in the trial wave function. A variational principle can be used to write a *variationally correct expression* for a scattering quantity in terms of a trial wave function. The quantity computed from the variationally correct expression depends quadratically (not linearly) on the errors in the trial function. For example, a variational principle for a reactance matrix element allows us to calculate the first-order correction (first-order extrapolation toward the exact value) for an approximate reactance matrix element in a trial function. This correction is called a *variational improvement* in the matrix element. The corrected value is sometimes called the stationary value. A *variational functional* is a functional that appears in a variational principle. A *variational method* is a method that uses a variational principle to choose the parameters in a trial wave function. A variational method should involve as the final step the use of the variational principle and the trial function to calculate stationary values for the quantities of interest. However, the variational principle may also be applied to calculate first-order corrections to scattering quantities from trial functions which were not obtained by variational methods.
27. T. Kato, *Phys. Rev.*, **80,** 475 (1950). Relationships between the Kohn, second Hulthén, and Schwinger variational methods are pointed out.
28. K. L. Williams, *Proc. Phys. Soc.*, **91,** 807 (1967). The Rubinow and the second Hulthén method are shown to be equivalent.
29. H. Feshbach and S. I. Rubinow, *Phys. Rev.*, **88,** 484 (1952). A variational principle is proposed and applied to *s*- and *p*-wave scattering from a potential well.
30. S. I. Rubinow, *Phys. Rev.*, **98,** 183 (1955). Reformulation of the variational principle of Ref. 29.
31. T. Kato, *Progr. Theoret. Phys.*, **6,** 295, 394 (1951). Upper and lower bounds on phase shifts with application to the static approximation for electron-hydrogen atom scattering.
32. I. C. Percival, *Proc. Phys. Soc.*, **76,** 206 (1960).
33. M. J. Seaton, *Proc. Phys. Soc.*, **89,** 469 (1966). Percival's method is extended to the treatment of many-channel scattering.
34. Yu. N. Demkov and F. P. Shepelenko, *Zh. Eksperim. i Teor. Fiz.*, **33,** 1483 (1957) [English transl.: *Soviet Phys. JETP*, **6,** 1144 (1958)]. The Hulthén and Kohn methods are compared, the Kohn method result is shown to satisfy an integral identity, and phase shifts are calculated for an electron scattered by the static potential of the hydrogen atom.
35. Yu. N. Demkov, *Variational Principles in the Theory of Collisions*, Macmillan, New York, 1963.
36. Additional demonstration proofs of the correct sign may be found in Ref. 13, equation 60, Ref. 16, p. 164, and Ref. 35, pp. 60–61. The last argument is discussed further in Ref. 16, p. 162.
37. H. S. W. Massey and B. L. Moiseiwitsch, *Phys. Rev.*, **78,** 180 (1950); H. S. W. Massey and B. L. Moiseiwitsch, *Proc. Roy. Soc.* (*London*) *Ser. A*, **205,** 483 (1951). The elastic scattering of electrons by hydrogen atoms using the Kohn and Hulthén methods in the static approximation and also including exchange and polarization effects. See also G. A. Erskine and H. S. W. Massey, *Proc. Roy. Soc.* (*London*) *Ser. A*, **212,** 521 (1952) where the computed wave function is used and see (a) H. S. W. Massey and R. O. Ridley, *Proc. Phys. Soc.*, **A69,** 659 (1956), (b) H. S. W. Massey and A. H. A. Moussa, *Proc. Phys. Soc.*, **71,** 38 (1958), (c) A. H. Moussa,

*Proc. Phys. Soc.*, **74,** 101 (1959), and (d) L. Spruch and L. Rosenberg, *Phys. Rev.*, **117,** 143 (1960) for further variational calculations.

38. T. B. Staver, *Arch. Math. Naturv.*, **51,** 29 (1951). The Hulthén method is used to calculate phase shifts for the scattering of electrons by hydrogen atoms.
39. L. Hulthén and S. Skavlem, *Phys. Rev.*, **87,** 297 (1952).
40. B. H. Bransden, A. Dalgarno, T. L. John, and M. J. Seaton, *Proc. Phys. Soc.*, **71,** 877 (1958).
41. C. Malinowska-Adamska, *Acta Physica Polonica*, **28,** 909 (1965). The Kohn method is used for electron scattering from several atoms in the static approximation. An approximation to the static potential is used.
42. T. L. John, *Proc. Phys. Soc.*, **92,** 62 (1967). The methods of Kohn, Hulthén, and Malik are shown to be members of a larger group of methods which are introduced. Three new methods are applied to electron collisions with hydrogen atoms in the static, static-exchange, and exchange-polarization approximations.
43. K. R. Brownstein and W. A. McKinley, *Phys. Rev.*, **170,** 1255 (1967). An investigation of the spurious singularities that occur in the Kohn method.
44. H. Morawitz, *Ann. Phys. N.Y.*, **50,** 1 (1968). Discussion of the original Harris method for the case of only one square-integrable basis function.
45. H. Morawitz, *J. Math. Phys.*, **11,** 649 (1970). Application of the Harris, Hulthén, and Kohn methods to electron scattering off an attractive exponential and a Yukawa potential.
46. M. Heaton and B. L. Moiseiwitsch, *J. Phys. B*, **4,** 332 (1971).
47. B. L. Moiseiwitsch, *Variational Principles*, Interscience, New York, 1966.
48. R. K. Nesbet and R. S. Oberoi, *Phys. Rev. A*, **6,** 1855 (1972). Introduction of the optimized minimum-norm and optimized anomaly-free methods and their application to a two-channel model problem.
49. R. K. Nesbet, "Matrix Variational Methods in Electron-Atom Scattering," preprint RJ1059 (unpublished); R. S. Oberoi and R. K. Nesbet, *Phys. Rev. A*, **8,** 215 (1973).
50. F. E. Harris, private communication.
51. F. B. Malik, *Ann. Phys. N.Y.*, **20,** 464 (1962). The Malik method is introduced and compared to Hulthén's first and second methods.
52. L. Hulthén and P. O. Olsson, *Phys. Rev.*, **79,** 532 (1950).
53. K. Ladányi, V. Lengyel, and T. Szondy, *Theoret. Chim. Acta*, **21,** 176 (1971); see also K. Ladányi, *Nuovo Cimento*, **56A,** 173 (1968).
54. H. A. Bethe and R. Bacher, *Rev. Mod. Phys.*, **8,** 111 (1936).
55. R. G. Newton, *Scattering Theory of Waves and Particles*, McGraw-Hill, New York, 1966, pp. 313–315; see also Ref. 59, pp. 176–178.
56. Ref. 16, p. 165.
57. In Ref. 56 it is stated that $M^{01}$ will "ordinarily be slightly nonzero" when $M^{00} = 0$. The case discussed here is the extraordinary case.
58. L. Spruch and M. Kelly, *Phys. Rev.*, **109,** 2144 (1958); L. Spruch, *Phys. Rev.*, **109,** 2149 (1958); L. E. Rosenberg and L. Spruch, *Phys. Rev.*, **120,** 474 (1966).
59. L. Spruch, in *Lectures in Theoretical Physics*, Boulder, Colo., 1961, Vol. 4, W. E. Brittin, B. W. Downs, and J. Downs, Eds., Interscience, New York, 1962, p. 161.
60. J. Nuttall, *Ann. Phys. N.Y.*, **52,** 428 (1969). A theorem is proved that states that for $s$-wave collisions with a well-behaved potential and given enough basis functions the Kohn method can yield phase shifts to any desired accuracy.
61. H. J. Kolker, *J. Chem. Phys.*, **53,** 4697 (1970). The Harris method is shown to be

capable of providing upper and lower bounds on the phase shifts under certain conditions.

62. I. Shimamura, *J. Phys. Soc. Japan*, **31,** 852 (1971). The Feshbach projection operator technique is used to describe the resonances and pseudoresonances of the Kohn method. Examples of the difference of zeroes of determinants of $\underset{\sim}{N}(M \times M)$ and $\underset{\sim}{N}[(M+1) \times (M+1)]$, defined in Section III. *E*, are given and these figures are closely related to our remarks in Ref. 136.
63. I. Shimamura, *J. Phys. Soc. Japan*, **25,** 971 (1968). A method for calculating upper and lower bounds on phase shifts is discussed and applied to *s*-wave elastic electron-hydrogen collisions.
64. N. Anderson, A. M. Arthurs, and P. D. Robinson, *J. Phys. A*, **3,** 587 (1970).
65. J. N. Bardsley, E. Gerjuoy, and C. V. Sukumar, *Phys. Rev. A*, **6,** 1813 (1972).
66. K. J. Miller, *Phys. Rev. A*, **3,** 607 (1971).
67. P. L. Kapur and R. Peierls, *Proc. Roy. Soc.* (*London*) *Ser. A*, **166,** 277 (1938).
68. E. P. Wigner and L. Eisenbud, *Phys. Rev.*, **72,** 29 (1947); A. M. Lane and R. G. Thomas, *Rev. Mod. Phys.*, **30,** 257 (1958).
69. See also A. M. Lane and D. Robson, *Phys. Rev.*, **178,** 1715 (1969).
70. C. Mahaux and H. A. Weidenmüller, *Phys. Rev.*, **170,** 847 (1968).
71. O. H. Crawford, *J. Chem. Phys.*, **55,** 2571 (1971).
72. R. S. Oberoi and R. K. Nesbet, IBM Research Report No. RJ1049 (unpublished).
73. L. Schlessinger and C. Schwartz, *Phys. Rev. Lett.*, **16,** 1173 (1966); L. Schlessinger, *Phys. Rev.*, **167,** 1411 (1968).
74. W. P. Reinhardt, D. W. Oxtoby, and T. N. Rescigno, *Phys. Rev. Lett.*, **28,** 401 (1972); T. S. Murtaugh and W. P. Reinhardt, *J. Chem. Phys.*, **57,** 2129 (1972).
75. G. Doolen, M. Hidalgo, J. Nuttall, and R. Stagat, in *Atomic Physics*, Vol. 3,. S. J. Smith and G. K. Walters, Eds., Plenum, New York, 1973, p. 257; F. A. McDonald and J. Nuttall, *Phys. Rev. C*, **6,** 121 (1972).
76. J. Schwinger, unpublished lectures on nuclear physics, Harvard University, 1947, referred to in J. Schwinger, *Phys. Rev.*, **78,** 135 (1950) and J. M. Blatt and J. D. Jackson, *Phys. Rev.*, **76,** 18 (1949). See also Ref. 27; E. Gerjuoy and D. S. Saxon, *Phys. Rev.*, **94,** 478 (1954); C. J. Joachain, *Nucl. Phys.*, **64,** 548 (1965); and D. A. Micha and E. Brändas, *J. Math. Phys.*, **13,** 155 (1972).
77. See, for example, N. F. Mott and H. S. W. Massey, *op. cit.*, pp. 118–120; B. H. Bransden, *op. cit.*, pp. 60–61.
78. S. Altshuler, *Phys. Rev.*, **89,** 1278 (1953). The algebraic solution of the Schwinger method for electron scattering off the static ground state potential of the hydrogen atom.
79. C. Schwartz, *Phys. Rev.*, **141,** 1468 (1966).
80. S. K. Knudson and B. Kirtman, *Phys. Rev. A*, **3,** 972 (1971).
81. W. Kohn, *Phys. Rev.*, **84,** 495 (1951).
82. B. L. Moiseiwitsch, *Phys. Rev.*, **82,** 753 (1951).
83. H. S. W. Massey and B. L. Moiseiwitsch, *Proc. Phys. Soc. A*, **66,** 406 (1953).
84. R. J. Huck, *Proc. Phys. Soc. A*, **70,** 369 (1957).
85. N. F. Mott and H. S. W. Massey, *op. cit.*, pp. 372–379, 544–546; see also Refs. 8, 38, and 88.
86. N. F. Mott and H. S. W. Massey, *op. cit.*, Chapter XII*ff*. Refs. 87 to 106 are only a representative selection of references for the eigenfunction expansion method in the coupled-differential-equations form as solved by numerical integration techniques.
87. H. S. W. Massey and C. B. O. Mohr, *Proc. Roy. Soc.* (*London*) *Ser. A*, **136,** 289 (1932).

88. H. S. W. Massey, *Encyclopedia of Physics*, Vol. 36, Springer, Berlin, 1956, p. 232.
89. G. F. Drukarev, *The Theory of Electron-Atom Collisions*, Academic Press, London, 1965.
90. R. Marriott, *Proc. Phys. Soc.*, **72,** 121 (1958); T. L. John, *Proc. Phys. Soc.*, **76,** 532 (1960); K. Smith, W. F. Miller, and A. J. P. Mumford, *Proc. Phys. Soc.*, **76,** 559 (1960); K. Smith, *Phys. Rev.*, **120,** 845 (1960); K. Smith and P. G. Burke, *Phys. Rev.*, **123,** 174 (1961).
91. P. G. Burke and H. M. Schey, *Phys. Rev.*, **126,** 147 (1962).
92. P. G. Burke and K. Smith, *Rev. Mod. Phys.*, **34,** 458 (1962).
93. P. G. Burke, H. M. Schey, and K. Smith, *Phys. Rev.*, **129,** 1258 (1963); P. G. Burke, S. Ormonde, and W. Whitaker, *Proc. Phys. Soc.*, **92,** 319 (1967).
94. P. G. Burke, "Theoretical Background and Achievements of the Close-Coupling Approximation," in *Physics of Electronic and Atomic Collisions: Invited Papers from the Fifth International Conference*, L. Branscomb, Ed., Joint Institute for Laboratory Astrophysics, Boulder, Colo., 1968, p. 128.
95. P. G. Burke, "Low Energy Electron-Atom Scattering," in *Atomic Collision Processes*, Vol. XI-C of *Lectures in Theoretical Physics*, S. Geltman, K. T. Mahanthappa, and W. E. Brittin, Eds., Gordon and Breach, New York, 1969, p. 1.
96. P. G. Burke, "Electron-Atom Scattering," in *Scattering Theory: New Methods and Problems in Atomic, Nuclear and Particle Physics*, A. O. Barut, Ed., Gordon and Breach, New York, 1969, p. 193.
97. P. G. Burke, J. W. Cooper, and S. Ormonde, *Phys. Rev.*, **183,** 245 (1969); P. G. Burke and M. J. Seaton, *Methods Comput. Phys.*, **10,** 2 (1971).
98. N. F. Lane and S. Geltman, *Phys. Rev.*, **160,** 53 (1967).
99. R. Marriott, *Proc. Phys. Soc.*, **83,** 159 (1964).
100. R. Marriott, in *Atomic Collision Processes*, M. R. C. McDowell, Ed., North-Holland, Amsterdam, 1964, p. 114.
101. A. C. Allison and A. Dalgarno, *Proc. Phys. Soc.*, **90,** 86 (1967).
102. S.-K. Chan, J. C. Light, and J.-L. Lin, *J. Chem. Phys.*, **49,** 86 (1968).
103. A. F. Wagner and V. McKoy, *J. Chem. Phys.*, **58,** 2604 (1973).
104. S. Yoshida, *Proc. Phys. Soc. A*, **69,** 668 (1956).
105. W. Laskar, C. Tate, B. Pardoe, and P. G. Burke, *Proc. Phys. Soc.*, **77,** 1014 (1961).
106. F. J. Bloore and S. Brenner, *Nucl. Phys.*, **69,** 320 (1965).
107. In some cases the eigenfunction expansion (216) has been made for the wave function but the problem has been numerically integrated for the coefficients $\alpha_{ip}^0$ and $\alpha_{ip}^1$ in a form other than coupled differential equations for the functions $X_{ip}(r_i)$; see, for example, K. Smith, R. P. McEachran, and P. A. Frasier, *Phys. Rev.*, **125,** 553 (1962); A. Degasperis, *Nuovo Cimento*, **34,** 1667 (1964); D. Secrest and B. R. Johnson, *J. Chem. Phys.*, **45,** 4556 (1966); B. R. Johnson and D. Secrest, *J. Chem. Phys.*, **48,** 4682 (1968); M. E. Riley and A. Kuppermann, *Chem. Phys. Lett.*, **1,** 537 (1968); W. N. Sams and D. J. Kouri, *J. Chem. Phys.*, **51,** 4809, 4815 (1969); D. Secrest, *Methods Comput. Phys.*, **10,** 243 (1971).
108. Y. Hahn, T. F. O'Malley, and L. Spruch, *Phys. Rev.*, **134,** B397, B911 (1964).
109. M. K. Gailitis, *Zh. Eksperim. i Teor Fiz.*, **47,** 160 (1964) [English transl.: *Soviet Phys. JETP*, **20,** 107 (1965)]; see also the discussion in A. K. Bhatia, A. Temkin, R. J. Drachman, and H. Eiserike, *Phys. Rev. A*, **3,** 1328 (1971).
110. F. S. Levin, *Phys. Rev.*, **140,** B1099 (1965); **141,** 858 (1966).
111. See, for example, M. L. Goldberger and K. M. Watson, *Collision Theory*, Wiley, New York, 1964, pp. 138–160; L. S. Rodberg and R. M. Thaler, *Introduction to*

*the Quantum Theory of Scattering*, Academic Press, New York, 1971, pp. 206–210; W. A. Goddard, *Phys. Rev.*, **157,** 73 (1967); and Ref. 120.
112. M. Rotenberg, *Ann. Phys. N.Y.*, **19,** 262 (1962).
113. J. F. Perkins, *Phys. Rev.*, **173,** 164 (1968).
114. P. G. Burke, D. F. Gallaher, and S. Geltman, *J. Phys. B*, **2,** 1142 (1969); S. Geltman and P. G. Burke, *J. Phys. B*, **3,** 1062 (1970); and P. G. Burke and T. G. Webb, *J. Phys. B*, **3,** L131 (1970).
115. N. Feautrier, H. van Regemorter, and Vo Ky Lan, *J. Phys. B*, **4,** 670 (1971); Vo Ky Lan, *J. Phys. B*, **5,** 242, 1506 (1972).
116. M. Gailitis, in *Effective Cross Sections for the Collision of Electrons with Atoms: Atomic Collisions III*, V. I. Veldre, Ed., Latvian Academy of Sciences, Riga, 1965 [English transl.: M. Gailitis, "Computing Lower Bounds for Phases of Electron-Hydrogen Scattering," in *JILA Information Center Report No. 3*, University of Colorado, Boulder, Colo., 1966, p. 129].
117. P. G. Burke and A. J. Taylor, *Proc. Phys. Soc.*, **88,** 549 (1966).
118. A. J. Taylor and P. G. Burke, *Proc. Phys. Soc.*, **92,** 336 (1967).
119. P. G. Burke and A. J. Taylor, *J. Phys. B*, **2,** 44 (1969).
120. W. H. Miller, *J. Chem. Phys.*, **50,** 407 (1969).
121. See also J. C. Y. Chen and K. T. Chung, *Phys. Rev. A*, **2,** 1892 (1970); J. C. Y. Chen, K. T. Chung, and A.-L. Sinfailam, *Phys. Rev. A*, **4,** 1517 (1971).
122. Y. Hahn, *Phys. Letters B*, **30,** 595 (1969); Y. Hahn, *Phys. Rev. C*, **1,** 12 (1970).
123. H. Feshbach, *Ann. Phys. N.Y.*, **5,** 357 (1958); **19,** 287 (1962); **43,** 410 (1967).
124. B. L. Moiseiwitsch, *Proc. Roy. Soc. (London) Ser. A*, **219,** 102 (1953); see also H. S. W. Massey and B. L. Moiseiwitsch, *Proc. Roy. Soc. (London) Ser. A*, **227,** 38 (1954), where the computed wave function is used.
125. R. L. Armstead, *Phys. Rev.*, **171,** 91 (1968). Elastic electron-hydrogen atom calculations using correlation functions; see also R. L. Armstead, *Univ. Calif. Tech. Rept. UCRL-11628*, 1964.
126. H. H. Michels and F. E. Harris, *Phys. Rev. Letters*, **19,** 885 (1967); H. H. Michels, F. E. Harris, and R. M. Scolsky, *Phys. Letters A*, **28,** 467 (1969).
127. S. K. Houston and R. J. Drachman, *Phys. Rev. A*, **3,** 1335 (1971).
128. Some other examples of applying multichannel variational principles to trial functions that were not obtained by variational methods are V. M. Martin, M. J. Seaton, and J. G. B. Wallace, *Proc. Phys. Soc.*, **72,** 701 (1958); and M. R. H. Rudge, *Proc. Phys. Soc.*, **85,** 607 (1965); **86,** 763 (1965).
129. R. K. Nesbet and J. D. Lyons, *Phys. Rev. A*, **4,** 1812 (1971). These authors also discuss using a dimensionless form of $M^{\alpha\beta}$ obtained by dividing the trial function by $(E - E_i)^{1/2}$.
130. K. T. Chung and J. C. Y. Chen, *Phys. Rev. Lett.*, **27,** 1112 (1971).
131. K. T. Chung and J. C. Y. Chen, *Phys. Rev. A*, **6,** 686 (1972).
132. Some of the formulas in Refs. 14, 15, and 16 are valid only for the case where all $\mu_i$ are equal. They may be generalized by substituting $v_i$ for $k_i$.
133. M. Baer and D. J. Kouri, *J. Chem. Phys.*, **56,** 4840 (1972). For applications, see M. Baer and D. J. Kouri, *J. Chem. Phys.*, **56,** 1758 (1972); **57,** 3441 (1972).
134. D. J. Kouri, *J. Chem. Phys.*, **58,** 1914 (1973).
135. W. P. Reinhardt, *Phys. Rev. A*, **2,** 1767 (1970), **4,** 429 (1971). See R. K. Nesbet, *Comments Atomic and Molecular Phys.*, **3,** 143 (1972) for a discussion of the relation of Reinhardt's Fredholm-determinant method to the work of Refs. 13, 14, and 107.
136. There is still confusion on this point in the literature; for example, Matese and Oberoi, Ref. 143, loosely say that the Kohn and Rubinow methods introduce

spurious resonant behavior in the cross sections at eigenvalues of $\mathbf{H}^{\eta\eta}$. Although the distinction between the pseudoresonance positions and the eigenvalues apparently has no bearing on the results presented or the conclusions drawn in that paper, it is often important to distinguish these. Following Ref. 13, the text discusses how the Kohn and Rubinow methods introduce spurious resonant or pseudoresonant behavior in the cross-sections at paired zeroes of det $\mathbf{M}^{00}$ and det $\mathbf{M}^{01}$ and at paired zeroes of det $\mathbf{M}^{11}$ and det $\mathbf{M}^{10}$. These paired zeroes tend to occur near eigenvalues of $\mathbf{H}^{\alpha\beta}$ since all four det $\mathbf{M}^{\eta\eta}$ have odd-ordered poles at such eigenvalues and since the det $\mathbf{M}^{\alpha\beta}$ generally have the shapes as functions of energy of the $\mathbf{M}_{11}{}^{\alpha\beta}$ shown in Figs. 6, 9, and 10. In this article we do not identify the resonance or antiresonance positions as the places where tan $\eta$ or cot $\eta$ is $\infty$, but rather as the midpoint of the region where $\eta$ is changing rapidly by $\pi$. (For problems with $P > 1$ one identifies the resonance or antiresonance as the midpoint of the region where the sum of the eigenphases is changing rapidly by $\pi$.) This is another important point that has caused confusion. For example, the discussion and formulas for positions and widths for pseudoresonances on p. 165 of Ref. 16 often yield the wrong position and too narrow a width. The error in position occurs because the discussion identifies the resonance with the place where tan $\eta = \infty$. The width of the resonance was incorrectly assumed to be the width of the region where cot $\eta \simeq 0$ rather than the width of the region where the phase shift increases rapidly by $\pi$. The width of the latter region depends on how close the zeroes of the numerator and denominator of the formula for tan $\eta$ are, not merely on the width of the region over which the denominator is about zero.

137. J. C. Y. Chen, *Phys. Rev.*, **156,** 150 (1967).
138. Nesbet and Lyons, Ref. 129, p. 1817, say that resonance calculations by the Kohn and Rubinow method are not valid near zeroes of $M_{11}{}^{11}$ and $M_{11}{}^{00}$, respectively. But even for resonances, the zeroes of $M_{11}{}^{\alpha\beta}$ tend to be near the eigenvalues of $\mathbf{H}^{\eta\eta}$. Nevertheless, both the Kohn and Rubinow methods are accurate near a real resonance.
139. G. Payne, *Phys. Rev.*, **179,** 85 (1969).
140. Y. Hahn, *Phys. Rev. A*, **4,** 1881 (1971).
141. E. R. Smith, R. S. Oberoi, and R. J. W. Henry, *J. Comp. Phys.*, **10,** 53 (1972).
142. J. Callaway, R. S. Oberoi, and G. J. Seiler, *Phys. Lett. A*, **31,** 547 (1970).
143. J. J. Matese and R. S. Oberoi, *Phys. Rev. A*, **4,** 569 (1971). See also J. Callaway and J. J. Matese, *Int. J. Quantum Chem.*, **S6,** 79 (1972).
144. I. Shimamura, *J. Phys. Soc. Japan*, **31,** 217 (1971).
145. A.-L. Sinfailam and R. K. Nesbet, *Phys. Rev. A*, **6,** 2118 (1972); see also R. K. Nesbet, *Phys. Rev.*, **156,** 99 (1967), where a Hulthén-like procedure was suggested.
146. R. L. Smith and D. G. Truhlar, *Phys. Letters A*, **39,** 35 (1972).
147. S. E. Wakid and R. W. LaBahn, *Phys. Rev. A*, **6,** 2039 (1972).
148. J. Stein and R. Sternlicht, *Phys. Rev. A*, **6,** 2165 (1972); J. W. Humberston and J. B. G. Wallace, *J. Phys. B*, **5,** 1138 (1972).
149. E. M. Mortensen and L. D. Gucwa, *J. Chem. Phys.*, **51,** 5695 (1969).
150. F. E. Harris and H. H. Michels, *J. Comput. Phys.*, **4,** 579 (1969); J. D. Lyons and R. K. Nesbet, *J. Comput. Phys.*, **4,** 499 (1969), **11,** 166 (1973); C. Bottcher, *J. Comput. Phys.*, **6,** 237 (1970); R. S. Oberoi, J. Callaway, and G. J. Seiler, *J. Comput. Phys.*, **10,** 466 (1972); D. E. Ramaker, *J. Math. Phys.*, **13,** 161 (1972); R. L. Smith and D. G. Truhlar, *Comput. Phys. Commun.*, **5,** 80 (1973).
151. M. J. Seaton, *Proc. Phys. Soc.*, **77,** 184 (1961); J. Lawson, W. Lawson, and M. J. Seaton, *Proc. Phys. Soc.*, **77,** 192 (1961).

152. I. H. Sloan and E. J. Moore, *J. Phys. B*, **1,** 414 (1968).
153. S. P. Rountree and R. J. W. Henry, *Phys. Rev. A*, **6,** 2106 (1972).
154. M. Datzeff, *Ann. Phys. Paris* (11th series), **10,** 583 (1938); R. G. Gordon, *J. Chem. Phys.*, **51,** 14 (1969); **52,** 6211 (1970); see also E. L. Hill, *Phys. Rev.*, **38,** 1258 (1931); E. R. Davidson, *J. Chem. Phys.*, **34,** 1240 (1961).
155. D. J. Wilson, *J. Chem. Phys.*, **51,** 5008 (1969); D. J. Wilson and D. J. Locker, *J. Chem. Phys.*, **57,** 5393 (1972); J. Canosa and R. G. de Oliveira, *J. Comput. Phys.*, **5,** 188 (1970); L. G. Ixaru, *J. Comput. Phys.*, **9,** 159 (1972).
156. R. J. Herbold, Ph.D. Thesis, Case Western Reserve University, Cleveland, 1968.
157. E. M. Mortensen and K. S. Pitzer, *Chem. Soc.* (*London*) *Spec. Publ.*, **16,** 57 (1962); D. J. Diestler and V. McKoy, *J. Chem. Phys.*, **48,** 2941, 2951 (1968).
158. V. P. Gutschick, V. McKoy, and D. J. Diestler, *J. Chem. Phys.*, **52,** 4807 (1970).
159. D. G. Truhlar and A. Kuppermann, *J. Chem. Phys.*, **56,** 2232 (1972).
160. See, for example, E. Vogt, *Rev. Mod. Phys.*, **34,** 723 (1962); W. M. McDonald, *Nuclear Phys.*, **54,** 393 (1963); M. Danos and W. Greiner, *Phys. Rev.*, **138,** B93 (1965); W. Tobocman and M. A. Nagarajan, *Phys. Rev.*, **138,** B1351 (1965); M. Nagarajan, S. K. Shah, and W. Tobocman, *Phys. Rev.*, **140,** B63 (1965); M. Danos and W. Greiner, *Phys. Rev.*, **146,** 708 (1966); A. M. Lane and D. Robson, *Phys. Rev.*, **151,** 774 (1966); P. J. A. Buttle, *Phys. Rev.*, **160,** 719 (1967); H. G. Wahsweiler, W. Greiner, and M. Danos, *Phys. Rev.*, **170,** 893 (1968); L. Garside and W. Tobocman, *Phys. Rev.*, **173,** 1047 (1968); J. Hüfner and R. H. Lemmer, *Phys. Rev.*, **175,** 1394 (1968); B. Pöpel, *Nucl. Phys. A*, **119,** 325 (1968); L. Garside and W. Tobocman, *Ann. Phys.*, **53,** 115 (1969); P. P. Delsanto, M. F. Roetter, and H. G. Wahsweiler, *Z. Physik*, **222,** 67 (1969); R. F. Barrett, L. C. Biedenharn, M. Danos, P. P. Delsanto, W. Griener, and H. G. Wahsweiler, *Rev. Mod. Phys.*, **45,** 44 (1973).
161. See also B. C. Eu and J. Ross, *J. Chem. Phys.*, **44,** 2467 (1966) and N. S. Snider, *J. Chem. Phys.*, **51,** 1489, 4075 (1969).
162. P. G. Burke, A. Hibbert, and W. D. Robb, *J. Phys. B*, **4,** 153 (1971).
163. P. G. Burke and W. D. Robb, *J. Phys. B*, **5,** 44 (1972).
164. See also W. D. Robb, *Comput. Phys. Commun.*, **4,** 16 (1972).
165. O. H. Crawford, *J. Chem. Phys.*, **55,** 2563 (1971).
166. J. J. Matese and R. J. W. Henry, *Phys. Rev. A*, **5,** 222 (1972). This article proposes that the Degasperis technique (Ref. 107) be used in the external region in derivative-matrix calculations.
167. For a review of such treatments of resonances see J. N. Bardsley and F. Mandl, *Rept. Progr. Phys.*, **31,** 471 (1968); see also D. T. Birtwistle and A. Herzenberg, *J. Phys. B*, **4,** 53 (1971). If calculations are to be done only for resonances, a greater variety of methods than are discussed in this article may be used. See, for example, the stabilization method discussed by H. S. Taylor, *Advan. Chem. Phys.*, **18,** 91 (1971) and M. F. Fels and A. U. Hazi, *Phys. Rev. A*, **5,** 1236 (1972); the Feshbach QHQ method discussed by T. F. O'Malley and S. Geltman, *Phys. Rev.*, **137,** A1344 (1965), Ref. 137, and D. A. Micha and E. Brändas, *J. Chem. Phys.*, **55,** 4792 (1971); and the multichannel Fano approach of U. Fano, *Phys. Rev.*, **124,** 1866 (1971), U. Fano and F. Prats, *Proc. Natl. Acad. Sci. India A*, **38,** 553 (1963), C. Block and V. Gillet, *Phys. Letters*, **16,** 62 (1965), P. L. Altick and E. N. Moore, *Phys. Rev.*, **147,** 59 (1966), and D. E. Ramaker, Ph.D. Thesis, University of Iowa, Iowa City, 1971. The multichannel Fano method is related to the algebraic close-coupling method when the latter is stated in terms of the Feshbach formalism as in Refs. 130–131. The references in this footnote are merely representative.

168. L. Schlessinger, *Phys. Rev.*, **171,** 1523 (1968).
169. F. A. McDonald and J. Nuttall, *Phys. Rev. Lett.*, **23,** 361 (1969).
170. G. Doolen, G. McCartor, F. A. McDonald, and J. Nuttall, *Phys. Rev. A*, **4,** 108 (1971).
171. F. A. McDonald and J. Nuttall, *Phys. Rev. A*, **4,** 1821 (1971). Also the method of complex coordinates has been applied to collisions of composite particles in J. Nuttall and H. R. Cohen, *Phys. Rev.*, **188,** 1542 (1969) and F. A. McDonald and J. Nuttall, *Phys. Rev. C*, **6,** 121 (1972).
172. J. H. Weare and E. Thiele, *Phys. Rev.*, **167,** 12 (1968).
173. G. Payne, *Phys. Rev. A*, **2,** 775 (1970).
174. W. R. Garrett, *J. Phys. B*, **4,** 643 (1971).
175. S. K. Knudson and B. Kirtman, *Phys. Rev. A*, **8,** 296 (1973).
176. H. Rabitz and R. Conn, *Phys. Rev. A*, **7,** 577 (1973).
177. Ref. 55, pp. 319–320.
178. I. Wladawsky, *J. Chem. Phys.*, **58,** 1826 (1973).
179. H. J. Kolker, *J. Chem. Phys.*, **58,** 2288 (1973).
180. F. H. Read and J. R. Soto-Montiel, *J. Phys. B*, **6,** L15 (1973).
181. A. U. Hazi, *Chem. Phys. Lett.*, **20,** 251 (1973).
182. E. J. Heller and H. A. Yamani, to be published.
183. J. W. Humberston, *Bull. Am. Phys. Soc.*, **18,** 709 (1973).
184. A.-L. Sinfailam and R. K. Nesbet, *Phys. Rev. A*, **7,** 1987 (1973); J. N. Bardsley and R. K. Nesbet, *Phys. Rev. A*, **8,** 203 (1973).
185. G. Banerji, A. S. Ghosh, and N. C. Sil, *Phys. Rev. A*, **7,** 571 (1973).
186. P. G. Burke and J. F. B. Mitchell, *J. Phys. B*, **6,** 320 (1973).
187. L. M. Koppel and J. Lin, *J. Chem. Phys.*, **58,** 1869 (1973).
188. The authors are grateful to Drs. Richard J. Drachman, Jerrold Kolker, John J. Matese, David A. Micha, B. L. Moiseiwitsch, and Earl M. Mortensen for sending helpful comments on the original manuscript.

# AUTHOR INDEX

# SUBJECT INDEX